THEY F*** YOU UP

THEY F*** YOU UP

How to Survive Family Life

Revised and Updated Edition

OLIVER JAMES

BLOOMSBURY

LONDON • NEW DELHI • NEW YORK • SYDNEY

First published in Great Britain 2002

This revised paperback edition published 2007

Copyright © 2002, 2007 by Oliver James

The moral right of the author has been asserted

Bloomsbury Publishing Plc,
50 Bedford Square, London WC1B 3DP

Bloomsbury Publishing, London, New Delhi, New York, Sydney

A CIP catalogue record for this book is available
from the British Library

ISBN 978 0 7475 8478 0

19

Typeset by Palimpsest Book Production Limited,
Grangemouth, Stirlingshire

Printed and bound in Great Britain by CPI Group (UK) Ltd, Croydon CR0 4YY

www.bloomsbury.com

To my mum and dad,
the principal cause of this book

This Be the Verse

They fuck you up, your mum and dad.
They may not mean to, but they do.
They fill you with the faults they had
And add some extra just for you.

But they were fucked up in their turn
By fools in old-style hats and coats,
Who half the time were soppy-stern
And half at one another's throats

Man hands on misery to man.
It deepens like a coastal shelf.
Get out as early as you can,
And don't have any kids yourself.

                                        Philip Larkin

Acknowledgements

Only Clare, with whom I share my life, knows the full extent of what has been involved in writing this book. She has not only put up with an often irritable and introverted partner, wreathed in increasingly thick clouds of cigarette smoke, but has read the various drafts and made unsentimental demands for improvement to the text.

Second place for vital observations from friends and family must be shared between my late mother, who provided unsparing diagnoses of the prose and its lack of linearity even from her deathbed, and Jemima Biddulph, who got involved late in the day but to very telling effect despite her busy life as a teacher.

Others who have allowed themselves to be sucked into reading and commenting on sections of the text and who have all made useful suggestions of one kind or another are: Lucy Astor, Penny Garner, Anna-Louise Garner, Teddy St Aubyn, my sisters Lucy, Mary and Jessica, and James Sainsbury.

At Bloomsbury, I thank Rosemary Davidson for taking the book on and subsequently giving me very clear steers on how to salvage it when, after almost a year's work, it was miles away from being right. Bill Swainson and Pascal Cariss made suggestions, both structural and detailed, which immeasurably improved the text.

At Gillon Aitken Associates, Gillon's austere insistence on the right proposal set me off on the right track, and subsequently I have been very grateful for the unwavering support provided by his colleague Lesley Shaw.

Thanks are due to the Larkin Estate and to Faber and Faber, for permission to reproduce 'This Be the Verse' from *High Windows* by Philip Larkin. Thanks also to Jane Beresford for permission to reproduce in Chapter 1 part of the interview by Feargal Keane with Rufus May, broadcast on BBC Radio 4 as part of the *Taking A Stand* series, 2001.

Contents

Prologue to the Revised Edition   xiii

Preface   1

Introduction   4

Chapter 1: Our Genes   12

Chapter 2: Scripting Our Family Drama   35

Chapter 3: Scripting Our Conscience, Aged Three to Six   93

Chapter 4: Scripting Our Relationship Patterns in
Our First Three Years   151

Chapter 5: Scripting Our Sense of Self in Our
First Six Months   200

Chapter 6: Be Your Own Scriptwriter   249

Conclusion   283

Appendix 1: The Dubiousness of the Minnesota Twins
Reared-Apart Study   312

Appendix 2: Twin Studies – A Warning   315

Appendix 3: Estimates of the Environality of Human
Psychology from Twin Studies   321

Notes and References   325

Index   373

PROLOGUE TO THE
REVISED EDITION

Short of marrying my wife and the two sparky nippers that have ensued from this union, it is one of the most gratifying events in my life (although not quite as pleasing as the two goals I scored in the inter-house cup final at school in 1972) that this book has been sufficiently widely purchased as to have justified a second edition because its subject is the one that has been closest to my heart for over thirty years. If readers have found it helpful, both in sorting themselves out and in becoming better equipped to do the business for their own children, then I am thrilled to have been able to promote these outcomes, surely the most important there are after the relief of starvation and medical illness.

A remarkable amount of new scientific evidence that strongly bears out the argument of this book has been published in the five years since the first edition in 2002, requiring a second one. In just that short period, the main proponents of 'genes for' mental illnesses like schizophrenia or depression have wholly recanted their former position. Having pored through the genes identified by the Human Genome Project they have been forced to admit that it is extremely unlikely that there are any single genes for any mental illnesses. The new position is that it must be a question of clusters of genes. We shall see, but thus far there is little evidence for this theory.

Meanwhile, the overarching importance of parental care continues to find confirmation. One of the most convincing demonstrations is studies of its effect on patterns of brain electricity and chemistry, and even, on the very size of different bits of brain. For instance, it is becoming increasingly clear that early nurture sets the thermostat for our levels of cortisol or patterns of brainwaves in the left frontal lobe. It is also becoming clear that subsequent good experiences, like therapy, can reset the levels to healthier ones and that bad ones can do the opposite. One of the most significant scientific events of the century so far occurred this year when a whole volume of a highly respected psychiatric journal was devoted to the overwhelming evidence that schizophrenia is often caused by sexual and physical abuse. At least half of people given that diagnosis suffered this experience.

I do not remotely imagine that the readership this book achieved is purely due to its sparkling prose and brilliant insights. In the 1980s I was fortunate enough to become acquainted with Nic Roeg, the film director. Although he had been the creator of numerous outstanding films, *Don't Look Now*, the one starring Donald Sutherland and Julie Christie, was by far the most watched and commercially successful. In Nic's view, it was no better than the others, it was purely a matter of luck that it happened to ring bells for people in the Developed world at that particular point in cultural history.

If this book had come out at the end of the 1980s, when genetic determinism was in its heyday, I doubt it would have chimed so well with the times. As I have argued in some detail in my latest book, *Affluenza*, it took the rise of what I have characterized as Selfish Capitalism (akin to market liberalism – Thatcherism, Reaganomics) for Richard Dawkins' book *The Selfish Gene* – published in 1976 – to become a bestseller, providing a rationale for Right Wing political beliefs. This book happened to coincide with a government which, at least in theory, regards the social environment as crucial, especially the role of parents. These things go in cycles. In my youth, in the Sixties and early Seventies, nurturism was all the rage, nature hardly got a look-in. The pendulum swung the other way in the Eighties, now it has swung back again. By 2002, I suspect that many people had become sick of hearing how their volition was largely illusory and that their genes were crucial.

However, the new openness to the importance of care in the early years was soon put to work in the service of Selfish Capitalism. Applying behaviourist theories, originally based on studies of reward and punishment in rats and monkeys, the likes of Gina Ford's *Contented Little Baby Book* and Jo Frost's Channel Four series *Supernanny* became popular. The true title of Ford's book should be 'The Contented Parent' and of Frost's series, 'Taming the Beast in the Nursery'. The needs of the parents are everything, the emotional needs of the small child are nowhere. Although there are countervailing popular alternatives, like the books of Steve Biddulph, and Jean Liedloff's *The Continuum Concept*, they were up against a government and national culture which was overwhelmingly concerned with parents working ever-longer hours in order to be able to afford ever more, bigger, consumer snacks, now. Blatcher and his Nouveau Labour acolytes proved perfidious not only in the matter of Weapons of Mass Destruction. It turned out that their personal lives were gruesomely dominated by a love of rubbing shoulders with celebrity (Blair), of chasing the sexually attractive (Blunkett, Prescott) and, worst of all, of chasing money (Cherie Blair's lecture tours and property investments) or associating with it (loans for peerages, hobnobbing with the rich).

Both their education and childcare policies reflected this rampant materialism, placing the creation of good little producer-consumers far ahead of the providing of the love and security which are so central to both real education and mental health. They wanted to addict future British children to their own Americanized values as much as possible.

In education, examination vied with university tuition fees as methods for locking children into a dog-eat-dog, commercially-driven existence. Playing fields continued to be flogged off as rapidly as they had been under the Tories, any curricula activity which would not directly contribute to the economy was downgraded or deleted and businessmen were allowed to buy academies in order to propagate a worrying mixture of Selfish Capitalist ideology and religion.

In childcare, just as hardly any of the Nouveau Labour elite or their partners actually cared for their babies and toddlers them-selves, the downgrading of the maternal role (whether executed by a man or woman) proceeded apace, since only paid work attracted status. Notably, the SureStart scheme for poor parents was largely converted into group daycare provision on the grounds that only through paid work could mothers have self-esteem and dignity. It just so happened that such a policy also swells the available pool of low-paid workers to be hired by the companies contributing 'loans' to Nouveau Labour's campaign funds. Interestingly, little or none of the Nouveau Labour elite opted for group daycare as the method of substitute care for their own children, only one-on-one nannies would do for the likes of little Leo. Not a year went by without one or other of the Nouveau cabinet ministers seeking to mine political capital by announcing measures to crack down on parents and their 'feral' children. Television screens were filled with out of control sink-estate children and parents being taught discipline. It has been desperately disappointing that the opportunities which arrived in 1997 have been squandered in these ways.

The core of the Nouveau Labour apple has proven itself rotten in all sorts of ways (Iraq, the rich getting obscenely richer, the various scandals involving almost every one of those who were closest to Blair). But it is the failure to create the conditions wherebye the emotional needs of small children are met which I regard as their greatest iniquity – they should have known better.

A less contentious, more down-to-earth level is that of the individual seeking understanding of how their childhoods is affecting their adult personality, beyond what is offered in this book, since writing it I have discovered a further method which deserves special mention. Although I have not done it myself, I know many have found the Hoffman Process highly effective in bringing alive the impact of their childhood and offering very practical ways to move on from blaming parents and repeating the past (see hoffmanprocess.co.uk).

When I wrote this book I was childless, now I am father to a four-and-three-quarter-year-old daughter and a twenty-one-month-old son. People sometimes ask if parenthood has changed my views. My answer is that no, it has not; if anything it has only served to increase my confidence in the research described herein. Doubtless, my wife and I will make a horlicks of bringing our children up, and to some extent this is inevitable. Like the psychoanalyst Donald Winnicott, I believe we can only ever hope to be 'good-enough' parents. My plea is that all of us take a long, hard look at how our histories have affected us because that way we have a better chance of not recreating our own problems in our children. I would like to think that this book can foster such change.

Oliver James
Idbury
October 2006

PREFACE

My earliest memory is of falling off the end of my parents' bed, aged eighteen months. Incensed by the arrival of my younger sister, I had thrown a wobbly on seeing her being breastfed. I skulked around for several weeks afterwards, grunting and looking nasty. Whenever anyone tried to pick me up I would push them away, and for three weeks my parents assumed I was in a bate because of my sister's arrival. Both my parents had trained as psychoanalysts so they were not averse to psychologizing, but, to be fair, any parents might have concluded that my bad temper was due to sibling rivalry. It was only when they took me to the doctor that they discovered I had cracked my collar bone in falling off the bed. Even though my dad was also a medical doctor, they had confused the psychological with the physical.

This, and countless later experiences working in and around the world of 'shrinks' and the mentally ill, has led me to the conclusion that over-intepretation of human psychology can be inadvisable. My favourite Freud joke has him sitting in his gentlemen's club in Vienna after dinner, enjoying a cigar. A hostile colleague wanders up and says, 'That's a big, fat, long cigar, Professor Freud', to which Freud replies, 'Sometimes a cigar is just a cigar.' I am all for the healthy scepticism with which ordinary people regard tricksy theories about what makes us tick, so it would be as well if I explained why I think I'm qualified to propound them.

After doing a degree in social anthropology, I trained as a child clinical psychologist and then worked as one for six years, part-time, in a mental hospital. I was fascinated by the way different kinds of parental care in early childhood could affect what sort of adult we become, and I wrote a lengthy, wordy theoretical treatise on the subject. I developed what has been an enduring interest in the scientific evidence, as well as the theories, regarding the great 'nature versus nurture' controversy, becoming a jackdaw for such studies.

1

I would probably still be rewriting my thesis today if in 1982 I had not been rung up, on the off-chance that I knew something about child development, by a university friend who worked for Granada television. As a result I got involved in producing documentaries, including several that entailed interviewing violent men. As will be seen in Chapter 3, they are a particularly interesting group because the links between their crimes and their childhood histories are almost unavoidable. They provided me with an invaluable insight into the role of the past in the present, because all too often violent men do to their victims what was done to them by their parents.

But equally fascinating grist to my mill was provided when I diversified into conducting psychological interviews with famous people on TV. Again, I was astonished by the transparency of the links between the patterns of early experience of these high achievers and their need to achieve. As with the violent men, in many cases I only had to ask simple, straightforward questions, such as 'Which parent were you closest to?' or 'How did your parents punish you?', for obvious connections between past and present to force themself upon the listener.

As well as using conventional clinical cases, I have included in this book examples of famous people whom I have met and psychobiographies of some whom I have not. Television and newspapers have made many of them a virtual part of our social circle, so when I use Prince Charles or Woody Allen or Paula Yates to illustrate a point it is easier to make the connection because we feel we already know them.

R.D. Laing, the creator of what is known as Radical Psychiatry, introduced his wild and wonderful polemic *The Politics of Experience*, published in 1967, with the words 'Few books today are forgivable.' I hope this one is. Laing continued that 'we are all murderers and prostitutes . . . we are bemused and crazed creatures, strangers to our true selves, to one another and to the spiritual and material world'. His book ends with the alarming sentence: 'If I could turn you on, if I could drive you out of your wretched mind, If I could tell you I would let you know.'

How I would love to be able to claim as my ultimate qualification for writing this book that I have achieved some kind of transcendent mental health, a state of emotional hygiene so superior and spectacular that I am now ready to share it. Alas, I cannot pretend to any such higher state. As you will see throughout the book, I too am struggling to make sense of the past in my present, just like everyone else. However, I will make one tentative assertion as to what reading this book could do for you.

Finding out how your parents cared for you when small, whether through your own memory or by asking those who witnessed it or by analysing the way you relate to others today, could lead to a more fulfilled life. Although this is not a self-help book as such, I do offer some practical methods for performing an emotional audit – a taking stock – of the past in your present and an exercise by which you can apply it to bring about insight and change.

Much of the book is devoted to making accessible the scientific evidence that early parental care is crucial in forming who we are, complete with footnotes to signpost those who wish to read the studies themselves. But whilst I have no desire to drive you out of your wretched mind I do share Laing's desire to turn you on.

INTRODUCTION

The actress Mia Farrow was born the fifth out of eight children. Aged nineteen, she expressed regret about this, telling a newspaper reporter that 'A child needs more love and affection than you can get in a large family.' Aged twenty-five, she gave birth to her own first children (twins), and soon afterwards a son followed. Now she could give them that life in a moderately sized family which she had never had herself. Yet within a year she had adopted two Vietnamese infants. Over the next twelve years she produced one other child and adopted a further six. In all, twelve children called her Mother. Today, she says: 'The benefits of large families are enormous. I want to re-create my childhood environment.'

Somewhere along the way she had blanked out what it is like to be lost in a crowd of siblings, but the scientific evidence is that the nineteen-year-old Mia was right. It's easy to see why meeting the needs of very large families can be difficult. Offspring of families with five or more children are significantly more likely to be delinquent and to suffer mental illness. Already, two of her adoptees have been convicted of shoplifting. Trying to meet the needs of twelve children without any help from nannies, as Farrow proudly reports, is liable to produce children who crave love and attention and who lack individuality.

When combined with multiple divorces, it's especially likely to produce girls who use their nubility to get undivided adoration and security from men old enough to be their father . . . girls like Soon-Yi, the adopted daughter now married to Mia's ex-partner, Woody Allen . . . and the conspicuously childlike twenty-year-old Mia herself, whose first husband, Frank Sinatra, was fifty when she married him (her second, the musician André Previn, was sixteen years older than she was). If anyone knew all this, it was Farrow, so what on earth was she up to? How could she re-create a childhood environment that, at least when she was nineteen, she knew in her case had been depriving? Perhaps she had inherited a gene which made her want a large family. But it is also possible that her behaviour is explained by the saying: 'Those who forget the past are condemned to repeat it.'

It is not only the likes of Mia Farrow who find themselves reliving the past – all of us do. In fact, the extent to which we repeat our childhood experiences is quite extraordinary. How we react to our friends as well as who we pick as a lover, our abilities and interests at work, in fact almost everything about our psychology as an adult is continually reflecting our childhood in our day-to-day, moment-by-moment experience.

It has been shown in recent experiments, for example, that when we meet someone new we impose preconceptions upon them based on our childhood relationships. Unwittingly, we confuse them with characters from the drama that was once our family life. The stranger's name, their way of talking, how they look, any one of hundreds of tiny details can trigger memories from the original family script, which we then impose on to that new person.

Not only do we relate to our intimates guided by childhood narratives and roles, the experiments suggest that we even get these people to behave in the ways we were used to back then. Whether we were seen as the sweet, lovable one or the black sheep of the family, we go out and find people who see us that way. If they stray from doing so we manipulate them into treating us like that, or else simply assume that this is how they see us regardless of the truth. No wonder it's so hard for us to find the right lovers and close friends. We first require them to fit our childhood scripts, and then in order for the relationship to work we must fit into their unconscious storylines as well. This evidence is proof of the truth behind the joke with which Woody Allen ended his film *Annie Hall*. A man visits a psychiatrist and tells him that his wife thinks she is a chicken. The psychiatrist asks why the man does not leave her, to which the man replies, 'I need the eggs.' That pretty much sums up all the marriages that I have ever known: each person needs the other's madness.

There is another striking piece of new evidence that it was her upbringing, and not her genes, that must be held responsible for Mia Farrow's behaviour. The pattern of electricity and chemistry which makes the thoughts and feelings in each person's brain unique is hugely influenced by the way that person was related to in early childhood. For example, if one's mother was depressed, the thoughts and feelings that this engendered become established as measurably different electro-chemical patterns in the frontal lobes of the right side of the brain. Psychologists know that these patterns are not inherited because they are absent at birth, and only show up if the mother behaves in a depressed fashion when relating to the child. Whilst not immutable, the earlier these patterns are established the harder they are to change. Unless something radically different has happened in the interim, they are still present years later. Dysfunctions in the right brain have now been linked to numerous mental illnesses.

As well as brainwaves, the chemistry of the body is tremendously influenced by early upbringing. For instance, the hormone cortisol is secreted as a response to threats or other demands for action from the environment. In normal people its levels go up and down according to what is happening at any given moment, but if we were living in a highly stressful family in our first six or so years of life this acts like a thermostat, setting our cortisol levels too high or too low in adulthood. If an adult was under constant threat as a child, with aggressive or neglectful or intrusive parents, that person's system can either close down (low levels) or become jammed on permanent alert (high). He or she may have low, blunted cortisol levels because he has simply got too used to the stimuli that demand fight-or-flight, so that they no longer trigger the hormone; or he may have constantly high levels, always ready for a rapid response to danger. Adults who suffered childhood maltreatment have specific cortisol patterns that reflect the type of maltreatment. Those who were sexually abused as children have high levels as adults, whereas those who suffered coldness and lack of love have low levels.

So profound is the impact of early care on our psychology that even the size of different parts of our brain is affected. Studies of the volume of the hippocampus, a region of the lower part of the brain which plays a crucial role in our emotional functioning, show that it is 5 per cent smaller in women who were sexually abused as children, the earlier the abuse, the greater the reduction. Sexual abuse is one of many childhood experiences which can cause adult depression and depressed adults also have reduced hippocampus volume. So the very size and shape of our brain depends significantly on how parents related to us in early life and whether or not we suffer adult mental illness.

In accord with this, the earlier and more severe the maltreatment suffered in childhood, the more profound its effect. For instance, in a study of 800 children aged nine years, the ones who had suffered severe maltreatment before the age of three were more disturbed than the ones who had suffered it aged three to five (but not aged nought to three). The latter group in turn were more disturbed than children who had only suffered it aged five to nine. The specific form of the maltreatment also predicted the type of later disturbance, so that the children who were physically neglected had different outcomes from those who had been physically abused, for example. Furthermore, their cortisol levels were abnormally high, chronically jammed on fight-or-flight, if they had suffered maltreatments of several different kinds, whereas their levels were abnormally low if they had only suffered occasional physical abuse. Had these children's brains been tested, the earlier the maltreatment the more severe would have been the abnormalities in electro-chemistry and structure. Our first six years play a critical role in shaping who we are as adults, physically and psychologically.

Whether dysfunctional or not, we bring our brain patterns and structures to bear in seeking friends or lovers or occupations, choosing people or activities which match what our electro-chemistry expects. For example, if maltreated as a child we are more likely to suffer severe traumas, such as being raped, as an adult (most victims of rape were not maltreated as children, but overall, if we were it probably makes us more likely to get into situations where abuse may happen). Of course, experiences in the teenage years and afterwards, such as finding the right lover or having therapy or taking antidepressant drugs, can change our pattern of brainwaves and chemistry. But for most of us, in most respects, our first six years significantly explain what sort of adult we are.

One of the commonest retorts to such evidence is to ask the question: 'How do you explain the fact that I'm so different from my siblings? We had the same parents, were raised in the same house, had the same upbringing – so how come our brains don't have similar patterns? It must be due to different genes.' The answer is the remarkable fact that siblings *don't* have the same parents. Each parent treats each child so differently that they might as well have been raised in completely different families. Believe it or not, our uniqueness has far more to do with that than with our genes.

In the 1990s there appeared a stream of books, newspaper articles and television documentaries based largely on a single study by Thomas Bouchard, an American psychologist, of identical twins reared apart from each other. His results suggest that a great deal of what we are like is caused largely by genes. But there are many reasons to be wary of the findings of this study (see Appendix 1) and even more to reject the way it has been presented in the media. The truth is that, with the exception of a handful of extreme and rare mental illnesses, such as manic depression and schizophrenia, the way we were cared for as children is, in most respects, far more influential on who we are today.

Realizing this can be a source of liberation in our lives, and it has huge implications for our society. If we are as heavily dominated by our genes as is frequently claimed, then neither personal nor social change is possible. If the poor are poor, the mad mad and the bad bad largely because of their genes, there is little point in increasing spending on education for the poor, or in talking therapies or enlightened prison regimes. It would be just as fruitless to seek to change the colour of a person's eyes by these methods. But the evidence, as I shall show, suggests that genes don't do much to explain our individuality, nor are they the reason why siblings are different from each other.

Knowing this, and applying the knowledge to ourselves, can help achieve profound improvements in our lives; applied to society as a whole, it could achieve a significant change for the better. To take an obvious example, an astonishing 90 per cent of people in prison are suffering from a mental illness, in most cases caused by their upbringing: as will be apparent by the time you read the conclusion to this book, if the childcare of the next generation could be improved, it would lead to a very substantial reduction in the amount of crime. More generally, the UNICEF *State of the World's Children* report for 2001 stated that 'for a government that wants to improve the lot of its people, investing in the first years of life is the best money it can spend. But tragically, both for children and for nations, these are the years that receive the least attention.'

A major impediment is that many of us are reluctant to consider the true causes of what we are like, let alone the childhood origins of criminality. Although in much of the developed world we live in what is known as a culture of complaint, unprecedentedly liable to fashion ourselves as victims, most people are still very protective of their parents. One of our greatest problems is our reluctance to accept a relatively truthful account of ourselves and our childhoods, as the polemicist and psychoanalyst Alice Miller pointed out. She wrote, 'Not to take one's suffering seriously, to make light of it or even to laugh at it is considered good manners in our culture . . . many people (at one time including myself) are proud of their lack of sensitivity to their own fate and above all to their own childhood.' But the implications of confronting the truth about the care we received are not as destructive as is often feared. If we identify our parents as having been damaging as well as constructive, it should not entail a futile blaming of them or Woody Allen-style introspection and pointless acrimony. Insight is not the same as self-pity.

I don't propose that, having realized the childhood origins of your tendency to anger bosses or always to fall for the wrong sexual partners, you should lambast your parents for their awful behaviour. Most do their very best by their children. It is up to us to take what they did to us and fashion it to suit our own purpose, to rewrite the script of our lives.

Of course, that isn't easy. T.S. Eliot was spot on when he wrote that 'Human kind cannot bear very much reality'. We are insulated from it by a rose-tinted bubble of positive illusions, believing that friends like us more than they really do and that nasty things are less likely to happen than is actually the case. We dress up the past to suit the present. For example, when university students are asked to recall their pre-university grades, nearly all of them slightly inflate what they scored whereas hardly any remember doing worse than was truly the case. Also, parents of high-achieving children put this success down to their enlightened nurture, and blame genes for characteristics of their offspring that they dislike. We put a positive spin on the future, too, to a remarkable extent. In one study, men who had tested positive for HIV believed they were less likely to go on to develop full-blown AIDS than men who had tested negative. A congenial version of reality keeps us sane.

Please believe me when I say that the last thing I want to do through this book is stir up trouble between you and your family, to burst the bubble of illusions you have about your own childhood or to add to the burden of anxiety that parents already carry. Try to read it from the standpoint of a child rather than that of a parent or potential parent. I think of my reader as the product of a family, rather than as a person responsible for one. That way, when the evidence seems to be implicating parental care as a key cause, you will be less likely to feel defensive. The aim of this book is to make you understand better the way your childhood past is operating in your present, and how this knowledge can help you.

In fact, one of the implications in these pages is that it is normal to be screwed up to some extent, that everyone suffers problems in childhood and that all of us will be the better for changing our notions of what is normal. For example, the statistics on actual mental illness show that one fifth of us are afflicted at any one time, and one third will be at some point during their lifetime. Depending on which studies are believed, between 20 and 40 per cent more of us have serious symptoms without actually being mentally ill. It is time for us to stop believing that it is only 'me' who has problems and to realize that it comes with the territory of being human. As Sigmund Freud put it, 'Neurosis is the rule, not the exception', and grasping this can help us to see that we are not alone. It is also the starting point for understanding what went wrong and learning that we have a choice: we can simply re-enact the past, or we can rewrite the script.

This book makes possible an emotional audit. Just as an accountant makes a painstaking annual examination of a business's transactions to see how it is doing financially, you can use this book to audit what happened in your childhood and discover how it is affecting your present. At the end of each chapter you will find a simple, brief procedure to help apply its particular findings to yourself, enabling you to audit the aspect of yourself that has just been discussed.

10

You can do this as soon as you finish each chapter, while its ideas are still fresh in your mind, or wait until the end and come back to the audit then. Keep a pen or pencil with you at all times and make a mark in the margin if something rings a bell, so that it will be there to remind you when you look back later. This audit will help define who you are and why you are like that, and when you come to Chapter 6 you will be able to take advantage of some suggestions I make for using it to change yourself.

You can continue to be like an actor, endlessly repeating the same role in a family drama whose script was written long ago by others. Or you can become your own scriptwriter. But before you start work on this task, you need to understand a little more about the fact that your destiny has not been programmed into your genes.

A WORD ON NOTES AND REFERENCES

Wherever there is a statistic stated in the text, or an indication that a study has been done (for instance 'studies show that . . .') there is a source note beginning on page 313. I have not used numbering within the text, to avoid distraction for general readers.

Chapter 1
OUR GENES

They fuck you up, your mum and dad.

On New Year's Eve 1980, a young American called John Hinckley made an audiotape of himself singing one of John Lennon's songs, 'Oh Yoko'. Hinckley gradually gets drunker and more melancholic as the new year approaches until, filled with sadness, he begins to strum his guitar, plaintively, with the familiar opening chords of the song. 'In the middle of the night,' he sings, 'in the middle of the night I call your name.' So far so good, sung pretty tunefully, but then comes a shock. Instead of the chorus 'Oh Yoko, Oh Yoko', he changes the words to 'Oh Jodie, Oh Jodie', referring to the actress Jodie Foster.

Hinckley had developed an elaborate fantasy that he was courting Foster. As he became more delusional, he devised a scheme by which he hoped to prove his worthiness of her love: he would shoot President Reagan. On 30 March 1981 he did so and, after a lengthy court case, was judged to have been suffering from schizophrenia and sent to a secure mental hospital for treatment.

In 1985, his parents published an account of their life with their son, setting up a charity to help other parents with children who have the same illness. They believe that it was caused by his genes, that his pathology was marked out in DNA from the moment of conception. But are they right?

Presented with such a question, clever people usually answer, 'It's not nature versus nurture, it's a bit of both', but even clever people can be wrong. In this case they are very wrong indeed. The truth about what makes us different from each other is that only very occasionally is it a case of 'entirely or largely genes'; mostly it is 'largely environment'; and only in a small minority of the psychological characteristics commonly found in us is it genuinely 'a bit of both', a fifty-fifty split between nature and nurture.

The problem with any nature-versus-nature debate is that we all bring to it a great deal of personal baggage, making it hard not to ignore uncomfortable evidence and not to exaggerate that which supports our prejudices. Although most people say it is the combination that explains us, not very deep down, perhaps after a few glasses of wine, a heartfelt preference for one or the other is soon encountered. Not surprisingly, this is because we are attracted to the theories which support the story we tell ourselves to keep our illusions rose-tinted.

For example, when mothers are asked what makes children tick they plump for theories that help them feel they are doing the right thing. Those who work full-time tend to believe that small children are resilient little things, capable of coping with what life throws at them, with no need for constant personal attention. They may feel that their toddler will actually benefit from being cared for by others when they are at work, and consequently be unworried about leaving him or her. So long as their child's basic needs are met, its genetic uniqueness will flourish. These beliefs are wholly reasonable, bolstering what will make the mother feel most comfortable with her arrangements. By contrast, when full-time mothers who have stayed at home are asked this question they tend to express the opposite view. Their infants need constant one-to-one care from their biological mother, and without it they will be damaged. These mothers are much more anxious about leaving their children with someone else, regarding them as fragile and in need of high-quality nurture. They place far more emphasis on mothering than on genes, and, again, these beliefs merge seamlessly with their choice to stay at home.

If either group reads evidence that contradicts their position, it threatens to prick the bubble of their illusions. The same goes for all of us on innumerable topics. For example, when the publics of several developed nations were surveyed, homophobes tended to see homosexuality as a choice and the result of upbringing, whereas homosexuals and their supporters saw it as a genetically inherited preference. Believing in genes removes any possibility of 'blame' falling on parents. It becomes an unchangeable, natural destiny, not a choice that homophobes could portray as an illness which might be treated with therapies.

What we feel about the nature–nurture debate is likely to knit up with our political beliefs. As long ago as 1949, a survey showed that politically right-wing people tended to see genes as critical whereas those on the left favoured the environment, and this appears to be true today. The right will tend to argue that the hierarchies of society reflect genetically given talents, so that the rich are there because they have better genes, the poor are poor because they come from less good genetic stock. Likewise, women should be at home caring for children because genetic evolution has equipped their sex better for this role. For the left, these things are seen as the effect of society, something that can be changed; as a recent review of the matter demonstrates, the right wants to uphold the status quo, so genetic theories keep their bubble of illusions intact.

Welcome to my bubble

It makes no sense at all for us to emerge from the womb predetermined to react to our particular bit of the world with specific personalities or talents or mental illnesses. It would be far more logical, in evolutionary terms, to be born flexible, wide open to the influence of parents and upbringing, because each family setting, each social class and each society requires a different response in order for the individual to thrive. The child must attract the interest and love of its parents, and genes could not anticipate the precise traits best suited for achieving this any more than they could prefigure the particular demands of class and culture – demands that can rapidly change, as the social trends of the last fifty years illustrate.

To make an analogy with card games such as bridge or poker, knowledge of what a playing card is, the rules of the game and the hierarchy of different hands would be our species-wide inheritance. This knowledge is essential in order to be able to play – akin to all of us starting life with a range of emotions, like humour or sadness, and with basic mental equipment, like the potential for thought and speech. But just as it would be unhelpful for us to have preordained responses to our parents and siblings laid down genetically, with

14

particular thoughts and feelings (like a love of opera or a habit of interrupting conversations) already prescribed before we have even met our family, so it is with our response to particular card hands. In order to make the best of the hand we are dealt, we need to be highly adaptable to signals from other players about what cards they hold and we must base our judgements about how to play our own hand on their past performance, nurtured by our experience of them. To preconceive our response to each configuration of cards, so that we always bet if we have a pair of kings or four cards towards a flush in poker, or always bid a grand slam if we hold twenty-three points at bridge, would be as unsuccessful as having genetically preconditioned ways of reacting to our parents.

Yes, genes do establish a basic repertoire of traits in nearly all humans, but the subtle differences between us in their expression are largely determined by our upbringing. We got where we are today, the only species on earth able to survive in all ecosystems, by being born pliably plastic to our particular family. Strong support for this reading of evolution has come recently from the completion of the Human Genome Project, the map of our genes. It was expected that humans would have at least one hundred thousand different genes but it turns out that we have only thirty to forty thousand at most, just twice the number found in the common fruit fly. That we have so few genes may well mean that we simply do not have enough for them to be specifying the minutiae of differences between us as individuals. Craig Venter, the head of one of the two groups conducting the study, concluded that genes cannot play more than a minor role in determining differences between us. In his words, his work proves that 'the wonderful diversity of the human species is not hard-wired in our genetic code. Our environments are critical.'

In fact, considerable evidence to suggest this predated the Genome Project. The best comes from studies of identical twins. Taken overall, the results do not support the oft-repeated claim that differences between us in our psychology are half caused by genes ('a bit of both'). The truth is far more interesting.

Because identical twins have identical genes, any psychological differences between them must be environmental in origin. In twin studies this degree of difference is compared with that between non-identical (known as 'fraternal') twins, who share only half of their genes. Fraternals make a better comparison group than non-twinned siblings, because they were born at the same time and have all the additional features particular to being twins. The critical point is that, if a trait is influenced by genes, identicals will be more similar in that trait than fraternals because their genes are 100 per cent similar, whereas fraternals have only half their genes in common. For example, 90 per cent of identicals have a similar height whereas this is true of only 45 per cent of fraternals. This greater degree of similarity of the identicals is assumed to be due to their greater genetic similarity. That they are so much more similar in their height than fraternals suggests that height is heavily influenced by genes.

The findings of twin studies

For the purposes of this book, with the exception of Thomas Bouchard's study, I shall assume that the results of twin studies are reliable, although most studies of adopted children produce much lower estimates of heritability than those of twins and there are many technical reasons to doubt that twin studies are very accurate (see Appendix 2). The fascinating fact is that, even when they are taken at face value, what they reveal is not how important genes are but how relatively unimportant. That is the view of Robert Plomin, the world's leading authority on the subject, when he writes that the main finding from twins is that 'most behavioural variability among individuals is environmental in origin'.

Whereas much of our physical make-up is strongly genetic, hardly any psychological differences are as predetermined (see Appendix 3 for a summary and references to the following statistics). The vast majority of the characteristics that have been tested by twin studies are less than half heritable; and, indeed, a great many crucial ones – like the propensity to violence or our romantic preferences or degree of masculinity-femininity – show little or no heritability. Whilst some personality traits are quite heritable, for instance extroversion and emotionality (both 40 per cent), many others, such as sociability (25 per cent), are not. Scores on intelligence tests suggest they are the most heritable general cognitive capacity (30 per cent in childhood rising to 52 per cent in adulthood), but many crucial mental abilities are not very heritable. Memory is 32 per cent heritable, creativity is 25 per cent and exceptional high achievement, up to and including genius, is largely if not totally environmental in origin.

There is no better way to illustrate this than through the lives of twins who were born with identical genes. Gayle and Gillian Blakeney, a pair of dark-haired, pretty, identical twins, starred in the Australian TV soap opera *Neighbours*. Shortly before I interviewed them in 1993, they had paid their first ever visit to an optician. By their age, then twenty-seven, everyone's vision is less precise but the amount of distortion varies from person to person. The optician had been amazed to find that the degree and type of deterioration were exactly the same in both twins. But it was not just their eyes that were identical – so were their faces and bodies. Within five minutes of meeting them, despite the fact that Gillian's red blazer made a visible contrast to Gayle's cream shirt, I had confused them several times. Yet, even though they looked so similar, their psychologies were extraordinarily different. This difference could not be due at all to their identical genes but only to differences in upbringing, which were, indeed, striking. I asked them about their personal histories, and an occupational psychologist provided an independent assessment through formal tests of personality and intelligence. Together, we arrived at the following profiles.

As a child Gillian had been assertive and aggressive, a rebel who liked to play with boys and preferred masculine toys. She had a short fuse, was liable to get angry with her parents and had run away from home on two occasions. Although born just nine minutes after her sister, in their family script she was portrayed as the baby of the family, a self-avowedly manipulative 'Daddy's Girl' as a result. Yet in her personality Gillian took after her mother, developing her handwriting, manner of holding herself, facial expressions and mental attitude. Extremely forceful towards me, she interrupted and refused to give way if she had a point to make. Gillian was suspicious and more secretive than her sister. She described their relationship as having been like a marriage in which she was the man and Gayle the woman. When sharing a flat with her sister, she did the man-about-the-house DIY.

She was going to hold out for the Dream Marriage with her

Perfect Man and would be heartbroken by infidelity. Her sex life began at seventeen with a boyfriend whom her father deemed inappropriate. She was attracted by very different traits in men, saying that she preferred 'the good-looking, ultra-masculine type whereas Gayle likes intellectuals'. She wanted to have children in Britain, to take them to Australia for a few years and return to Britain for their secondary education. Her husband would have to fit in with her plans, although she expected him to be successful and very masculine and that this would lead to lots of arguments, which she believed they would enjoy. Although independent and assertive in her dealings with men, she used feminine wiles. She was more overtly sexy and flirtatious than her sister.

Gayle was so different in all these respects that one might imagine she came from a different family. If judging her only on her personality, one would never guess that she had exactly the same genes as her sister. Markedly less assertive, as a child she had occupied the niche of being helpful around the house, offering compliance as a way of winning her parents' approval. Unlike her sister, she had a doll which was 'my baby'. There were no signs of the tomboy or rebel, and she would never have thought of running away. She took an elder-sisterly, protective role towards 'baby of the family' Gillian. She did not have Gillian's short fuse, and only showed aggression under extreme provocation. In adulthood, her handwriting, manner of holding herself, facial expressions and mental attitude were said to be very like her father's, in contrast to Gillian's similarities in these respects to her mother. In ordinary siblings, at least in theory, this could be explained by one having inherited genes from the father, the other from the mother, but that cannot be the case with genetically identical twins.

Gayle was more reflective and listened more carefully when I spoke to her. She was also more forthcoming. Whereas Gillian concealed a significant fact about their father, Gayle freely volunteered it to me, saying, 'Dad's business collapsed in the early 1970s. He's not the same person as he was – a broken man.' Much more open and trusting, none the less Gayle dammed anger up; and just occasionally the dam burst: 'If I do blow my fuse I'm dangerous. I don't like arguments at all.' She was the cook when sharing a flat with her sister, and took the role of housewife. She believed that her sister was 'much prettier – it's the structure of her face', although their faces are indistinguishable to a stranger. Gayle's sex life began five years later than Gillian's, at twenty-two. With her differing preference for brainy types, she expected to marry a man who would introduce her to a new social world and to live where he needed her, whereas Gillian expected her man to fit in with her plans. Whilst Gayle had no intention of devoting herself totally to the role of housewife, she believed that in a marriage 'you have to act The Wife – I will have to be accommodating'. She felt it would be wiser to see marriage as a practical pact between two adults when she settled down; unlike Gillian, Gayle believed it quite possible that her husband would be unfaithful, but felt that this should not necessarily mean divorce.

Whatever else caused these two women to be so different, it could not have been their identical genes. Indeed, even within the limits of a single meeting it emerged that differences in the way their parents related to them were crucial. Their mother exhibited a completely different attitude to the aggression of each daughter, and both parents clearly favoured Gillian.

Unaggressive Gayle recalled a telling incident. 'I had this angelic doll which became completely precious to me. That doll was my baby – that doll was my life. So of course if Gillian was going to inflict any emotional pain on me it was to rip off the doll's head, throw it down the hill and into the jaws of a dog – which she did.' Retaliation followed swiftly. 'I grabbed her Snoopy doll, scratched its head on the ground and ripped its ear off.' Their mother's response to the doll massacre provides a perfect illustration of what has made the two girls so different. 'Mum got angry at me but not at Gillian. She said, "You can always put a doll's head back on but you can never take the scratches out of Snoopy's nose" – not very fair comment. I said, "But that's my baby – she's ripped the head off my baby!" But Mum was always more on Gillian's side.'

In recent years it has become fashionable to claim that such different treatment is caused by the child's genetic temperament, so that 'difficult' children make themselves unpopular whilst lovable ones attract favouritism. But this is simply not possible in the Blakeneys' case, since the two were born the same. Although it can be difficult to get to the truth of the matter, since parents rarely want to admit to having treated their children differently and children are often highly protective of their parents, such differences in treatment caused by the parents' projections rather than the child's supposedly inborn temperament are the norm, not the exception, and are found to some degree in all families.

The hundreds of differing reactions like this, in which their mother supported Gillian's aggression and suppressed Gayle's, day in, day out, were bound to have an effect. They may have something to do with Gayle's tendency as an adult to dam up aggression and with why she occasionally felt so disempowered that she became violent. Perhaps, when words did not work, only physical force could make her point. It may also explain why she was generally a more obedient child whereas Gillian, being encouraged to express her aggression, was rebellious. Gillian developed a habit, which she displayed over lunch with me, of always leaving some of her food on the plate – much to her parents' vexation. Gayle said she was more compliant: 'I gained respect and individual worth by being extremely helpful about the house and being bubbly, because Mum and Dad were very busy and hard-pressed. I was into being rewarded. Gillian was into being cute.'

Gillian incurred the anger of her father by having a delinquent first boyfriend, but her mother 'adored' this boy. She managed to divide and rule her parents. One of the ways in which she achieved her extra liberty, she reported, was to 'play on being the little baby, the youngest-of-the-family routine. I knew how to be cute and I was quite a manipulative little thing. If I wanted something I knew how to get it.' Despite the fact that they had been born identical and only nine minutes apart, their father took a special shine to her. Gayle recalled that Gillian used to snuggle in and be the 'cutesy bubs' on their father's lap. 'Dad used to go, "How's my baby?" to Gillian.' Gillian confirmed that 'I was definitely my dad's little girl, of the two of us. I was his "Gilly-Gum".' This may have given her confidence in dealing with men, and may also explain her father's anger, possibly jealous, towards her first, 'inappropriate', boyfriend.

Moving from Gillian and Gayle as specific examples to more general findings, twin studies show that only a small handful of characteristics, all of them rare mental illnesses, have more than 50 per cent heritability. Even in these cases the way we are cared for in childhood and afterwards significantly affects how ill we become. For the vast majority of the one fifth of us who are suffering from a full-scale mental illness at any one time, genes play only a small part. This is because the commoner that an illness is, the less heritable: only the rare ones are very genetic.

Huntington's Chorea, which causes brain degeneration, seems to be almost completely heritable, so that virtually everyone who has the relevant gene develops it; but it affects just one in twenty thousand people and is the only severe mental illness for which a single specific gene has been identified as the cause. Where genes do play a part in severe mental disorders it is a variety of interacting ones, although, as yet, not a single one of the principal mental illnesses has been proven to be the result of possessing a particular gene or genes. The next most heritable mental illness is autism, thankfully only affecting a tiny number of children (about 0.2 per cent). It may be as much as 80 per cent heritable but there are still large differences in how the child turns out, depending on the age at which the problem is diagnosed and starts to receive treatment. After autism comes manic depression (affecting 0.5 per cent of the population at any one time), which may be as much as 60 per cent heritable. Major depression (4 per cent of people) is less so, about 50 per cent, similar to schizophrenia (1 per cent of people). Since these are all illnesses that affect very few of us, the much commoner neuroses (15 per cent of people) and so-called 'minor' depression (18 per cent) are of far greater concern, as are alcoholism and other increasingly common addictions. All of these are much less heritable, ranging from 30 per cent at most to no genetic influence at all. Thus genes play only a minor role in most cases of mental illness, since it is neuroses, minor depression and addictions that are by far the commonest of these.

Taken overall, even in those rare cases where genes do account for half of a trait the environment remains crucial. For example, some (but not all) twin studies suggest that the propensity to smoke tobacco is as much as half heritable. (This is something I am very inclined to believe: I gave up in 1988 but have been using nicotine substitutes on and off ever since. Whilst writing this book, I am ashamed to say, I have succumbed to a 10-a-day habit which I would love to be able to blame on genes.) What this means is

21

that differences between individuals in their proneness to smoking addiction are half caused by genes, but that these differences can only be fulfilled under certain environmental conditions. There was not a single smoker in Europe before the sixteenth century, because until then there was no tobacco there. Very few women smoked until the twentieth century; now the habit is more popular among young women than among young men. To say that individual differences in smoking are half caused by genes is simply not true. It all depends on the environment in which the individual is living.

Another dramatic recent example is divorce. One twin study found it to be about 50 per cent heritable, but, again, what does this really mean? The number of divorces in England in 1857 was just 5, because at that time a divorce required a specific Act of Parliament. The law was changed the following year, but there were still only 590 in 1900 and a paltry 4000 by 1930. How come the 50 per cent of people supposedly with 'divorce genes' were remaining married? Not until after the Second World War did divorce really rocket, from 12 per cent of all marriages to today's 40 per cent. These facts prove beyond question that social forces are the principal cause of divorce, so in what sense is divorce 50 per cent heritable? How can genetics accommodate the fact that divorce is not permitted at all in some countries, or account for the widely varying rates between developed nations? What has happened to the 'divorce gene' in these cases? It is nonsense to suggest that divorce is half heritable unless it is also specified that the environmental conditions that have emerged in developed nations since the mid-twentieth century (changes in the law, greater affluence and so forth) must be present to cause this. Also very telling are the huge differences in rates of mental illness between nations. Americans are six times more likely to suffer than citizens of Shanghai (China) and Nigeria. In general, citizens of English-speaking nations are twice as likely to suffer as those in mainland Europe. This is extremely unlikely to be anything to do with genes since they share the same genetic stock, and I have argued elsewhere that the reason is Selfish Capitalist governance. Likewise, if you compare rates in Singapore and China, the populations of which also share genes, they are far higher in the (Americanized-Anglicized) Singapore.

Very recently, as a result of the Human Genome project, a New

22

Zealand study found that certain genetic variations created the vulnerability to depression, cannabis use and violence: if you had the wrong genes, good early care meant they were not likely to find expression in dysfunction but bad care meant you developed those problems in later life; likewise, if you did not have those genes, even if you had bad care, you were much less likely to develop the problems. This accords with the 'bit of both' model. However, the jury is very much out on the role of this gene. On the one hand, four studies have partially replicated the original finding of the effect of the gene on depression. Unfortunately, all these studies had small samples, categorized the gene in different ways and in two of them the replication was only for women, not men. Three other small studies also provide partial support. All of this evidence is widely accepted to be very far from conclusive as replication. On the other hand, three have not replicated the effect of the gene on depression. One of these had a much larger sample than all the replication studies, and in this case, people with the genetic variation which should have made them vulnerable to depression in the event of adverse early care were actually less vulnerable, an important contradiction of the original study. Furthermore, in large studies of samples of depressed patients, their genes do not vary in the predicted manner compared with undepressed people: the depressed are not more likely to have the variant which increased the likelihood of depression when coupled with childhood maltreatment.

On a wider scale, there are reasons to question the role of this genetic variation. Two of the strongest predictors of who gets depressed in a developed nation are being of low income and being a woman – the poor and females are twice as likely as the rich and males to be depressed. Two studies have found no greater occurrence of the supposedly depression-inducing genetic variation in low-income people. Apart from the two replication studies mentioned above, in all the others the gene is not found more often in women than men. At best, the implication of this line of research so far is that the genetic variation may play no part in causing depression and even if it does turn out to play a role, the key factor is whether the environment is adverse – especially, whether there has been childhood maltreatment.

* * *

Similar studies of Attention Deficit Hyperactivity Disorder have also been done, with mixed outcomes. In the coming years, it is possible that other studies will show that we are born with potentials for all kinds of problems which are only fulfilled if our parenting is poor, but as yet this has not been proven and may never be.

Even in extreme mental illnesses, where genes seem to be most at work, the environment can be crucial. The *cause célèbre* for this issue amongst scientists has been schizophrenia. It is frequently asserted by psychiatric specialists that the disease entails a brain abnormality caused primarily by genes. Because sufferers from the illness seem so dramatically different from the norm, this is easy to believe – surely upbringing could not make someone so severely disordered? The rest of this chapter will look at the causes of this illness as an illustration of how important nurture can be, even if a significant genetic basis does seem to exist.

Schizophrenia as an example

Whilst I was working at a therapeutic community for the mentally ill I witnessed a fairly typical schizophrenic breakdown. Julie had studied politics and graduated with the best first in her year. Her relations with the university authorities were very stormy but she completed a postgraduate degree and published a book, a scholarly Marxist analysis of the American electoral system, before dropping out of academia. Her analysis of current British politics and economics was exceptionally well informed. When I met her she was in her early thirties and had been working full-time and very hard in left-wing politics for three years. She was admitted to the hospital because she had suffered from a general mental malaise, though nothing approaching schizophrenia. She was a warm, unassuming woman, well liked by many. A vegetarian and a heavy user of marijuana, she was easy-going if quick-witted and intense. She was short and slight in build and full of nervous energy, forever adjusting her spectacles with twitchy movements.

Her relationships with men followed a pattern. Either she was on friendly, asexual, sisterly terms, or in a few instances she fell in love with an idealizing gush. On the occasions that active sexual relations were established, they were broken off by the man very rapidly. She could be highly critical of men in general and sometimes voiced the idea that she might be a lesbian.

About a year after we had first met, I arrived back from holiday to find Julie and a group of other patients sitting round the kitchen table about to share a pot of tea. It was poured, but before anyone could sample it Julie insisted, 'Don't drink it, it's poisoned.' We ignored her with the fluid skill that groups develop for these occasions, but it proved to be the first note in a symphony of symptoms which were to grow into a cacophony over the next fortnight. She believed there were two transmitters at either end of the city which were sending a signal through the house, driving her mad. She spent hours crouched naked in the bath, scrubbing her clothes 'to get them clean'. A fear of 'dirty', poisoned food precluded eating. Some of what she said made no sense to me, but at other times she was full of dazzling insights, whether about my motives for sitting with her or about society in general. Like the last minutes of a dying light bulb, these insights illuminated a great deal; but they were followed by 'darkness' – days of psychotic incomprehension. Eventually we called her parents, who lived far away, and it was decided to transfer her to a hospital with locked wards because she had become a danger to herself.

Conventional psychiatric wisdom is that Julie's breakdown was caused primarily by the impact of her genes on her brain, causing it to malfunction. Emil Kraeplin, a nineteenth-century German doctor, laid the faulty foundations upon which psychiatrists have built their highly questionable edifice regarding her symptoms. He claimed that symptoms of madness cannot be understood in terms of psychology. It was a biological and largely genetically caused ailment, ultimately visible by directly examining the patient's diseased brain. This model still prevails within psychiatry. Textbooks divide mental illnesses into the discrete categories Kraeplin devised. The diagnostic system reflects his basic assumptions about genetic causes and the need to explain to patients that they are ill with a physical sickness no different in kind from tonsillitis. In 1978, the American psychiatrist Gerald Klerman updated Kraeplin, asserting that 'psychiatry treats people who are sick and who require treatment for mental illness . . . there is a boundary between the normal and the sick . . . there are discrete mental illnesses . . . the focus of psychiatric physicians should be particularly on the biological aspects of mental illness.'

But there is a stack of evidence that none of these claims stand up when scientifically tested. Considerably more of the population than expected have at least one symptom of madness, stretching its supposed boundary to breaking point: hearing voices, with whom they may have conversations; hallucinating sightings of familiar people; delusions, such as that they have travelled in a UFO; and severely paranoid thoughts. Take the mania of manic-depressives: it seems to be a psychological defence against feeling depressed. When asked to name words in a test, people who have recovered from manic-depression find it harder to say ones that are related to depression than euphoria. Although they say they are fine, in fact they often seem to be suffering from low-grade depression. Further studies suggest that depressive words are emotionally troubling to them and they have large fluctuations in their self-esteem. When followed up, it turns out that the greater the number of hidden signs of depression they have when ostensibly well, the greater the risk of the illness recurring. During their manic phase, despite euphoria, patients are very similar to depressives in their scores on some tests. The frantic activity of mania is a desperate attempt to distract the self away from a depressive core and can take the form of a frenetic battle to replace low self-esteem with achievement or activity.

Paranoia is a common symptom of schizophrenia which seems to be anything but the purely mechanical malfunction that conventional psychiatry deems it. Paranoids the world over are expert at avoiding blame for negative events and taking the credit for positive ones. When they were given a game to play on a computer over which they had no control of the true result, they claimed all the credit for winning and claimed it was rigged (quite rightly, in fact!) if they lost. By contrast, depressed patients barely took more credit for winning than losing.

Another key schizophrenic symptom is speech disorder – incoherent or confusing or downright weird language. Recent studies have shown clearly that this symptom is much more likely if the subject of conversation is emotionally charged. When asked to talk about sad rather than happy memories, the speech becomes measurably

26

more disordered, and the more personal the subject-matter, the more so. Nor are hallucinations meaningless. If a patient is poor or of low status they are more likely to hear voices to whom they feel subordinate. Emotional stress makes seeing or hearing things much more likely: fully 13 per cent of bereaved spouses have heard their dead spouse's voice.

Studies of twins are the cornerstone of the genetic argument. Yet, if they are to be believed, they actually prove that in half of cases genes are not the main cause. The reason that this can be said with such confidence is that if a hundred schizophrenics have an identical twin, on average only between one third and one half of their siblings will be schizophrenic too. That is higher than the rates found among the siblings of schizophrenic non-identical twins (around 15 per cent), suggesting a significant genetic component. But the extraordinary fact is that, in the case of the identical pairs where only one of them is schizophrenic, genes could not be the cause of the illness. Because the twins have exactly the same genetic code, if one of them were being made schizophrenic by their genes then so would the other. The only possible cause of the difference between the twins must be exposure to differing environmental influences.

In theory, these could be physical, like differing positions occupied in the womb or differing exposure to dangerous chemicals in childhood. But as we shall see in the next chapter, there is abundant evidence that the main reason is that they experienced very different childcare, even though they had the same parents. We shall also see that the most likely explanation for schizophrenia is that, whilst in some cases it is very largely caused by genes, in others it is largely the result of upbringing and in others still, it truly is a case of 'a bit of both'.

Apart from twin studies, the fact that schizophrenia is found throughout the world suggests it does have a genetic component – is part of the package that goes with being human, so that some members of any community will suffer it. However, that still does not mean that every case is largely genetic in cause. A survey of 2 million Danish people found that the more of your childhood you had spent in urban environments, the greater the risk. An international study even showed that the kind of symptoms you

27

have depend on this. In rural areas, symptoms like loss of interest in appearance and cleanliness were commonest whereas in cities you were more likely to hear voices and fear persecution – hardly likely to be anything to do with genes. Poor people are about twice as likely as the rich to suffer schizophrenia. As long ago as 1939, American sociologists showed that the nearer you lived to the slums at the centre of Chicago and the poorer you were, the greater your risk. That its rate varies considerably between social classes (twice as common among the poor) and between races tends to be dismissed by psychiatrists, yet it is three times more frequent in American Afro-Caribbeans and up to 16 times more common in children of West Indian immigrants to Britain. This latter is almost certainly nothing to do with genes because relatives of immigrant schizophrenics from the same genetic stock who remain living in the West Indies do not have these elevated rates of illness – emigration and the response of the host nation to this minority cause the increased rate. Non-white people living in white neighbourhoods are more likely to become ill than ones living in non-white areas. Alongside racism, there are huge pressures on Afro-Caribbean parents, leading to extended periods apart from their children. Of 38 Afro-Caribbeans who had become ill in one study, a third had been separated for more than four years from their mothers during childhood, and half from their fathers. Psychiatrists also play down the uncomfortable fact that the illness tends to last much longer and to be more severe in rich, industrialized nations compared with poor, developing ones (as strong a proof that modern life is deranging as you could hope to find). In fact, if you go mad in a developing nation you are fully ten times less likely to have any recurrence of the illness – a huge difference, also nothing to do with genes.

Many psychiatrists also argue that because the illness tends to run in families, it must be heritable. Whereas only 1 per cent of the general population are likely to get the illness in their lifetime, 17 per cent do so if one of their parents had it and this rises to 46 per cent if both parents did so. That this could be partly explained by the fact that being raised by mad parents could be maddening is rarely even considered.

Given the conviction among psychiatrists that the illness of

28

schizophrenia is no different in kind from a bacterial infection or cancer, the main treatment is drugs. About one quarter of patients are not helped at all by them, and around 15 per cent of all schizophrenics eventually commit suicide. Where the drugs do work, the decrease in symptoms is not large – only a 15–25 per cent improvement; and set against this are terrible side-effects, commonly hand tremors and other neurological problems accompanied by a sense of disorientated emptiness. Most treatments reject the idea that what the patient is saying makes any kind of sense. Nurses and family members are very actively encouraged by psychiatrists to regard the delusions as nonsense.

As we have seen, the idea that schizophrenia is always caused by genes is proven to be untrue by the fact that half of people who have an identical twin with the illness do not themselves have it. This is borne out by a particular fact that psychiatrists who believe the illness is largely genetic find hard to explain: at least 20 per cent of schizophrenics completely recover, most of them able to live their lives without any drug treatment at all. A telling example of a recovered schizophrenic is Rufus May. Not only did he become wholly sane but, having done so, he trained as a clinical psychologist and now treats schizophrenics in a community project in a deprived part of London: a poacher turned gamekeeper, a lunatic who has taken over the asylum. If his illness had been as genetically determined as, say, the colour of his eyes, this would have been impossible.

May was eighteen years old when a psychiatrist told him he was a paranoid schizophrenic and would have to take medication for the rest of his life. There followed seven months in a mental hospital and the forcible administration of drugs. Yet today, he has not taken any medication in thirteen years. He feels that the assumption that he had an incurable illness, and the exclusive use of drugs to treat it, actually impeded his recovery. Now mentally healthy, he told his story in a remarkable interview with Feargal Keane for the BBC Radio 4 programme *Taking A Stand*.

This is how he described the build-up to his breakdown:

> Only a few weeks before, I'd just gained a new job. My girlfriend left me a couple of months before, so I was struggling with some sense of abandonment and struggling with just becoming eighteen,

feeling that I suddenly had to achieve something and be somebody. So there were those pressures and I think the job was actually very boring.

I was a trainee draughtsman, given very little to do, and rather than seeing myself at the beginning of a dull career, it was easy to move to playing with ideas that I might be really an apprentice spy.

I turned up at my job one day and they asked me to deliver a parcel at short notice from Kings Cross to Manchester, and so they gave me some money for the ticket. I got down to Kings Cross, and just before the train was leaving I noticed that I had lost my ticket and a man had brushed past me.

I wondered if he'd pick-pocketed me.

With no time to think, the train whistle went and I decided, impulsively, to run round the barrier and jump on to the train. I went into the toilets because I thought I might have been spotted and I thought, 'Well, I need to change my appearance.' I got some water to wet my hair and I put my tee-shirt over my shirt.

This was very exciting for me. It reminded me of when I was a child reading about spy stories, and then I thought, 'Well, what if I really am a spy? What if this is all set up to see if I've got the ability to deliver a parcel under difficult circumstances?' And this idea really appealed to me, until on the way back I came down to earth when a diligent ticket collector knocked on the toilet door where I had lodged myself in to avoid him. He only let me off when I acknowledged that I was just an office junior.

I think that was the time when I was gradually beginning to drift more and more into this fantasy world. It had a lot of appeals to it. It gave me a sense of importance and I could use my creativity to really escape from a dull reality. A depressing reality, because I had to some extent messed up, in academic terms. I drifted for many years really.

The more I invested in these exciting ideas, the more I got from them. I started to enter my own world more and more, and the intense excitement meant that I found it more and more difficult to sleep. I think that sleep deprivation played a key part and in a way I started to dream while I was awake.

If you notice, in your dreams you're always the central figure.

Whatever happens around you is related to you and that's what my life became like. Street signs became personal messages for me. A person scratching their head was a special sign that I had to decode. Newspaper articles had special meanings. Everything revolved around me, just like in a dream.

I think there is no doubt I was very confused and needed someone to make sense of those experiences with me, but crucial in the next few months was that decision [by his doctors] to dismiss everything I was going through as a meaningless product of a carnivorous illness, a disease called schizophrenia, which I think is a very contentious idea . . .

Q: What was the doctors' response to you when you tried to say, 'Look, I need something more than being locked up here. I need something more than drugs'?

A: I think we got into a battle about that. It was seen as me lacking insight into the fact that I had mental illness, and therefore I needed medication. It was very difficult to bring up embarrassing side-effects, like impotence, in a ward round with maybe fifteen people, lots of student doctors. It felt like appearing on stage, a very important five minutes once a week, to go in and show that you were ready to have your medication reduced or ready to be discharged. There seemed to be a battle of wills going on.

Q: How were the rest of the patients treated?

A: Different staff were different. There were staff who treated you with disrespect, who were quite bullying and quite dismissive, and there were staff who treated you as an equal, but they were trained not to talk to you about your unusual ideas. If you mentioned an idea that you had, somebody would get out a game of chess, or something, and suggest you play that.

Q: So there was no therapy?

A: There was no therapy.

Q: How often a week did you see a psychiatrist?

A: Once a week.

Q: For how long?

A: A few minutes.

Q: So that was what your therapy amounted to, a few minutes a week?

A: Therapy was drug therapy . . .

Q: How did you make the transition from the point where you successfully weaned yourself off medication to becoming a psychologist?

A: I did a range of jobs. I even worked in Highgate cemetery as a night security guard. I started doing drama at a community centre, and they started asking me to do work with adults with learning disabilities, and I asked them why and they said, 'Because we think you're good at it.' I was moved by this and thought I could develop that role as a care worker, and as I started to do this I started to develop the idea: what if I managed to infiltrate the mental health system and change things from within? But this took many years. I did ten years of care work and psychological training where I kept quiet about my psychiatric experiences because I wouldn't have been allowed to do that training, probably, I wouldn't have been allowed to do that work if people had the knowledge that I had a diagnosis of schizophrenia . . .

Q: In your process of training you were very concerned to keep the fact of your past illness quiet. Yet you did encounter people who had been nurses while you were a patient. What was their reaction to you?

A: It was strange because, in a sense, I'd become the undercover spy of my delusions and there I was being spotted by someone. One time I was in a meeting and the nurse had recognized me and he was saying, 'Don't I know you from somewhere?' And he was trying to work out how he knew me, and I was kind of grinning broadly with my teeth slightly gritted, saying, 'Yes, but let's not go into that right now.' I met with him afterwards and he was very good actually and said, 'Don't worry, this is confidential between you and me' . . .

If you look at the six or seven years before I actually had a psychotic episode, I was kind of struggling, I was blocked. The actual psychosis allowed me to come out of myself and move on. I very nearly became a long-term mental health patient – I strongly believe that. But nevertheless, through the struggle that I went through, it has given my life a sense of meaning and I want to create better mental health services that are more enabling. I want to change the way we think about human experience.

Having spoken with Rufus May, I would not be at all surprised if the book that he is currently writing about his vital journey does, indeed, change the way we think. In the meantime, the question remains: in the half of cases of schizophrenia that are not primarily the result of genes, what does cause them?

Whilst schizophrenia is not the main focus of this book, it is a *cause célèbre* in the nature–nurture debate. My answer to this specific question is distributed throughout the book. Towards the end of Chapter 2 I present a wide variety of evidence that the role in which we are cast within our family drama can be highly influential. In later chapters, I show how early infantile experiences may create a potential for the illness. Whether this potential is realized may depend on subsequent childcare, in particular whether our parents give mixed messages and are unsupportive, and whether they are abusive. Only in some cases are genes the main reason; in others it may be largely or wholly environment, and in others still it may include elements of both nature and nurture – and schizophrenia is one of the most genetic human characteristics there is.

As this book proceeds, you will see that the real tension in explaining why we are as we are in most respects is between past and present, not nature and nurture. My principal question is this: to what extent is the way we were cared for during our first six years more influential on the kind of adult we have become, compared with the second or third period of six years, or present-day adult experiences such as unemployment or divorce? In particular, how is the kind of care received during different periods within the first six years linked to specific adult outcomes?

After presenting the influence of our role in the family script in Chapter 2, I introduce what I think of as three key psychic characters who are differently assigned lines in the unique drama of our personal history, at different ages.

- Our 'conscience' (Chapter 3), an internal policeman, is formed primarily by the way we were curbed or encouraged when aged three to six years old.
- Our 'pattern of attachment' (Chapter 4), the set of assumptions about how we will be treated by others which we bring to all our

33

relationships, is heavily influenced by the way we were cared for as a toddler, up to the age of three.

- Finally, our 'sense of self' (Chapter 5), the fundamental sense that we exist, and exist in our body, is caused, at least in part, by our care in early infancy.

Each chapter delves ever further back into the murky recesses of childhood. You will be able to remember and apply to yourself much of what I cover in Chapter 2, but probably not with Chapters 3–5 because none of us can remember much, if anything, of our earliest years and we may have only fleeting awareness of the way we were treated then. None the less I hope that, in joining me on this journey into your past, you will begin to see that, for the great majority of us and in most respects, it is our unique relationships within the family, not our cocktail of genes, that determine the gradual emergence of volition, of the capacity for choice, as we struggle during childhood to convert what has been done to us into something that is our own. The battle continues in adulthood. For the whole of our lives we are grappling to become the organ grinder rather than the monkey, to explore and bring alive the past in our present so that it works for us rather than against us, to shift from being actors in a play whose script was written by our early childhood history to becoming the author of our own authentic experience.

Chapter 2

SCRIPTING OUR
FAMILY DRAMA

They may not mean to, but they do . . .

Families are similar to a theatrical drama. Like fictional characters, we are each assigned a scripted role, tightly directed in its performance, clothed in psychological costumes and required to sing and dance to our family tune. This is proven whenever there is a family gathering – at Christmas, for example.

From the moment we gather on Christmas Eve or the day itself, our parents and siblings demand that we enact our appointed role. Never mind that we may have long since ceased to be the clever one or the fatty, the attention-seeker or the moaner, our family treats us just as they always did and within minutes of walking through the door we are back in the nursery. The achievements and independence of adulthood are swept away and we find ourselves performing a role that we thought was long obsolete.

A simple method for testing this idea is to become the author of our own lines in next Christmas's drama by deliberately deviating from the script. As you will soon see, the other characters will be taken aback, proving the script's existence. Like all good writers, we need to have done our research. To establish what our new lines should be, we need to establish what will be expected of us by the other actors who, in the case of this drama, were also the writers of our original role.

Begin by listing all the characteristics you believe your family routinely attributes to you. Some of them will be projected on to you by all the family members. Not only, perhaps, do both your parents treat you as terminally lazy or untidy or selfish, so do your siblings. But other attributions will be specific to particular family members or to coalitions of them. Perhaps your older brother and your mother are convinced that you are a fussy eater, but your father and other siblings don't share that view. By the time you gather for Christmas you should have a clear picture of the part that each will be expecting you to play. Now comes the fun. Instead of acting in the predictable manner, you can do the opposite.

If you are known for being stingy when buying presents, give ostentatiously expensive ones. If you are notorious for never doing any washing up, be the first at the dishwasher or the sink after every meal. If you are known as a late riser, be there to greet the family at breakfast. Part of the fascination of this game will be their denial of any change. Just as actors in a real drama would be amazed, indeed scared, if one of their fellow players suddenly started coming out with lines from a different play, initially pretending that it was not happening, so it will be with your family. When the difference becomes undeniable they may gang up and try to impose your old role upon you, probably by mocking your attempts to wash up or by making a joke of the fact that you were out of bed before them. They may keep referring to past occasions, like the time you overslept and did not make it in time for Christmas lunch. But instead of being infuriated, gently remind them of the new evidence to the contrary. If nothing else, it could be quite amusing. And who knows, next Christmas you could be starring in a role which is, this time, one of your own creation.

Scripting the difference between you and your siblings

As R.D. Laing put it, when it comes to families, 'We are acting parts in a play that we have never read and never seen, whose plot we don't know, whose existence we can glimpse, but whose beginnings and end are beyond our present imagination and conception.' Your particular family script had a major influence on why you are so different from your siblings. The fact that you are so different, despite having been raised by the same parents, is frequently the opening salvo from supporters of the 'genes' theory during nature–nurture debates in the pub or over the supper table. Parents with small children, especially after the birth of their second, are liable to explain the uniqueness of each by genes (however passionately they may have argued the opposite before they had children). 'They *are* born different,' asserts the harassed parent. 'They have distinct characters from the beginning. How else could they be so different, having had the same parents and been raised in the same family?' The answer, of course, is that they were treated very differently. If you are extrovert or neurotic or stupid, your siblings are almost as unlikely to share these characteristics as children to whom you are genetically unrelated, and, upon further thought, this actually makes a lot of sense. If you compare the treatment you received at the hands of both your parents to that received by your siblings, you will soon realize that it was very different. I can offer an example of just how radically different this treatment can be from my own family history.

During the early 1970s it was not uncommon for young people to consume hallucinogenic drugs, and in some cases to have panic attacks and paranoid episodes as a result. There was an occasion when one of my sisters smoked some exceptionally potent marijuana and became briefly convinced that she was the victim of a plot that everyone who spoke to her was in on. When my parents were made aware of her disturbed state, their reactions were as different as can be. My father was furious, insensitively scolding my sister for having been so foolish as to have taken drugs. My mother was a model of understanding and sympathy.

A few years after this incident I took some LSD, a potent hallucinogen, and became filled with doom and panic. I instructed a friend to drive me to the nearest mental hospital, but he took me home. In the sitting room with my parents I explained that I was thinking of plunging out of the window to a messy end. Their reactions were the polar opposite of those they had demonstrated to my sister's plight. This time the condemnation came from my mother: 'You bloody fool! How could you be so stupid as to take drugs?' My father was as concerned and sympathetic with me as he had been angry and unsympathetic with my sister. For both of them there was one rule for my sister on drugs, quite another for me. This is typical of the dramatically different relationships parents

37

have with each of their different offspring. There are a whole variety of reasons why.

Parents are at different stages in their lives when their different children are born, and very often so is the state of their marriage. At the start of parenthood, having a firstborn has a radical effect on most women, usually entailing the transition from worker to full- or part-time mother. Meanwhile, the father may feel excluded by the new arrival or may be delighted; he may feel under tremendous pressure to be a breadwinner or that his destiny has been fulfilled. When baby number two comes along, the family has already been altered by baby number one. Perhaps Mum and Dad are not getting along so well in the face of the new pressures, or perhaps it has brought them closer together. Perhaps they are on the up financially, or perhaps money has become a huge bone of contention. Perhaps they have moved to a new area, upsetting one or both of them. In any of thousands of ways, the state of emotional contentment and preoccupations of each parent are likely to have changed considerably from those present at the birth of their first child.

On top of all this, each parent dumps on to each offspring a quantity of baggage from their own childhood history. Parents always re-enact how they were cared for as children, to some degree, and they also often go to great lengths in other respects to do the opposite of what was done to them. If firstborn themselves, perhaps the arrival of a second or third child triggers memories of being displaced when they were small, making them extra-sympathetic to their own firstborn. If lastborn, perhaps they are concerned to ensure that the new child doesn't get the same 'raw deal' they felt they had. If it is a boy and the father had a bullying brother, if it is a girl and the mother had a sister who was prettier than her . . . the variations on birth order and relations with siblings are innumerable and the effect on how the parent relates to the newborn profound.

The child's gender also combines with its place in the family in creating the script for that child as parents rehash their own childhood. Perhaps the mother was hoping for a boy and got a girl; perhaps the father was. The mother may have had an intrusive mother herself and consequently be determined to give her daughter space. Perhaps the father feels that his parents were not strict enough and is anxious not to make the same mistake with his son. More profoundly still, perhaps a parent was treated cruelly or neglectfully and re-enacts the maltreatment with a particular child by inflicting the same mental cruelty, violence or sexual abuse.

The changes in the emotional state of each parent, and in the marriage between one child and the next, combine with the biographical bric-a-brac that each parent loads separately on to each child, creating a unique psychological environment every time. Whilst a few family-wide traits, such as a concern for punctuality or a desire for good exam results, will be demanded by both parents from all their children, the remarkable fact is that most of each individual child's nurture is particular to that child and she develops her own niche in the family as a result of receiving radically different care.

Scripting our gender

From the moment that your mother learnt that she was pregnant you became a character in your family drama: your parents wondered if you were a boy or a girl. Gender is a common hook on which parents hang their biographical baggage and social stereotypes. In general, boy infants get more reactions from parents and are given toys that require active problem-solving and parental involvement. Later on, boys are encouraged to explore and play alone whereas girls are kept closer through fear for their safety, making them more used to having to conform to adult expectations. Fathers play more roughly with sons and expect higher standards of achievement, whereas they encourage dependence and passivity in daughters. Both parents interrupt daughters and listen to sons more. Subsequently, at school, the sexes tend to segregate themselves along gender lines so that opposite-sexed ways of behaving are less likely to be learnt. Teachers expect boys to be more aggressive and girls to be cooperative.

Of course, these different reactions to children are partly caused by the fact that the sexes are probably born different. Doubtless to some degree men really are from Mars and women from Venus as a result of producing different amounts of the sex hormones, testosterone and oestrogen. Before they have children parents may assert that gender differences are all a matter of sexism, but afterwards most change their tune. Little boys and girls are, indeed, different from birth, triggering different reactions to them. But there is no doubt that whether adults think a child is male or female hugely affects how they relate to it as well, regardless of what it is actually like. For example, when adults are told a girl baby is a boy they are liable to interpret its cries as angry. If the same baby is presented to other adults who have been told it is a girl, they interpret the cries as expressing fear. Numerous other clever studies like this show that we relate to children according to a set of sex stereotypes, regardless of their actual behaviour. Hence, even today, there is still a strong tendency for both parents to believe sons are better at maths and sport, regardless of the evidence, and likewise to believe daughters are better at English. Since parents' beliefs about their children's capacities have a considerable effect on what they actually achieve in later exam results, these stereotypes continue to influence performance in the real world. Again, I can offer an example from my own experience.

I was the third of four children, the others all being girls. Neither of my parents ever denied that they were delighted to have a son after having had two daughters, and none of my sisters had been backward in pointing to the preferential treatment I enjoyed in some respects purely as a result of my gender (they are rather less vocal on the subject of the many additional problems I encountered because of this). The difference is most conspicuous to me in the matter of my father's response to my atrocious academic performance throughout most of my childhood.

Having had permissive early care, very much encouraged to do my own thing, I was not pleased to be sent to school at the age of four. It seemed preposterous that I should have to obey rules. Frankly, I was delinquent, fighting a lot with other boys and paying little attention to my school work. Soon I persuaded my long-suffering mother to withdraw me from school and let me wait a year before starting again. When that time came I was sent to liberal schools, so that my reluctance to buckle down to work and to behave did not create too many problems for me – although it did to those around me, and my reports were a catalogue of despair. Then, when I was eight, came a big change. My father insisted that I move to an extremely strict, conventional preparatory school (although my parents were not stinking rich, they had enough money for private education). As a result of his own childhood experiences it was important to him that his son should pass the common entrance examination to gain entry to public school, and to achieve this aim it was obvious that I would need to attend a more academically taxing institution. At this prison (as I saw it), for two long years I and two equally recalcitrant pupils shared the bottom three places in every subject. Beatings from the headmaster became a regular occurrence, sometimes for misbehaviour, sometimes for laziness. Despite the accumulating evidence to the contrary, my father never ceased to tell me that I was 'very clever' and that I could do 'brilliantly' if I chose to. He commissioned an IQ test which proved that I was above average and made heroic efforts to teach me Latin, geometry and so forth during the holidays – alas, to no effect. When I was ten the headmaster gave up on me. He told my startled parents that, since I was mentally subnormal, they should look for a place in a 'special school'.

At my next, more easy-going, preparatory school I continued as before and duly failed common entrance. But my father refused to accept what seemed obvious to everyone except him and me, namely, that I was thick. He broke off a family skiing holiday to deliver me to a crammer which would prepare me to retake the exam. Having left it to the last moment, I repaid his faith with due diligence and after fourteen weeks of unbroken sweat and tears (only one beating, but at this place the main punishment was to be sent on a run; I covered several hundred miles) managed to pass. But the boom-and-bust pattern did not end there. I failed the internal exam at the end of my first term at public school and had to stay down a year. I mooched and misbehaved my way up to 'O' levels in my mid-teens, ignoring the steady stream of letters from my father attempting to kickstart me all over again. After I had done predictably badly, scraping eight low-grade passes, my father came to the school and sat me down with a glass of Pimms on the terrace of a hotel on the banks of the Thames. In this pleasant setting he presented me with three rather less pleasing, if memorable, alternatives. I could stay at school, continuing not to work at my books and scrape some bad 'A' levels, but when I left would have to take a job in the City, like becoming a trainee stockbroker. In our family nothing was more despised than 'a job in the City', so I gave that idea short shrift. The second option was to leave school and take a menial job mending railway tracks. Although I was quite attracted by this, fancying myself to be the next George Orwell (I did read novels; it was school books I objected to), I was finally won over by his third choice: go to Cambridge.

When my 'decision' to go to Cambridge was discussed with my mother she nearly lost her temper. All the evidence was that I would have my work cut out to get into any university at all. Living, as she did, in the real world rather than sharing my father's and my fantasy that anything was possible, she used the adjectives 'absurd' and 'ridiculous' to describe the plan. But the odd thing is that after that conversation with my father I put my nose to the grindstone and managed, after a great many more ups and downs, to do quite well.

What is interesting about this tale of infinite forbearance is that, in penning the soap opera of my life, one of the scriptwriters, my father, had decided that the character of his son would go to Cambridge and was prepared to go to tremendous lengths to make this fiction come true. As my three sisters frequently point out, he had no such aspirations for any of them. He was perfectly content for them to go to unacademic schools, even though they showed far more early promise than I (indeed, despite his attitude to them, all eventually achieved degrees at respected universities and have had successful careers, encouraged by my mother who had enjoyed a distinguished one herself). But, purely because of what was between my legs at birth, my father related completely differently to me in regard to academic performance. Had my younger sister been born before me I have no doubt that she would not have been the object of his aspirations. Only the arrival of a boy child would have triggered them.

Our gender also heavily influences the script our parents write for our sexuality. As I shall explain in greater detail in Chapter 3, families are awash with sexual desires and the way your parents responded to yours is important. Severely inhibiting reactions lead to inhibition, whilst appropriate enthusiasm means you grow up able to enjoy this instinct, as in the case of 'Daddy's Girl' Gillian

Blakeney and her father. But some fathers find their daughters' sexuality disturbing and suppress it, just as some mothers reject their sons' masculinity. Two extreme examples are found in the stories of the rock star Elton John and the actor John Gielgud.

According to Philip Norman, Elton John's biographer, his mother 'had really been hoping for a girl'. She would not let him play with other boys, keeping him tied to her apron strings. He describes himself as having been like a girl: 'I looked like Shirley Temple, with a mass of bubbly hair.' His adult enjoyment of stereotypically female pursuits, such as shopping and colourful clothing, are a matter of public record, as is his homosexuality. In the case of Gielgud, also a homosexual, both his parents were longing for a girl when he was born, having had two sons already. He immediately became the apple of his mother Katie's eye. Whereas his siblings were raised by nannies, Katie gave him her undivided attention for his first three years. Her relationship with him was so close that it was inevitable that his siblings would gang up on him. In his autobiography, Gielgud wrote that, 'Mother was inclined to spoil me, particularly as I was supposed to be delicate and "artistic".' This was part of her scripting for the future actor. He feigned illnesses to obtain special treatment from her and used her favouritism to ward off demands on him. He recalled that 'I would say "Mother wouldn't like me to do that. I might get sunstroke"' and that on his first day at kindergarten 'I sat down in the middle of the room and burst into tears.' He loved dressing up and hated manly activities, writing that 'I was very fond of an audience and I was dreadfully bad at cricket and football.' His mother's script for him was greatly influenced by her own history and he strongly identified with the women from his mother's family, a famous dynasty of theatrical women. His grandmother was a nationally celebrated actress, his aunt was the legendary Ellen Terry and two other aunts were distinguished actresses. He wrote that 'I have never been able to follow the various ramifications of my father's family and my mother's theatrical relations always had the stronger hold upon my interest.'

Of course, these are extreme cases but they illustrate the influence of parental expectations and attributions in shaping your sexual identity.

Scripting our place in the family

At least as big a determinant as gender on the role in which your parents cast you in the family drama is your place in the family, known as birth order. Although there are many exceptions to the following rules, firstborn children are more likely to be self-assured, assertive, competitive and dominant compared with lastborns. Firstborns are exceptionally common among leaders, such as prime ministers and presidents. They are more liable to be conscientious, to endorse conventional morality, to identify with their parents' values and to be obedient to parental authority. They achieve more in their careers, having done better at school, with parents pitching dinner-table conversation at their level rather than that of younger siblings. They are more anxious about their status, more emotionally intense, less quick to recover from upsets. They are more vengeful and prone to anger. They do not usually enjoy risk-taking in general and dangerous sports in particular, preferring swimming, tennis, golf and other non-contact sports. They tend not to favour backpacking and world travel.

Lastborns are generally very different in all these respects. Compared with firstborns, they are less self-confident. They are more altruistic, emotionally empathic and, when small, get more involved with other children. They are less conscientious and less prone to anger and revenge, more sociable and easy to be with. They are far more open to experience, being unconventional, adventurous and rebellious. They embrace risk-taking and contact sports, and are more likely than firstborns to indulge in such pursuits as rugby, football, boxing and parachute jumping.

A major reason for these differences is that, depending on our birth order, we adopt different strategies to get our parents' attention. Given that there can never be enough love to go round, we have to carve out a niche for ourselves to attract it, cultivating different skills and personality attributes from our siblings to mark ourselves out. That is why we tend to be more different from siblings born immediately before or after us than we are from next-but-one siblings: in order to seem distinctive to our parents, we go to greater lengths to establish how different we are from immediately adjacent brothers or sisters. Likewise, if one of us is strongly identified with one parent, the next one along is liable to identify with the other parent. Sucking up to one parent and having them on your side is better than having no one, even if you cannot have both.

If we are a firstborn, we tend to take the line of least resistance in establishing our niche by providing parents with what they want and siding with them. When the next child comes along we, the firstborn, are quick to assert our prior claims and do our best to dominate. Since we are bigger and stronger, this makes us tend to be more self-assured. However, depending on how successful our younger siblings are in grabbing attention, we may be angry and vengeful at being supplanted. Behind the veneer of confidence there is liable to be considerable fear. As the oldest we are likely to be roped into child-rearing tasks and to be required to act as the responsible one in the family. This makes us less keen on risk, more conservative, anxious to conserve the status quo which preceded the birth of our siblings. Being a conscientious pupil at school is a simple way to curry favour with parents if this is important to them.

By the time the lastborn comes along, many of the obvious niches for attracting parental love and investment are liable to have been taken. Older siblings have already grabbed the role of comedian or sporty one or sexy one (the appeal of the Spice Girls' individual nicknames – Sporty, Baby, Posh – must make instant sense to children engaged in the battle for family niches). Given limited options, as a lastborn we may find our best strategy is to innovate and reject the status quo. Travel may appeal as a way of escaping from a family which has less to offer us, whilst also providing useful ideas on how to be different. At the same time, we may find that being altruistic and supportive is a way to pre-empt hostility from older, stronger siblings. This makes us a good team player.

In seeking niches we are also governed by how our parents react to our birth order. Firstborns are more likely to attract high parental aspirations, especially in relation to academic performance, and they are also very liable to be vehicles for parental emotional baggage. As pioneers they are the first to go to school and to encounter the challenges of adolescence, and at least one parent will usually have a 'special' relationship with them. They get more attention, more child-centred care and greater intensity of engagement in infancy, but by the age of three, if a second child has come along, they are treated more coldly and with greater restrictiveness than their siblings will be at the same age.

As the family grows in size, the firstborn continues to express the core attitudes and convictions of the parents, but gets less and less positive attention. Fathers of firstborns talk to and touch them in their infancy more than they do with any subsequent children, but later on the father is also more likely to discipline them and interfere with their activities, demanding higher performance levels. Both parents attempt to accelerate the firstborn's achievement of developmental milestones, such as walking and reading, more than they do with subsequent offspring. In simple terms, the good news for firstborns is that their parents tend to be thrilled by their arrival and lavish undivided attention upon them. The bad news is that when a sibling comes along the relative neglect that results is more of a shock to the system, and the special demands to succeed are not served with extra helpings of love; indeed, if anything these tend to go to laterborns. Most of what is true of firstborns is also true of only children, with the difference that they do not have the blow of losing their parents to a sibling. Onlies tend, therefore, to be even more high-achieving than firstborns because they never cease to be the conduit for all parental ambitions.

When applied to famous people, the evidence of the effects of birth order is illuminating. Whereas professional revolutionaries such as Leon Trotsky, Karl Marx and Fidel Castro are liable to be laterborns, establishment politicians tend to be firstborn. Scientists who are laterborn, like Charles Darwin, are more likely to develop radical theories than are firstborns. The later in birth order a scientist is born, the greater the diversity of their interests, perhaps because being multi-talented increases the chances of striking a parental chord. Scientists prone to acrimonious disputes about who has been first to invent a theory which they believe themselves to have originated are three times more likely to be firstborns, perhaps still anxious about siblings stealing their toys.

The size of the gap between us and our nearest sibling affects the degree of impact imposed by birth order. Children aged four are far less affected by the birth of a sibling than if it had happened when they were younger because they are better equipped to cope. Interestingly, in Sulloway's study of famous historical figures, the greater the gap between siblings, the less the likelihood that a younger sibling would be politically revolutionary or scientifically radical – a large gap created less need for the laterborn to reject the status quo in order to get parental attention. Perhaps the most startling evidence for the idea that a small gap seriously upsets an older child comes from studies showing that it affects which of us suffer from schizophrenia. (This, as explained earlier, is the most extreme mental illness of all and, like depression, is usually presented as largely genetic in cause, an assertion which I shall be frequently challenging throughout the book.) The studies show that adults who had a sibling less than two years younger are significantly more at risk than those with larger age gaps. This cannot be explained by genes. Nor can the fact that another study found that children who had been exceptionally gifted before developing schizophrenia were significantly more likely to be firstborn than normal gifted children. Whilst it is obvious that having a sibling who is close in age, or that being a firstborn, are not, on their own, causes of schizophrenia, they are an influence if the child is already vulnerable.

The eventual size of the family is also important in creating the family script. Large families of five or more offspring tend to stretch parental resources, both material and emotional, making the children more likely to be emotionally deprived. As a result, offspring from such families are at greater risk of most kinds of problem. They must work harder to identify a niche. Laterborn children from large families are especially likely to have a wide diversity of interests and to be attracted by travel. Girls in large families are more stereotypically feminine because they are more likely to be roped into childcare by their hard-pressed mother.

On the whole, firstborn traits such as leadership, assertiveness and competitiveness are associated with masculinity, whereas laterborn ones such as displaying affection, cooperation and flexibility are more stereotypically feminine. Where girls are firstborn they are prone to be more masculine, especially so if their siblings are all girls because they identify more strongly with their fathers. When boys are laterborn, they tend to be more feminine. One study actually found that homosexual men are significantly more likely to have

47

an older brother than heterosexuals; perhaps they were influenced towards more feminine traits by the fact that the conventionally masculine niche had already been occupied. Obviously, having an older brother does not in itself cause homosexuality, but as will be seen in Chapter 3, the way our parents related to us when we were children has far more influence on our sexuality than genes.

An example of how birth order is influential in general, and of how the firstborn child can be the vehicle for parental expectations and emotional problems in particular, is Prince Charles. Royal biographers have tended to suggest that he was born with an innate shyness, inhibition and lack of confidence. It is, however, far more likely that these traits derived from the treatment received at the hands of his father, Prince Philip – although, as will be explained in Chapter 3, Charles's early care from his mother and nanny were also influential. The journalist Jonathan Dimbleby, who was given full access to Charles and his personal papers in writing his biography, portrays the problem as a clash of innate personalities, with Charles as the shy, sensitive one and Philip as domineering and insensitive, if well-meaning. Whilst there is no doubt that both were like this, there is good reason to doubt that the causes were genetic.

Tommy Lascelles, a leading courtier, described Philip as 'rough, ill-mannered, uneducated and probably not faithful'. Effectively orphaned at the age of ten, Philip had a childhood of severe emotional deprivation, with separated parents, and he used his firstborn son as the prime carrier for this emotional legacy. There are innumerable well-documented examples of Philip's belittling, sometimes even cruel and hostile, attitude to his son from a young age, and it is highly probable that this contributed to a melancholic tinge to his psychology. Charles told Dimbleby that his father was 'unable or unwilling to proffer affection and appreciation', and the Queen bowed to all Philip's demands in regard to Charles's care and education. He regarded his son as over-sensitive and in need of toughening up before he could face the harsh realities of life, so he sent him to Cheam, the strict preparatory boarding school that he had attended, and subsequently to Gordonstoun, his harsh public school. Whilst all the princes were sent to these schools, it was particularly inappropriate to send Charles, who was known to be a fragile, shy boy. Philip explained his reasoning as follows: 'Children may be indulged at home but school is expected to be a spartan and disciplined experience.' Quite how he perceived his own fierceness as indulgence, when his son was at home, is hard to imagine. Not satisfied with this, Philip had even gone to considerable lengths to sack the most loving figure in Charles's early life, his nanny Helen Lightbody, for being 'too indulgent'.

At Cheam, Charles suffered terribly from homesickness. He later recalled that 'it was not easy to make a large number of friends . . . I'm not a gregarious person so I've always had a horror of gangs . . . I've always preferred my own company or just a one to one'. According to Dimbleby, he tried to be friendly 'but an emotional and impetuous nature was severely constrained by an outward reserve and formality – behind which he also concealed his painful insecurity'. If Cheam was 'loathed . . . a misery to him', Dimbleby's words, Gordonstoun was torture. The Queen would have preferred him to go to Eton because it was near their home at Windsor, but this proximity was precisely Philip's reason for preferring Gordonstoun, in the wilds of Scotland. According to Philip, a boy with such a 'shy and reticent disposition' needed 'something that would draw him out and develop a little more self-assertiveness in him'. Only someone closed to the emotional needs of others could have imagined that the vicious environment then found at Gordonstoun would have this effect.

For vicious it was. Accounts of the school at that time, such as that by his contemporary William Boyd, the novelist, confirm that mental cruelty was the norm. A gang of thugs roamed Boyd's house, beating up small boys and extorting food and money. Charles's housemaster was an intimidating disciplinarian with a capricious temper. Dimbleby writes that the other boys picked on him 'maliciously, cruelly and without respite'. They developed a culture in which talking to the heir to the throne led to instant social excommunication. Socially unskilled to start with, Charles was up against a deliberate campaign to isolate him. This was carried over on to the rugger pitch, where it was a matter of honour for thugs to crush him. William Boyd recalled hearing boys boast to each other, 'We just punched the future King of England.' Writing home, Charles stated that 'I don't like it much here. I simply dread going to bed as I get hit all night long . . . I can't stand being hit on the head by a pillow now.' His love of politeness and feeling for the spiritual side of life were deeply offended by the crude language and lack of sensibility of his fellow pupils. He sought consolation in pottery and music. It might be supposed that Philip, as a previous pupil at the school, would have realized that this would have happened to a future king – especially one with Charles's personality.

Charles's unique position as firstborn heir to the throne also wrote his part in the wider family script, that of a focus for mockery. Both Philip and his uncle, Lord Mountbatten, made Charles the constant object of banter and ridicule, while they doted on his immediate sibling, Princess Anne. A fanatical horse-lover, she rapidly acquired the language by which equestrian enthusiasts communicate and spent hours debating equine arcania with her parents. Time and again, Charles would be revealed as ignorant of the meaning of some highly technical term. At mealtimes he would be tested and found wanting, to general amusement and his humiliation.

It may be that Philip envied his son his role as heir to the throne, since he himself was sidelined as consort to the Queen. This would have made the humiliation satisfying to the father, but Charles's siblings joined in too. Where one's brother is given huge extra importance the desire to bring him down can be powerful – especially so if encouraged by one's father. Whilst Charles's brothers, too, were sent to Gordonstoun, he was the one upon whom Philip focused his emotional insecurities most strongly. Had Prince Andrew or Prince Edward been born first, they would have suffered the same fate and almost certainly have been very like Charles. Just as Anne managed to monopolize the family niche of horse-lover, Andrew adopted his father's Jack-the-lad attitudes to women and Edward, typical of the lastborn, was the most radical, deviating from the royal status quo with his career in theatre and television. Who knows, had their birth orders been reversed the thespian might have been Charles.

Of course, a special status is not conferred only because a child is firstborn. Physical beauty is another common peg for parental plotlines, especially in developed nations which are obsessed by looks. The beautiful reap a rich dividend from this. It should come as no surprise that newborns who have been independently rated as cute are held more by their mothers, who stare more into their eyes and speak more baby-talk to them. Less attractive newborns get less of this kind of interaction and greater attention paid to physical needs, like burping and mouth wiping, and their mothers are more easily distracted from attending to them. When mothers of twins are observed, they give less attention to one of them if he or she is of low birth weight and looks slightly sickly. (Interestingly, both parents are more likely to claim that a newborn resembles the father than the mother, a fact that has been attributed to anxieties about paternity; this is no small matter, it would seem, because at least 10 per cent of children are not genetically related to the person they believe to be their biological father. This rises to one third in some low-income communities.) Preferential treatment continues into later childhood. For example, adults are more likely to give attractive seven-year-olds the benefit of the doubt after they have been naughty, saying that they doubt the child has done the wicked deed. They are less likely to believe the child will repeat the wickedness if it is attractive.

In adulthood, when shown pictures of models or average-looking women men are more likely to say they would help the beauty, whether it be in terms of lending money or something more extreme, like donating a kidney or jumping on a terrorist hand grenade. In the context of practically any positive quality you can think of, we tend to assume that good-looking people do it more, do it better and enjoy it more. Despite the Dumb Blonde stereotype, we expect attractive women as well as men to be intelligent. We assume they have more, better and more varied sex. If badly behaved, they are more likely to get away with it, whether it be shoplifting or cheating at exams – although, interestingly, the one crime they are more liable to be convicted of is fraud, because we assume that good-lookers are smooth-talking charmers. We are so in awe of them that, in experiments, when asked to stand next to strangers we will keep an average of 23 inches distance from attractive people compared to just 10 inches for the less attractive. We cut them more slack, and they are more assertive as a result. Asked by a researcher to wait in a room, attractive people only last three minutes before complaining whereas the less attractive wait nine.

Because we accord them such special status, the beautiful do in reality have a better time in most respects. As children we want them to be our friend, and as adults they are more popular with the opposite sex, have more dates and do, indeed, report more sex of a more varied kind, starting younger. Good-looking men bring their women to orgasm more, doing so simultaneously more often. The only fly in the social ointment is that good-lookers, especially women, tend to be less popular with their own sex.

Scripts for exceptional achievers

Of course, in most cases both parents don't favour the same child equally. Whilst one parent may zero in on a child because he or she is beautiful, the other may have a special bond with a less attractive one. Whatever the reason, being deemed special is a major cause of exceptional achievement.

The pop star Michael Jackson was raised in a cruel regime of beatings, emotional torture and tyranny, as described in his sister La Toyah's autobiography. He was humiliated constantly by his father, who never missed an opportunity to tell his son, 'You're nuthin'.' What sets Jackson's abusive family apart is that his father used his reign of terror to train his children as musicians and dancers. The Jackson Five spent all the hours not at school being literally whipped into a top-class pop act. They were not allowed to socialize with other children, forced to return home as soon as school was over in order to practise their act until bedtime, and beaten with a belt if they showed any resistance. Michael was picked out by his father for special attention, required to achieve the highest standards and to practise the most. This is probably the reason he was also the most talented.

Although it is often assumed that musical ability is inherited, there is abundant evidence that this is not the case. It seems that all of us are born with perfect pitch, the capacity to match notes perfectly. Highly musical children were sung to more as infants (and in the womb), and were encouraged more to join in singing games as toddlers, than less musical ones; this, of course, was long before any musical ability could have become evident, which suggests that nurture is critical from the start. Studies of classical musicians prove that the best ones, the soloists, practised considerably more from childhood onwards than ordinary orchestral players; this is because

52

their parents urged them to put in the hours from a very young age. The best violinists at one music school were found to have averaged twice as many hours' practice by the age of twenty-one compared with the less good ones, and the same was true of children selected for entry to a specialist music school compared with those who were not accepted. All the children who gained entry to one school had parents who had very actively supervised music lessons and daily practice from young ages, giving up substantial periods of their leisure time to transport the children to and from lessons, attend concerts and so on. Early in life these children had been nominated by their parents as 'musical' and had internalized this label. Jackson's story, albeit unusually brutal, bears this out.

On top of his extra ability, Jackson also had more drive. This may have been the result of being the closest of his siblings to his mother. Just as his father picked him out for particular violence, his mother regarded him as unique. 'He seemed different to me from the other children, special,' said his mother of Michael. She may not have realized that treating him as special may have been part of the reason he became like that. Such special status is almost always found in the backgrounds of exceptional achievers, but it can be intimates other than the parents who confer it. The former soccer player Gary Lineker and the Conservative politician Kenneth Clarke are both examples of high achievers who occupied a special status in the eyes of their grandfathers.

Lineker's grandfather was a skilful footballer who had to forgo any professional ambitions in order to mind the family fruit stall in Leicester. Lineker recalls, 'My grandad would always come to watch me play from the age of about eight', and he was there when Lineker got his first big break. 'I was spotted at thirteen by a scout from Leicester City. The funny thing was that he knew my grandad, who asked him what he was doing at the match. Grandad was very pleased when he was told it was me who was being looked at.'

Whereas Lineker came from a stable home, Kenneth Clarke's mother Doris was an alcoholic who died from cirrhosis of the liver. His brother Michael recalls their childhood: 'My mother would go into enormous highs and lows. She would spend two days drinking alone in her bedroom.' Both sons may have compensated for their difficult childhoods by becoming successful, but they chose different domains in which to do so. Whereas Kenneth says, 'Money is not a high priority', for millionaire car salesman Michael it is so crucial that, he recalls, 'I had a Rolls by the age of twenty-three. Ken couldn't understand why anybody would want anything like that.' The difference was that Kenneth had a special status in the eyes of his grandfather, a highly politicized man who would have loved a career in politics. Kenneth admired him and sought to fulfil his dream for him.

These are exceptions, however, because unfulfilled parental ambitions, rather than those of grandparents, are the most common in childhoods of exceptional achievers. In general, we are more likely to pursue a similar profession to one of our parents and, if we are prodigiously good at a subject from a young age, it is almost always because this was one in which our parents were either accomplished or wished that they had been. These parents 'hothouse' their children, often from birth, creating a regime of accelerated learning. A typical example is John Adams, who passed 'O' levels at age eight (when the norm is sixteen) and 'A' levels a year later. His father Ken published a book, modestly entitled *Your Child Can Be a Genius*, giving a detailed account of the parenting by which this was achieved. There are numerous famous mathematical prodigies whose parents even went so far as to move with their pre-pubescent child to a university town so that their offspring's studies could be pursued at a higher level. Although outstanding early ability tends to be presented in the media as a genetic freak, this is hardly ever the case; the exception is a handful of isolated skills, like being able to calculate (there are children who for no apparent external reason to do with nurture are able to multiply and divide improbably difficult numbers without blinking). There are virtually no authenticated cases of prodigies who came from families in which they were not hothoused or otherwise helped. In the early years the parents go to tremendous lengths to nurture a particular talent, and subsequently no expense is spared to obtain the best possible teaching. Nearly all prodigious modern sportsmen and women, like the tennis players Venus and Serena Williams, have been fanatically coached from a young age, usually with their parents watching from the sidelines. In the case of the Williams sisters, for example, their father declared his intention of creating world-beaters from the moment of their birth.

But having exceptional early gifts written into our family script is not all accolades and subsequent eminence. Childhood prodigy is not necessarily the precursor of adult genius – in the vast majority of cases it is not. Nor is being able to score high marks in intelligence tests a guarantee of lifetime achievement. A famous study of 400 American children with IQs above 140 (the average is 100) found that they achieved no more than what would be expected from a person of their social class. None of them became geniuses. If anything, the capacity to pass exams or do well at IQ tests may be more a measure of our desire to please our parents and teachers than of originality.

Overall, our intelligence and academic performance have a great deal to do with our social background. If Afro-Caribbeans and children from poor homes score an average of 10 points lower than whites and affluent children, this is almost certainly more due to upbringing than to genetics. This is proven by the fact that children from poor homes who are adopted by middle-class parents score an average 12 points more than their biological parents. Furthermore, as nations become more educated their inhabitants' IQ scores rise, strongly suggesting that the test assesses the kind of training received, not innate ability.

The most interesting evidence of what causes outstanding achievement comes from the study of children whose family script was affected by the loss of a parent whilst they were young. It suggests that mental ability is hugely influenced by emotional motivation. One third of the six hundred people with more than one column devoted to them in the *British and American Encyclopaedia* suffered early parental loss. I refer mainly to males in what follows as nearly all the evidence refers to that sex. Before the advent of modern medicine this crucial advantage was given to 35 per cent of British prime ministers and 34 per cent of US presidents – double the average of 17 per cent in the general population who lost parents in childhood. In contemporary surveys of British entrepreneurs, 30 per cent had lost a parent before the age of fifteen (compared with an average 8 per cent in the general population today); Robert Maxwell lost both. In the world of the arts, Byron, Keats, Wordsworth, the Brontë sisters and between 40 and 55 per cent of other eminent British writers (depending on definitions) were bereaved in childhood. The great French writers who lost a parent early include Rousseau, Baudelaire, Zola and Molière. In science think of orphaned Isaac Newton or Charles Darwin, in popular music Lennon and McCartney. Striking modern examples of early bereaved high-achieving females include Madonna and the entrepreneur Anita Roddick. Whatever the field, where a study has been done it shows far higher than normal rates amongst outstanding achievers of all kinds. You may be able to hothouse a prodigy, but genius comes from early adversity.

Some of the emotions that early loss provokes are universal. Where boys are concerned, the bereaved child is almost always made insecure by the disruption of the family home and the temporary loss of love and attention while the remaining parent grieves. If he has lost his father, he feels inadequate and is always in danger of being assailed by deep self-loathing because he cannot replace the breadwinner, the source of stability and (if Freud is to be believed) the partner in the marital bed. The astonishing feats of diligence and intelligence which result are a lifetime struggle to repair damaged self-esteem and to prove his worth to all the world. The mastery of these feelings, especially of depression, is a springboard for immense compensatory energies, the source of strivings for accomplishment, status, power or wealth.

The battle for mastery can work in several ways. The most common is to develop a formidable determination to trust no one and to seek maximum control of our environment so that we will never be let down again. Where this is coupled with rage it often evolves into a comprehensive grievance against all social injustice. Of course, this can be the 'society's to blame' cry of paranoid, antisocial young people who feel they will never get a square deal from teachers, police and, eventually, prison officers. But it can also be transformed into the profound commitment of a Mahatma Gandhi or a Leo Tolstoy (who had lost both parents by the age of eight), shaming or cajoling the world into behaving more justly. Taken to their logical conclusion, the combination of rage and desire for mastery become totalitarian. They amount to a personal mission to wrest life's outcome from fate, to cheat destiny and at the same time to impose one on everyone else. Here is the charismatic dictator with his 'will to power' or the revolutionary who is prepared to use violence to force the world into his mould. As a child he believes that his orphaned status marks him out from his peers as special, rather than handicapped. He sees himself as 'chosen', and in later life nothing can stand in his way. Many of the most famous (and infamous) world leaders suffered early loss, including Hitler (who described himself in *Mein Kampf* as 'plunged into the depths of grief' at the premature death of his father), Stalin and Napoleon. The long list of such ruthless and effective individuals covers Lenin, Robespierre, Danton and, more recently, Ho Chi Minh, Eamon De Valera and post-colonial presidents Amin, Sukarno, Nasser, Kaunda and Kenyatta. Such men usually lack any insight into their personal despair and insecurity and project all their problems on to the body politic or into ideological systems of thought. Rather than feel depressed, they act depressingly; rather than feel helpless, they overpower and dominate their immediate subordinates and perhaps millions of fellow citizens; and rather than contain and understand their volcanic rage at having been the victim of fate, they direct it outwards with a terrifying destructive power.

Whether the urge for mastery expresses itself like this depends on how much the rage and the desire for revenge are tempered by guilt, and this in turn is a matter of the child's family script, before and after the loss. Guilt powered the morality and overwhelming sense of conscience of Gandhi and Tolstoy. As boys they felt they should have saved their father and that his death was their fault. Alternatively, a boy who felt little guilt is able to behave amorally in the pursuit of his adult goals. Whereas Tolstoy or Gandhi were not prepared to countenance violence, dictators and revolutionaries are usually conspicuous for their lack of sympathy for the thousands or millions of lives that they claim. What capacity for guilt they do possess is swept away by rage and vengefulness.

All these men used the control of fellow humans as the means of expressing their need for mastery, but the arts and sciences can serve this function too. The novelist or poet has total control over invented characters or the sentiments expressed. The composer dictates what sounds shall be made by which instruments (picture the intensity of the sense of power of George Frederick Handel, who lost his father at the age of eleven, while writing his masterwork *The Messiah* in a continuous burst of creativity lasting just twenty-eight days). The painter determines all aspects of the canvas, and the sculptor plays God with his clay. In the performing arts, the virtuoso musician makes his instrument do his bidding to perfection and the actor is in total command of his or her voice, facial expressions and body. Subjectively, their sense of control can be as great as that involved in dictating a nation's fate. On top of this, in all the arts there is an element of controlling other humans by engaging their emotions – often the very ones that the artist cannot cope with.

The effect of parental loss on artists and scientists is perhaps greatest of all on their creative imagination. Where the father was never known by the child it can create the sense of a gap to be filled. In his autobiographical work *The Words* the philosopher Jean-Paul Sartre discussed the impact of the loss of his father, Jean-Baptiste, at the age of two: 'A father would have weighted me with a certain stable obstinacy. Making his moods my principles, his disappointments my pride, his quirks my law, he would have inhabited me . . . the death of Jean-Baptiste was the big event of my life.' The absence meant 'I was not. I was not substantial or permanent. I was not the future continuer of my father's work . . . in short, I had no soul.' Filling this vacuum by writing about the nature of nothingness, and insisting on its centrality for the rest of us, in novels like *Nausea* and monographs like *Being and Nothingness* was Sartre's life work.

Sartre was too young to know his father, so the effect was not one of bereavement. But most high achievers suffering an early loss were old enough to have loved the deceased parent and to have felt the sadness and anger which such a death provokes. This can become a recurring theme in the work of artistic achievers as they continually try to rework the experience and make it more tolerable. Edgar Allan Poe (both parents dead by the time he was four) spent a whole night with his mother's corpse before neighbours discovered them together. The persisting confusion in his fiction as to whether the dead are living or the living dead is often traced back to this experience. Rather than becoming a horror writer, Tracey Ullman adopted comedy as her domain after a similar experience when she was five years old. In a TV interview in 1987, she told me that she was playing with her father when he died suddenly. In the months and years that followed she began performing comic routines to cheer up her bereaved mother.

Writers, particularly poets, are more likely to have lost their mother (as opposed to their father) than are other types of high achiever. This may help to explain the strong association between depression and writing. Writers tend to be depressives and they also have a high incidence of alcoholism. Their depression often takes the form of wistful longing, and they drink to drown their sorrows. In some cases their lives have been one long wake provoked by mother-loss. Historically, fathers were often remote and punitive, rather than present and nurturing, figures. Their loss would be far less likely to cause a sense of full bereavement than that of the mother. In addition, the child who loses a mother must usually adjust to a new mother-figure, unlikely to be as satisfactory as her predecessor. The result of losing the mother and perhaps suffering subsequent neglect is the intense sense of worthlessness, hopelessness and melancholic rumination which, if the child grows up artistic, creates circumstances favourable to writing as a way of expressing it.

But mother-loss is also found amongst other kinds of genius. Charles Darwin lost his at the age of eight and was scarred by a series of other deaths in childhood. Several authors have directly related this experience of a perfidious world to his attempt as a scientist to demonstrate that the cruelty has a purpose and is rule-governed rather than random. He proved that the natural world evolves because the maladaptive are less likely to reproduce their genes and only the fittest survive. This offered a comforting explanation for why his mental health had been subjected to the massive threat that the loss of his mother posed. Evolution demanded that he must adapt and become one of the fit.

However, these various theories only take us some of the way to explaining why loss spurs achievement, for more often than not it is a major reverse rather than a career boost. Quite apart from the short-term distress caused to the child and surviving parent, the loss usually creates long-term financial hardship and stretched childcare resources. It is hard to see how these help rather than hinder the child's progress; indeed, in the vast majority of cases the loss is damaging. Adult criminals (32 per cent), patients suffering from

59

depression (27 per cent) and juvenile delinquents (30 per cent) are three to four times more likely than average (8 per cent) to have lost one of their parents before the age of fifteen. It is only very rarely that genius flowers from loss (genius itself, after all, is a very rare thing); criminality or depression are far more common.

The line between success and emotional problems resulting from loss is very fine and there are close parallels between the psychology of the power-obsessed leader and that of the often rather mad, antisocial ordinary person. Both are extremely vulnerable and sensitive at times, yet at others remarkable for their thick-skinned ability to block the world out and go their own way. Both invest great energy in themselves and what they produce, and are hugely narcissistic. Both tend to have weak, remote or destructive relationships with their intimates. Both regress easily to childlike behaviour, but the successful personalities have learnt to control this to their benefit whereas in the disturbed ones it is out of control. Above all, both have a great capacity for suffering and exhibit profound dissatisfaction with their environment. They strive vigorously to change it, rather than themselves. They refuse to conform to the status quo and are often original and behave in ways that are not considered normal.

What makes the difference between two outcomes: the antisocial, disturbed character and the one who becomes a world-beater as a result of the same tragedy? There is, of course, no definitive answer to such a complex question, but family scripting plays a large role. The age of the child at the time of the loss, the nature of the relationship to the lost parent when they were alive and the circumstances of the family all make a difference, but one factor recurs as critical. After the bereavement, the main person who cares for the child needs to be unusually disciplined and loving. Geniuses are not made by childhood trauma or the deprivation of needs alone. Whilst they may adopt the RAF's motto of 'Per Ardua Ad Astra' ('Through adversity to the stars'), the adversity must be combined with just the right blend of love and discipline. It is essential that after the loss the mother or mother-substitute is passionate about her son (the evidence is only available for boys) and that makes him feel he is special. As Freud put it, 'A man who has been the indisputable favourite of his mother keeps for life the feeling of conqueror, that

60

confidence of success which often induces success.' There are many examples of this among high achievers, from George Washington, who said he was 'very close' to his mother, to Josef Stalin, who as a child was 'close to only one person: my mother'. However, it is not enough for the mother to be unusually devoted. She must also be exceptionally competent and disciplined. Washington's mother was 'active, capable and resolute' whilst Stalin's was 'a woman of severe and determined character, firm and stubborn, puritanical in her ideas, inflexible in her manners and very demanding towards herself.' Whilst few people in the Western world would regard her son's achievements as admirable, there are still many in the former Soviet Union who regard him as a hero, and there can be no doubt that what he achieved was extraordinary.

Apart from parental loss, several other environmental factors are strongly associated with high achievement. Being a boy is obviously one, being the eldest is another, having a family that is in some way socially marginal, such as an immigrant one, is yet another. But none of these plays nearly as powerful a role in the drama of a life as parental loss.

Depressing scripts for high achievers

Being scripted for exceptional achievement, whether as a result of losing a parent or simply because one or both of our parents demand it of you, is often a distinctly mixed blessing. It may result in a more successful career than those of our siblings, but there can be a heavy price to pay – depression. The feelings of low self-esteem ('I'm fat . . . I'm useless . . . I'm loathsome', when we aren't), hopelessness about the future and powerlessness to control our lives that we feel when depressed are often misrepresented as being largely genetic in origin. In most cases, the truth is that how we are treated in our family is crucial.

In general, when adults who have been depressed are asked about their childhoods, compared with those who are not depressed they are more likely to describe unsupportiveness, strictness and lack of love. They do so regardless of whether they are depressed at the time of being interviewed, so their account is not merely the result of feeling negative. Allied to parental coldness they often report a tendency to control the child too much, interfering with her unnecessarily and trying to dictate her every thought and feeling as well as her behaviour. This usually makes the child extremely anxious, as well as at risk of depression.

The more abuse and neglect that is reported, the greater the likelihood of depression. A study of 800 women from the general community found that 41 per cent of those who had been severely physically abused as children had been depressed during the last year, and this was true of half who had been severely sexually abused. Depression was found in one third of those who had been neglected. Depression was much rarer (only 8 per cent) in women who had suffered none of these adversities. When children are directly observed with their parents in childhood and then have their mental health assessed as adults, parental care also emerges as vital. A recent British study showed that of five-year-old children whose mothers had trouble coping and who lived in disharmonious homes were four times more likely to grow up to suffer depression at the age of thirty-three than children from undisturbed homes. Another study showed that when the parents of five-year-olds had been rejecting and overcontrolling their offspring were more likely to be severely self-critical at the age of thirty-one.

The early role of emotional deprivation in causing depression, especially during the first three years, is described in Chapters 3 and 4, but being scripted to be a high achiever in the family drama during the early years and afterwards is also important. Many very successful people are afflicted by self-criticism, by feelings of unworthiness, inferiority, failure and guilt. They are plagued by a fear of disapproval, criticism and lack of acceptance by others, and are prone to harsh self-scrutiny. They may set themselves impossible standards, strive for excessive achievement and perfection, and are often highly competitive and hard-working, making huge demands on themselves yet never feeling lasting satisfaction even if they succeed. Their best is never good enough, and one subtype of this depression is known as Dominant Goal.

In such cases, their self-esteem relies on the achievement of a lofty goal and they shun any other activities that divert them from this quest. When they were children only achievement was rewarded by their parents, so high marks or some outstanding performance were sought as a way to ensure acceptance. What love was on offer was conditional on doing well. In time such people select some distant goal for themselves which is pursued fanatically in the belief that it will transform their life. Attaining it, they feel, will mean that others will treat them in a special way, finally valuing them. Whereas the Dependent type of depressive uses fantasies of a relationship in which they are loved to derive feelings of self-worth, the Dominant Goal type obtains meaning and esteem from fantasies of achievement. A clinical example from the authors of this theory Silvano Arieti and Jules Bemporad, shows how it works:

Mr B was a middle-aged research scientist who came for treatment complaining of a lack of interest in his work, a sense of futility in all of his activities, difficulty sleeping, fatigue, various psychosomatic complaints of recent origin, and a subjective feeling of depression. The various symptoms had developed over the previous two years after he had not received a specific position that he had greatly desired. It became clear that his depressive episode was not simply a response to missing out on the coveted job but that his failure to obtain the position signalled to him that his career plans, which were meticulous, had been thwarted forever. He spoke of great aspirations of being awarded spectacular prizes and becoming director of a prestigious research institution. His being passed over for the job forced him to realize that he might never achieve his goals.

This man was obsessed with his work, putting in extraordinarily long hours (although he did not particularly enjoy what he did) and being so driven that some colleagues refused to collaborate with him. His marriage was a disaster. He expected his wife to plan her life around his work and expected special treatment from her because of the alleged importance of his work. He had no hobbies or interests but was consumed with fantasies of his glorious future, when everyone would respect him.

He was raised by parents who were poor and centered their hopes for upward mobility on their children. The patient's two siblings were also driven professionals. In his childhood, he had been made to understand that he had a special mission in life and that the pursuit of excellence (and prestige) was his way of repaying his parents for the sacrifices they had endured to send him to the best schools etc. Pursuits that were not 'productive' were forbidden. The belief in this messianic role was further fostered by the local minister who took the patient under his tutelage and, impressed by the boy's willingness to work and learn, also painted a glorious picture of his future if he applied himself. Eventually, Mr B lived only for this great goal that, once obtained, would bring all sorts of gratification and meaning. The remote possibility that he might simply remain a respected but not extraordinary scientist so discouraged him that he no longer found meaning in his activities. The goal and the quest were his motivation for being.

Expensive mental hospitals throughout the developed world are awash with high achievers like this, some of them very famous. Elton John, who has been treated for both depression and addiction, is one example, reporting that 'I get depressed easily, very bad moods.' He used success to bolster his self-esteem but, as with all Dominant Goal depressives, no amount of public recognition or wealth was enough. He says, 'Whenever I had a hit record or met someone, I always wanted more, too much. I never gave myself time to stop and smell the roses.' Both his parents insisted on his becoming a musician and entertainer as a child, expressing unfulfilled ambitions of their own. John recalls that 'When I was a kid I was always being told not to do things. I was suppressed by, and petrified of, my father. I dreaded it when he came home.' His parents divorced when he was fourteen and, although it was his mother who was the unfaithful one, John blamed his father. However, his mother also set him extremely high standards. From early childhood she forced him to take piano lessons, and his consequent skills became a pawn in his parents' disharmony. His father wanted him to be a classical pianist but his mother encouraged his interest in popular music and she it was who prevailed. Unhappily married and lonely, she got him to cheer her up by playing. John recalls that he was 'on stage' long before he became a professional and that he was 'always being forced to play. If I was at a wedding, for example, they'd say, "Come on! Give us a song!".' A famous photograph of him, aged seven, depicts him at the piano turning to face the camera. His biographer,

Philip Norman, comments that 'the smile has the resigned quality of the professional who must perform, whether he feels like it or not'. Typical of the Dominant Goal depressive, John has said that 'it's more or less impossible for me to have a personal relationship with someone. I like being alone but I crave for someone to love. It's really a tortured existence.' Unable to sustain real relationships as a child, he escaped into the theatrical, histrionic fantasy of his stage persona. It became his solace, so that, when singing the tragic lyrics of his songs, he may feel them more keenly than his actual existence. He, not Marilyn Monroe or Princess Diana, may be the 'Candle in the Wind'. In 1976 he reported that 'my life in the last six years has been a Disney film and I have to be a person in my own life'.

The author and actor Stephen Fry, whom I interviewed for a television programme in 1988, is another well-known example. A lifelong depressive, Fry used success as a buffer against a nagging, irrational self-hatred, explaining that 'my achievements have been driven by a fear of inadequacy and unpopularity'. He has the Dominant Goal self-loathing: 'As an adolescent I was shy and awkward. I had an appalling body image, thought of myself as a quite revolting specimen and still do to some extent. I don't think of myself as an oil painting – oil slick would be closer. The fact that I don't inflict myself on women is the greatest favour I can do them.' The principal cause of this low self-esteem appears to have been his father. A scientist and businessman, Alan Fry had a brilliant mind and used it to find fault with his son. Says Fry of his relationship with his father, 'There was a lot of tension and rivalry. He knew I was bright and therefore he was very irritated. He scared the living daylights out of me until I was twenty.' His father was hypercritical: 'He frowned at anything I did with any degree of competence.' This attitude was still detectable in comments that Alan made about his son in 1991: 'I sometimes feel like saying to him, "Stop doing this pappy and ephemeral stuff on the box and get down to some serious writing." Stephen spends a lot of energy doing things that aren't worthy of him.' Fry seems to have devoted himself to proving his father wrong by an impressive array of achievements. But after he finally suffered a depressive breakdown, narrowly avoiding suicide, he acknowledged that his Dominant Goal life had caused it. He had believed that all he had to do to be happy was write a few good novels and be a good comedian and actor. Yet, despite having made a lot of money and achieved a unique esteem amongst his peers, he now realized that this was not going to work.

More recently, in two documentaries on BBC2 in 2006, Fry presented a new interpretation of his problems: that he has manic-depression. An alternative possibility is that he suffers from the lack of identity, febrile moods and narcissistic self-preoccupation of someone with a personality disorder (see Chapter 5). He seemed to be reassured (as well as upset) by having found a label for his behaviour, one which was depicted as completely absolving him of any human agency because it was a physical, neurological malfunction. It also enabled him to ignore completely his earlier public statements about his father's attitude to him and its possible role in creating his disturbed frame of mind. He ended the programmes completely at sea, regarding himself as only having the choice of taking potentially toxic drugs to control the illness or soldiering on as he is.

Interestingly, Fry might have been able to avoid this misery had he read the substantial body of scientific evidence which proves that greater unhappiness comes to those who regard the pursuit of the approval of others and of external rewards more highly than self-acceptance and engagement with friends and families. In particular, dreaming that component of the American Dream in which the main goal is material wealth has been shown by many US researchers to be especially bad for you. People who put money before family or community or emotional fulfilment as motives are considerably more likely to report feeling depressed, anxious, to abuse substances and to suffer narcissistic personality disorder. A remarkable series of recent studies have shown that American children from affluent backgrounds are actually at greater risk than those from poor homes of suffering depression, anxiety and compensatory substance abuse, whether still at High School or at College. As I have argued in my book *Affluenza*, someone with Fry's childhood history is put at far greater risk of becoming ill by living in a Selfish Capitalist, English-speaking nation.

Many studies have proven that having highly critical, insatiably demanding parenting in your family script causes Dominant Goal depression. Adults who suffer from it are more likely as children to have been subjected to a torrent of negative words like 'bad', 'stupid', 'inadequate', 'useless' and 'unwanted'. At least one in ten of all children are exposed to such hypercriticism. In the past this attitude was more commonly tied to high achievement in boys but there is good reason to believe that girls are increasingly likely to be the object of it. Whereas only a handful of their mothers attended a university, today's young women are as likely to go as their brothers. Some mothers who were frustrated by not being allowed to attend university and discouraged from fulfilling their career potential have poured these unfulfilled ambitions into their daughters. In many cases this has simply righted the wrongs of previous generations, but in some the pressure on the daughter makes her prone to Dominant Goal depression. Although consciously the mother only wants the best for her daughter, in practice she tends to view the girl as an agent for satisfying her own ambitions, expressed by making love conditional on performance and by being excessively controlling.

A crucial method used to overstimulate achievement which is also liable to trigger depression is comparison to inappropriate models. For example, in a sample of disabled children, whether they were depressed was significantly affected by who their parents compared them with. Children of parents who used healthy, able-bodied children as the basis for comparing their disabled offspring's performance were more depressed than those whose parents used other disabled children as the standard. Daughters of pushy mothers tend to set impossibly high standards, with the curious result that many high-achieving girls actually have lower self-esteem than their less successful sisters. In America, over one fifth of affluent teenage girls suffer serious depression compared with only 7 per cent of girls from the general population. In Britain, a study of 5,000 fifteen year olds found that 38 per cent of the most affluent were at greatest risk of depression and anxiety. By contrast, 27 per cent of girls from low-income homes were afflicted. This trend only changes after the girls pass College-age, when being of low income predicts higher rates of mental illness. The pressures are highest in the world for American girls at ages 11 and 13, and second highest (with the British top) at 15. An in-depth British study compared academically high-flying middle-class girls with working-class ones, following them from the age of four. All the middle-class girls, without exception, were considerably more anxious and stressed by the age of nineteen. Despite mostly having done very well, they still felt they had not achieved enough. Elspeth Inch, the head teacher of a leading British girls' school (King Edward VI in Birmingham) confirmed this finding when she told me that 'Even our really clever girls do not realize how clever they are.' At its worst, this amounts to a soul-destroying perfectionism, something I encountered when interviewing a 32 year old woman called Karen for a television documentary in 1998. She had suffered a suicidal depression for several years.

'What I find surprising is that I've set myself goals and achieved them and yet I'm not happy. I've got a good job, very well paid, for which I work hard. I've got a lovely boyfriend, I've got a lovely two bedroom flat with a garden, in Chelsea, and a lovely little cottage in Somerset where we go for weekends. I'm sure that if I was to list my achievements and current situation, plenty of women would say "I'd give anything to be there" and that makes it worse because I think I've got all this and I'm not happy. So what's going to make me happy? Absolutely nothing. This is as good as it gets and I'm still not happy.'

Like all Dominant Goal perfectionists, Karen used achievement to bolster her self esteem.

'I love my job, it's been a lifesaver because I've got self-esteem in that area, it's the rest of my life I have self-doubts in. It's just total self loathing, that's the way I feel: "You don't deserve a job, you don't deserve friends, you don't deserve a flat. You're just the lowest of the low, you're the worst person in the world." People that like you are silly, they've been taken in by an act. The real you is this horrible, hateful person that doesn't deserve anything.'

At one point Karen's GP became so concerned about the risk of suicide that he sent her to a hospital. Karen told the nurse that she had to work the next day and the nurse asked 'what's more important, your job or your life?' to which Karen replied 'My job! Don't be so silly.' Her perfectionism was extreme. If she left her keys in the wrong place she would become furious with herself, she could not tolerate mistakes of any kind. When I challenged this, pointing out that she must know at some level that mistakes were inevitable and natural, she replied 'No, Oliver, there is no need for mistakes, it really is possible to be perfect. If only I was perfect then everything in my life would be fine.'

Although she did excellently at school, Karen's best was never good enough. If she returned home having come second in a subject, her mother's only response was to ask why she had not come first. This is typical of the perfectionist's childhood. Their parents are overtly and covertly hypercritical. They barely reward success and instead constantly find fault, urging ever better performance so that the child never senses that he or she deserves to feel pleased with his or her efforts. Subtle disapproval is conveyed by implicit disappointment. The parent may occasionally say 'Well done, darling', but with looks and intonation that suggest this praise is not sincere. Withdrawal of love is only a tiny mistake away, so that the child comes to think, 'If I try a little harder and do a little better, if I become perfect, my parents will love me.' In general, daughters are more likely to be perfectionists if their mothers, but not their fathers, are. Perfectionist daughters with perfectionist mothers are also at greater risk of depression, feeling suicidal and, not surprisingly given the degree of anxiety about 'getting it right', to being obsessive. The mothers intensely scrutinize their daughters' behaviour and try to prevent attempts by the girls to be independent and assertive. In most cases the parents are tremendously self-belittling as well, setting themselves equally unfeasibly high standards. When children are followed over time, those whose parents were rejecting and excessively controlling before the age of eight are more prone to be harshly self-critical, depressed and dissatisfied in both early adolescence and young adulthood. The children have learnt to be like their parents.

Of course, perfectionism can play a healthy role in high achievement. Indeed, features of perfectionism are often found in studies of high achievers. They derive pleasure from painstaking effort, they strive to excel, but they are also able to tolerate limitations on what is possible. They have the pathological perfectionists' love of being well-organized and their high standards, but do not share their low self-esteem and are able to feel pleased with a job well done. The comedian and writer John Cleese told me that if he had any special gift it was that he would refuse to accept that his work was good enough long after everyone else was satisfied – something that has been confirmed by others who know Cleese. I found a similar concern in John Lloyd, one of the most talented British comedy television producers of all time. After producing *The Hitchhiker's Guide to the Galaxy* for radio, his sole television credits were *Not the Nine O'Clock News*, *Spitting Image* and *Blackadder*. Having achieved these remarkable successes, he felt able to retire from television production. But most perfectionists are not like this. They tend to be hugely self-critical or fearful of criticism from others, or prone to demanding absurdly high standards from those around them, or a mixture of all these.

Whether our perfectionism is pathological depends on the precise role we have been scripted, but the kind of society beyond is also influential in how much the tendency is overstimulated. If there are more mothers using their daughters as vessels for unfulfilled ambitions today, we also live in a much more perfectionist culture. Academic pressure to succeed starts younger and there is far greater measurement of performance at school and work compared with, say, 1950. Girls are particularly vulnerable because they tend to be more compliant when small, more easily taken over by a system in which they are what they scored in their last test. Combine perfectionist mothering with this new academic pressure, and we have a prescription for depression.

On top of all this, girls also face an increasing demand to live up to preposterous physical ideals of beauty. Although it is frequently claimed that eating disorders are largely caused by genes, there is abundant evidence that parental care plays a major role. Perfectionist girls are at greater risk of suffering from bulimia (binge-eating followed by making oneself sick), but only if they have low self-esteem as well. This neatly illustrates the importance of the fine print in our family script. If our parents encourage us to feel good about ourselves as well as demanding high standards, we are spared bulimia. More generally, parents of bulimics tend to be at war with their daughters, belittling and blaming them, sulking a lot. Parents of girls who suffer from anorexia (self-starvation) are more liable to give out double messages, mixing nurture and comfort with suppression of their daughter's emotions and discouragement from independence. But again, even if vulnerability to eating disorders is

70

created by the family script, it is exacerbated by the wider society. That there are large fluctuations in rates, over time, within different societies and between genders, undermines the idea that genes play much of a part. Today, our unprecedented bombardment with wafer-thin beauties on advertising hoardings, on TV and in the cinema is at least part of the reason that women are ten times more likely than men to be dissatisfied with their current weight. It is hardly surprising that, in a recent British survey of 900 women aged eighteen to twenty-four, more than half of those who were of average weight wished they could be slimmer; 40 per cent did not feel comfortable being naked in front of their partner, and 20 per cent reported having stayed at home at least once a month because they were so dissatisfied with how they looked.

All of this helps to explain why so many academically able, outwardly successful and attractive girls suffer from eating disorders and depression. Like Karen, a slim and good-looking woman, they may outwardly appear to be Having It All (the title of Helen Gurley Brown's 1960s' bestseller) and yet feel utterly worthless. The modern equivalent of Brown's book might be Helen Fielding's *Bridget Jones' Diary*. It is no accident that this very different story, of ludicrously inflated expectations coupled with personal despair, has sold millions worldwide.

Whilst modern pressures are particularly acute for young women, they do affect men as well. There was a sharp rise in suicide amongst British young men between 1987 and 1993 which some commentators attribute to an increasingly competitive, winner–loser culture. New Zealand has the highest suicide rate in the world, with men six times more likely to kill themselves than women. This is thought to be because New Zealand men have to live up to an impossibly exaggerated model of masculinity, excelling at sport and concealing their feelings to such an extent that if they commit suicide their death comes as a complete surprise, even to close friends and immediate family. But overall, it is increasingly likely to be Bridget Jones, not her brother, who is at greatest risk of being allotted the character of family perfectionist.

Ugly duckling scripts

I have dwelt at some length on the hazards of being selected to fulfil parental dreams, of the travails of being allotted a Dominant Goal or perfectionist script, but it can be even worse not to have either parent gunning for us. As might be expected, children with a positive relationship to both parents fare better than those who get on with only one. Those with negative relations with both parents do worst of all.

Favouritism is found in all families, and inevitably there is only a limited amount to go round. Two-thirds of children claim their parents show some form of preferential treatment, and where there are small children no Christmas Day is without incendiary feelings about who has got the best presents. This is the norm, but in some families one of the children has to live with the consequences of actually being unwanted. A Czechoslovakian study identified 220 children whose mothers had given birth to a child they had twice been refused a request to abort. The mothers of these babies were less likely to breastfeed them compared with wanted children, and at the age of nine these children were doing much worse at school, where they were less diligent and more prone to explosive irritability and defensiveness. At fifteen the children were still academically worse, reported by their teachers to be less conscientious and obedient. When asked about their mothers' attitude to them, the children claimed she showed less positive interest. They felt either neglected by her or that she was interfering and overcontrolling.

For such extremely unfavoured children, or ones who are acutely aware that their siblings are getting an unfair share of the parental pie, the sense of having had a raw deal can endure for life. They attribute the preferential treatment to a predictable litany of traits: gender, as in 'My brother was the favourite because my parents wanted/preferred a boy'; age, 'I had it more difficult because I was the oldest'; physical attributes, 'My sister was more attractive than me and got more attention'; intelligence, 'My sister got more attention because she was brighter'; need, 'My brother got more attention because he was sickly'; and personality, 'My parents preferred my

72

sister because she fitted in with the family and its values better than me.' Parents may shape the unfavoured child's behaviour by adversely comparing them to models from outside the family, like successful public figures or children from other families (in my own childhood family we were always being told to emulate a particularly well-mannered pair of goody-goodies, but I suspect that my parents would have been horrified if we had actually been like them). Adults who report having had a favoured sibling are more at risk of depression, anxiety and feelings of hostility. They are prone to a sense of inferiority, feeling unattractive and unpopular. Accepting their parents' evaluation of them wholesale, they anticipate criticism and rejection from others, lack confidence and are easily cowed into a passive, defensive submissiveness.

Carol Thatcher, the daughter of Margaret Thatcher, the former British prime minister, suffered from being unfavoured. Her twin brother Mark was her mother's favourite, and Carol was the recipient of negativity. She recalls that 'Mum would say "Carol, those shelves are such a mess" if she visited my flat. Mum has never been keen on anything I have worn. "Ghastly" was a word frequently used.' As tends to be the case in homes where there is strong favouritism, there was little love for either child. Even Mark recalls that, although his mother did indulge him when she was present, there were 'only spurts of motherhood'. Carol states that 'Mum was perfect. There was nothing she couldn't do, but she didn't do it with enormous warmth. Mum and me weren't really that close. I don't remember her being physically demonstrative.' In fact, Margaret was scary. 'As a child I was frightened of her, and later I was conscious of talking to her knowing her mind was elsewhere. I used to console myself by thinking, "Carol, maybe it's because you talk such drivel."' Carol has a low opinion of herself. She compares her looks unfavourably to her mother's and doubts her attractiveness to men. 'Mum's better raw material than me. I still don't measure up. "Unloved" is not the word to describe me but I never felt I made the grade. I always felt I came out second' (compared with her brother). In some families such unfavoured status is made up for by love from the other parent, but Carol reports that she was not close to her father either: 'There wasn't much of a daddy–daughter relationship.'

Shaming is one of the most frequently used weapons in the parental armoury where a child has unfavoured status. Shame is an acute sense of being exposed, scrutinized and judged negatively. It often entails painful self-consciousness and feelings of awkwardness, inhibition and reticence, all attributes that Prince Charles suffers from. But it is also accompanied by anger directed at the parent who does the shaming. This may have something to do with Charles's very public advocacy of causes that his father must loathe, like ecology. Shame is most often induced by humiliating put-downs, belittling comments that leave the child feeling he has been too big for his boots, all of which Philip directed towards Charles. The shamed child is left with no dignity, no sense that he is in control of himself or of others' perceptions of him. Shame is also achieved by parental displays of disgust at the child's bodily activities, be it eating, such as repeatedly telling a child that he is revolting at the dinner table, or towards displays of sexuality, such as when a toddler flashes his genitalia at others. Where there has been physical or sexual abuse, chldren often feel a horrified shame of themselves. In general, adults who were devalued, demeaned or scorned by their parents are significantly more likely to be depressed, angry, anxious and alcoholic. Shaming in early childhood has been shown to affect brain development adversely.

At least one in ten children report that one of their parents 'really seemed to dislike me or have it in for me'. Having the role of the unfavoured child is usually accompanied by the additional misfortune of being on the receiving end of painful and unwanted emotions. Prince Philip seems to have used Charles in this way and to have encouraged his sister to do so as well, through mockery at his lack of equestrian knowledge. In recent years, alongside a fashion for exaggerating the effect of genes on us there has also been a trend towards downgrading the influence of parents and replacing it with that of siblings and peers. There are those who would argue that Charles's parents played a far smaller role in shaping him than did his brothers, sister and school contemporaries. But this is not borne out by the scientific evidence. In the first place, it is simply not true that peers have much effect on each others' enduring personality traits or values. For example, parents are far more influential on how religious a child is, how hard he works or what career he chooses. In the second place, the supposed power of peers to determine how each other turn out is easily exposed as being merely the result of choosing like-minded friends. For example, it is not peers who cause each other to take drugs or have sex at early ages, it is children choosing other children who are that way inclined. What makes a child inclined to pick such peers is the way she was cared for during her early years, so parental care, not peer pressure, is the key factor. Parents have a huge effect on what sort of peers their children spend time with, and subsequently on how susceptible they are to peer influence. Hence, children who routinely do not have a responsible adult waiting for them when they get home from school are more likely to get involved with antisocial other children, and are more easily influenced to become antisocial themselves. Children who have closer relationships with parents who supervise them more closely are much less susceptible to peers who pressurize them into self-destructive behaviour, such as drug-taking, or antisocial behaviour, such as stealing.

The commonest emotions that a parent projects on to family members are depression and aggression. Violent fathers tend to hammer their frustration and anger into their sons, but, as in the case of Michael Jackson, such fathers often do not hand out their cruelty evenly, usually focusing it on a particular character in the family drama. Depressed mothers are very liable to deal with their irritability and self-hatred by inducing it in their children, but usually in one child more than the others. The mother's vulnerability can induce the fearful child to take over the role of parent, so that two-year-old children of depressed mothers are more liable than those of undepressed mothers to act responsibly, showing distress and signs of guilt when confronted by worrying scenes. Precisely which child is the butt of the mother's aggressive depression and how much he is affected depends on when the mother becomes depressed and for how long. She may be fine for the firstborn and only depressed by the birth of the second, but even if she is depressed for most of her parental life she will relate differently to each offspring. Children will also vary in the amount of support they get from fathers and siblings, buttressing the impact of the mother. I encountered a fine example of this complex weave of

75

causation when I interviewed a family for a BBC 2 *Horizon* television programme which I co-produced in 1995.

Michael, then aged twenty-nine, had a history of violence and minor criminality in his teens and early twenties. He worked in casual jobs and was single. By contrast his brother David, aged thirty, was law-abiding, had had a steady job for thirteen years driving forklift trucks and had been happily married for two. Their parents, Maureen and Terry, had separated when Michael was eleven. Here is the rebellious Michael's account of himself and his family:

David's a total opposite. He took lots of exams at school, he stuck it out. He got himself married, stayed with the same girl, never strayed. I see him as responsible and reliable, all the things that I would like. I wish I could swap places with him. I never used to like him and he didn't like me. When my dad left, David stuck by my mum more than I did. He took on the father role and me and him used to fight nonstop then. He gets on better with my father than me.

I've no time for my dad. He was a joke, really. He'd put about four or five pints of strong beer down his neck really quickly, climb into his 3 litre Monza motor, piss off to Dorking and wonder why he pranged the motor! When he started seeing his bird in our house during the day, when my mum was at work, I'd think, 'He's upstairs shagging some old bird, and my mum is the most attractive woman on the planet.' Even at a young age I found it hard to understand.

When my mum sold the house she got my brother sorted out, and this is what hurt. All she said to me was, 'You've got to find somewhere else to live for yourself' about two weeks before we had to leave.

David's account was as follows:

Mike hangs around with a very different sort of crowd from me. He likes wide boys, flash cars, booze, girls. I like the intellectual side of life, books, films, that sort of thing. I save every penny and I enjoy my life with Linda [his wife].

Mike says to me, 'At least I'm not boring like you, sitting indoors all the time, doing nothing.' I say, 'But you'll be nothing if you carry on like this.' He's like the Trotters from *Only Fools and Horses* [a TV series about a pair of 'wide boy' brothers]: 'Ooh, I'll have made my fortune by the time I'm old.' It's always just around the corner.

He idealizes people he meets. Everyone's a 'Diamond Bloke' until he meets the next one – he loves all that cockney spiel. He sees a lot of people, drifts in and out of friendships and relationships.

I wouldn't say he is a violent man. During that skinhead craze in the eighties he used to go round beating up language school kids, foreigners, but that was just letting off steam. I went through some trouble at school as well as Mike, but I was more devious.

The divorce happened at the wrong time for both of us. I took on board what my grandpa said to me at the time: 'You've got to look after your ma now, be the head of the family.' Basically Mike looked for role models outside the family after the divorce and got in with a bad set. I was very lucky in the people I took as my models.

Mike is like my parents in all the wrong ways. He bottles things up and needs a drink to open up, like my dad. He's like my mum in that he's very highly strung.

Michael and David recognized that they were very different. Some of the explanation for these differences emerged from talking to their mother, Maureen.

I suffered postnatal depression after Michael was born. He wasn't a very good sleeper, so I was really down in the dumps a lot. When he was one I had to go out to work and I left him with a childminder. She kept Michael in a wet nappy all day, so that he was red raw when I got him home – I suppose nothing like that happened to David.

As a boy he was very mischievous. I had a job getting him to school and he didn't concentrate very much there – he liked to clown about. Punishing him was like banging your head against a brick wall. He played truant – I forever had the school inspector up. And he was involved with the police – I had to get him from the police station.

David was more serious. He thought he ought to be the man of the house after Terry [Maureen's husband] had gone, but Michael thought otherwise and there was a lot of fighting between them.

When Terry left, Michael really did want to go with him – I don't know why. Terry said, 'No, you'd better stay with your mum', and that did turn Michael against his dad. Terry had a different relationship with Michael than with David, and so did I. We always seemed to rub each other up the wrong way. We really clashed – I don't know why. Even now I'm afraid of what I say to Michael just in case we row. To me he's always been the baby, most vulnerable.

This story illustrates how divorce tends to have a negative effect if it is part of our family's plot development. Children of separated parents are about twice as likely as those from intact families to suffer a whole host of problems, from delinquency to depression to teenage pregnancy. But the story also shows that the impact of divorce depends heavily on our particular role in the family and its history. Like all of us, Michael and David had very different scripts. Having a positive relationship with one of our parents can innoculate us against becoming disturbed, if they are not getting on or if they divorce. Michael lacked such a relationship, whereas David got on well with both his parents. Responses to divorce also depend on the kind of care received prior to it. As we shall see in Chapter 4, Maureen's postnatal depression during Michael's earliest months may have weakened his identity; Chapter 3 explains that leaving him as a one-year-old with inadequate substitutes when she returned to work probably made him insecure in relationships and even more prone to delinquency. Both these childrearing experiences were unique to Michael – not shared by his more stable brother. This major difference probably partly explains why the two boys subsequently reacted as they did to their parents' divorce. However, as Chapter 3 shows, their very different relationships with both their parents after the age of three could also have played a big part.

If the impact of divorce is so different for different offspring, how much more dramatic is the example of schizophrenia. That one child from a family can suffer this very severe delusory mental state while his siblings remain normal and realistic suggests to many people that genes must have played a part: they had the same parents, did they not? Here is the ultimate example of how difference in family roles can make for radically different outcomes.

Scripting schizophrenia

There is a good deal of scientific evidence that the different ways in which parents relate to their children is a major cause of schizophrenia. However, these studies are very rarely referred to nowadays, partly because they do not accord with the core beliefs and wishes of the psychiatric establishment. Nearly all the work was done before 1990 and it has been buried deep in the journals, with much less new research done since that time.

The other great problem entailed in pursuing this line of thought is that implicating parents can easily be perceived as blaming them. There is no reason, though, why this should be so – it ought to be wholly a matter of scientific explanation, and explaining and blaming should be completely separate processes. My object in resurrecting the evidence that parental care can be the cause of schizophrenia is not to cause distress but to remind us of two major points. In those cases where, with the right sort of support and intervention, the illness could have been prevented, and in those where it has already developed, the kind of care provided is extremely important. Whereas parenting and management of the illness can be improved, the same is not true of our genes – at least not for the foreseeable future.

The starting point for an environmental explanation of schizophrenia is uncontested, even by those who support the 'genes theory'. There is now a large body of evidence to prove that the course of the illness in young schizophrenics is hugely affected by the way families react when released from hospital. In about half of the families from which schizophrenics come, the parents are negative at this stage. They may make a lot of critical comments, be hostile and get too involved in trying to control their offspring. Whereas 55 per cent of patients who return to a negative home like this relapse into schizophrenia, only 16 per cent do so if their parents are positive – a gigantic difference. If negative parents are trained to be less so, the relapse rate of their children falls. On top of that, when patients with negative parents are sent to a hostel instead of being sent home, they are less likely to relapse. Whether or not patients keep on taking their prescribed drugs makes far less difference to their relapse rate than parental reactions.

This evidence proves conclusively that parental responses are an important cause of schizophrenic relapses, but it tells us nothing in itself about the original reason for the illness. It could be that the parents are negative because they have been stretched to the limit by an incredibly difficult child with a genetic ailment, and doubtless that is sometimes the case. Compared to families with no schizophrenic members, studies of relationships in families which do have one show that the parents tend to be more dominant, perhaps attempting to quell an impossibly difficult child. They also tend to be more bewildering, perhaps because the parents are confused by their child's weirdness; and they are less respectful of the child's autonomy, perhaps drawn into controlling an otherwise deranged offspring. However, in other cases these patterns of care may be caused not by the child but by the parents.

It could be that the parental negativity has been there from before the illness began and was, in fact, its cause. In other cases still, both explanations could apply: a degree of genetic proneness to schizophrenia in the child could have meant that those parents who were ill equipped to cope with such a worrying offspring became negative, and therefore exacerbated the schizophrenia. So the critical question becomes this: did the parental negativity predate the schizophrenia? Despite the general reluctance of psychiatrists to investigate this possibility several sound studies strongly suggest that it often does.

In one, families with problem boys who had not yet developed any signs of schizophrenia were closely observed: the parents' negativity and tendency to be mystifying were measured. When the mental health of the boy was tested fifteen years later, in manhood, the more emotionally disturbing the parents had been in childhood the greater was his likelihood of being schizophrenic or of having symptoms of this illness. In all the cases of full schizophrenia, both parents had been both negative and confusing in their dealings with the boy of fifteen years earlier. In general, the more the parents had been like this, the greater were their son's subsequent schizophrenic symptoms. Where the parents had been neither negative nor mystifying, none of the men showed signs of schizophrenia.

Of course, it is possible that plenty of schizophrenia does have a strong genetic component; but even in these cases a relatively recent study of the role of environment shows how important it is. Fifty-six children adopted at a young age after being born to schizophrenic mothers, and who were therefore, in theory, at higher genetic risk of developing the illness, were compared with 96 adoptees who were at low risk because none of their biological parents had it. The families were extensively observed when the children were small and the adoptees examined regularly into adulthood. The results are dramatic.

On its own, simply having a greater genetic risk (because of schizophrenia in the biological mother) does not increase the likelihood that the child will develop schizophrenic symptoms. Genes alone do not cause the illness. However, if there is a high genetic risk and it is combined with confusing care during upbringing in the adoptive family, the likelihood is greater. So the genes do have an effect, but only if the family environment was of the kind that exacerbates schizophrenic potential. What is more, the environment is important in reducing the illness. If the child is put at high risk by his biological mother's genes and the adoptive parents provide positive care, he is protected from this potential being realized.

Recently, the psychiatric establishment experienced an earthquake that shook its intellectual foundations. In absorbing the contents of a recent edition of one of its leading journals, *Acta Psychiatrica Scandinavica*, it has had to rethink many of its most cherished assumptions. Not since the publication of R.D. Laing's book *Sanity, Madness and the Family* in 1964 has there been such a significant challenge to their contention that genes are the main cause of schizophrenia and that drugs should be the automatic treatment of choice.

With his colleagues, guest editor John Read (whose name I shall use as a generic term for this body of evidence), a leading New

Zealand psychologist, slays these sacred biological cows. The fact that some two-thirds of people diagnosed as schizophrenic have suffered physical or sexual abuse is shown to be a major, if not the major, cause of the illness. Proving the connection between the symptoms of post-traumatic stress disorder and schizophrenia, Read shows that many schizophrenic symptoms are directly caused by trauma.

The cornerstone of Read's evidence is the 40 studies that reveal childhood or adulthood sexual or physical abuse in the history of the majority of psychiatric patients. A review of 13 studies of the specifically schizophrenic found rates of sexual abuse varying between 51 per cent at the lowest and 89 per cent at the highest. Crucially, psychiatric patients or schizophrenics who report abuse are much more likely to experience hallucinations. The content of these often relate directly to the trauma suffered. At their simplest, the hallucinations involve flashbacks to abusive events which have become generalised to the whole of their experience. For example, an incest survivor believed that her body was covered with ejaculate. The visual hallucinations or voices often tyrannise and belittle the patients, just as their tormentors did in reality, creating a paranoid universe in which intimates cannot be trusted.

More research is needed to establish if abuse also heralds the other key symptoms of the illness, for example, incoherent, confusing or downright weird language. Studies already show that this is much more likely if the subject of conversation is emotionally charged. When asked to talk about sad rather than happy memories, speech becomes measurably more disordered, increasingly so the more personal the subject matter. We need research to investigate if verbal incoherence is abuse-related. Another symptom is dissociation (feeling disconnected from oneself and surroundings). This has been shown many times over to be more common in the abused. It remains to be seen if catatonia (complete withdrawal from the world) is so related, although it seems extremely likely that it will be because elements of this are often found in victims of trauma. The same may be true of depressive thoughts, which many schizophrenics suffer and which are rife in the abused. About 15 per cent of schizophrenics die by suicide, and depression is the norm in the suicidal. Suicide attempts are more common among people who have been abused as children, suggesting an abuse-schizophrenia connection, via depression and suicide.

Among the sexually abused there is a greater risk of developing schizophrenia the earlier the abuse happened, the closer the relation to the abuser and the more invasive the acts. For example, the mildly abused are twice as likely to be psychotic as the unabused whereas the severely abused are 48 times more likely. Of course, not all schizophrenics suffered trauma and not all abused people develop the illness. What makes the difference? Apart from the possible role of genes, less overtly cruel early childhood maltreatment, other than actual abuse, may be important. In a review of the 33,648 studies conducted into the causes of schizophrenia between 1961 and 2000, Read found that less than 1 per cent was spent on examining the impact of parental care. What is more, the amounts have decreased steadily, from 1.6 per cent during the 1960s, to four times less than that during the 1990s.

As noted earlier, more general evidence of the crucial role of environment comes from sociological studies. Schizophrenia is around 12 times more common in children of West Indian immigrants to Britain. Increased rates have been found for immigrants to other countries. Poor people are several times more likely than the rich to suffer schizophrenia, and urban life increases the risk.

Rates of schizophrenia vary as much as 16-fold around the world, as does its course. It is less common in developing nations, and the illness tends to last much longer and to be more severe in rich, industrialised nations compared with poor, developing ones (even so, about 20 per cent of schizophrenics in developed nations recover completely and many are able to function normally, without taking anti-psychotic drugs). In fact, if you become ill in a developing nation where hardly anyone is treated with drugs, you are 10 times less likely to have any recurrence of the illness – a huge difference, also nothing to do with genes.

What it may have a lot to do with is the administration of drugs. They have been shown to impede traumatised people from understanding their voices or visions and recovering from them. There is a close relationship between the drug companies and the psychiatric establishment. While it may not be the intention, the establishment explanation of the causes of and solutions to schizophrenia are crucial components in the process of selling drugs. If patients can be persuaded their illness is an unchangeable genetic destiny and that it is a physical problem requiring a physical solution, drug companies' profits will grow. Read shows those who buy this genetic fairytale are less likely to recover, and that parents who do so are less supportive of their offspring.

The huge importance to drug company profits of the bio-genetic refrain becomes apparent when you learn that most people do not hum along with it. Surveys find that the majority of people mention such environmental factors as trauma, stress and economic hardship as the commonest causes of schizophrenia. It may be seen that the drug companies have an uphill struggle to persuade them otherwise, for which they badly need the help of the psychiatric establishment's towrope. In Read's analysis, letting go of that rope will prevent it strangling the many schizophrenics whose illness has been caused by abuse. Genes may still emerge as a major cause of vulnerability to schizophrenia, as may problems during pregnancy. There is already no question that illicit hallucinogenic drugs are a major reason some vulnerable people become ill. But even if this is true, following Read's important work, it will be hard to ignore its implications.

Some of the evidence for environmental causes of schizophrenia was borne out when I met the parents of Julie, the woman described in Chapter 1, whose breakdown included an obsession with dirt. Although I don't have a detailed clinical knowledge of the case, her father's behaviour was certainly negative. When he entered the bedroom where his daughter had endured her nightmare he immediately began to empty the wardrobe and chest of drawers. He made three piles: one for burning ('impossibly filthy'); one for throwing away ('filthy, but not a health risk'); and one for washing. It turned out that Julie's father worked as an environmental health officer for the local council. This made some sense of her preoccupation with cleanliness and poisoning. He was strikingly aggressive towards the dirt and mess, as he saw it, and equally remarkable was his total lack of respect for Julie's personal autonomy. He seemed to have no awareness that in throwing away clothes, treasured objects and various items that she had written during her illness he was trespassing on her personal life.

There were grounds for speculation as to his role in causing her hostility towards men, authority and the British economic and political system. Her relations with men and with any formal authority were unsatisfactory to her. Whenever she got close to either she began to talk in a frenzied way about their 'contradictions', dragging her knowledge of Marxist theory into an explanation of why a man had jilted her or why a policeman should not have stopped me for speeding on the way to the hospital. It was difficult to avoid the thought that she had chosen Marxism as a way of coping with her father. It could be that the extent of her understanding of Marxism's application to British society was in direct proportion to her difficulties in coping. As they increased, she was forced to apply her defensive Marxism in ever greater detail and in ever widening contexts.

In the end, it was Julie's 'contradictions' and not those of the capitalist system which led to a collapse, not of the existing economic order but of Julie's capacity to cope. Perhaps she tried to save herself by identifying with what she was trying to oppose: her father and his concern with cleanliness. Scrubbing her clothes and refusing 'dirty' food may have been a last-ditch attempt to save herself, by 'becoming' her tormentor.

There are several other studies which suggest that parental negativity and confusing behaviour predate the onset of schizophrenia. There are also at least seven which demonstrate that it is not characteristics of the sick offspring which provoke the negativity. Equally telling are those studies of identical twins where one is schizophrenic and the other is not (known as 'discordant' twins). In these cases genetics could not be the cause of the discordance; only an environmental explanation is possible. They show that from a young age the twins were markedly different because they received very different care. Sometimes one twin is born lighter and weaker as a result of differences during the pregnancy. The parents often use this as a peg on which to hang their own preferences, showing a tendency to prefer the stronger one over the other from birth. As the children develop, the future schizophrenic is more submissive, shy, neurotic and obedient, tending to depend increasingly heavily on his stronger twin, who is liable to be more outgoing, academically successful and lively. Remember, these differences cannot be caused by genes but must be the result of nurture. The disturbed twin is more likely to identify with the parent who has the most emotional problems, and this parent also projects more negative feelings on to the future schizophrenic. This is liable to be the mother, who may be disturbed in a number of ways, such as prone to depression or to incoherent thought patterns. Where the mother is like this and the father is heavily involved with one of the twins, if he is sensitive and loving it can protect that twin from disturbance.

The parents in these families tend to project their positive feelings on to one twin, whilst the future schizophrenic becomes the recipient of negativity and confusing behaviour. For example, when the identical Tim and George were born, Tim was felt to have his father's ears, to be more difficult and weaker. George was seen as more energetic and exploratory, more outgoing and sociable. As their mother, Jill, put it, 'Each of your children are individuals even when they're identical . . . You can love them all equally but that doesn't mean your reaction to each of them is the same.' This was especially true of their father, Terry. A crucial issue for him had been a competitive struggle with his older brother Kevin, compared with whom he had always felt weaker, less successful and more plodding. Terry saw all the attributes that he felt his brother had and would have liked for himself in his son George from birth, and the negatives that he felt about himself were all loaded on to Tim. Unfortunately for Tim, Jill also projected all her negativity on to him and she was a very confusing mother liable to say one thing and do another, criticising behaviour today that had been applauded yesterday. He was the one to become schizophrenic.

The fact that these studies show how schizophrenics are more likely to have been the object of negativity and confusing behaviour in the period before they develop the illness provides support for the controversial theory of R.D. Laing. Although Laing is loathed by the psychiatric community the public do not share its hostility, for his books remain in print over thirty years after first publication. They suggest that schizophrenia can be the result of being treated in totally confusing ways by parents. In a famous example, a mother visited her schizophrenic son in hospital. As they approached each other, the son made as if to kiss her on the cheek. But his mother froze and turned her head away, so he backed off. She then said, 'Darling, don't you want to show that you're glad to see your mother

86

by giving her a kiss?' Laing provides numerous examples of this kind of 'Double-Bind' in families of schizophrenics, where parents give two contradictory instructions so that whatever the child does is wrong. Eventually, according to Laing, the only option for a child in this situation was to shift to a different order of meaning, to avoid feeling permanently furious and confused. Instead of attributing their overt meaning to people's words or gestures, such as saying hello or shaking hands, the child would begin to assume that they, and everything else, had a deeper, concealed, more symbolic meaning. Since this was true of so many of the Double-Binding communications of their parents, in order to make sense of his family a child of this sort would learn to over-interpret the world.

Incest is the ultimate Double-Bind. The child is often told that what the parent is doing with them is natural and normal, so that, for example, in father–daughter incest he may tell her that she needs to be taught about sex, with him as the instructor, before attempting it with outsiders. But alongside these injunctions comes a contradictory one, accompanied by severe threats, that the child must not reveal to anyone else what has happened. Coupled with a growing awareness that incest is prohibited in the wider society, the child begins to feel deranged by the huge inconsistencies. Sexual abuse often leads to a symptom known as 'dissociation'. Rather than endure the reality that the sexual acts are happening, the child simply removes himself mentally from his body, perhaps even watching from the ceiling. It is only a short step from this to believing you are someone else altogether, someone in a position of control or power, someone whom it is far preferable to be – like a spy. As Rufus May's story illustrated in Chapter 1, reinterpreting the world according to a cranky personal theory gives the schizophrenic a good feeling. It also removes him from a world of confusing Double-Binds, negativity or sexual abuse.

Whilst Laingian theory does help to explain schizophrenia and is supported by scientific studies, there are many children who have been in Double-Binding or sexually abusive families who do not develop the illness. A weak sense of self resulting from early childhood care that lacks empathy, as will be described in Chapter 5, may be an additional necessary condition, creating a greater vulnerability to adverse later care. As we shall see, early deprivation greatly increases the likelihood of dissociation when combined with later abuse. A strong hint that early deprivation, with a consequent weak sense of self, is important comes from yet another fact that psychiatrists find uncongenial: children whose mothers are schizophrenic are twice as likely to develop the illness as children with schizophrenic fathers. Since it is mothers who care for infants in most families, and since schizophrenic mothers are much more likely than normal ones to be depriving, this could suggest that early deprivation can be critical.

What can be true of schizophrenics applies to the rest of us, too. We are constantly living out the consequences of early childhood years.

The past is in your present

It has become clear that Sigmund Freud was wrong about several important things, like the exaggerated significance he attached to the supposed sexual attraction of small children to their parents (known as the Oedipus Complex). However, recent studies have begun to show that in most fundamentals he was right. They prove that he was right in asserting that we have an unconscious and that it governs much of our thought, feeling and behaviour. He was right to state that we repress into this unconscious what we cannot bear, and that much of our mental life is devoted to elaborate defensive activity designed to keep it there. He was right when he said that our inner lives are immensely complex, with conflicting wishes and paradoxical impulses coexisting and fighting for expression. But his single most important discovery to have been confirmed by the evidence of the last forty years is the sheer extent to which our childhood scripts govern the way in which we interpret the adult present.

When at university I shared living space with five other students, and it was striking how specific and passionate were our individual reactions to the inevitable mess in the kitchen. To one it was so intolerable that he set up a cleaning rota that was, alas, ignored. Another did not care about the lack of washing up, but felt that if it was done it should be done properly, imposing upon himself the unpleasant task of redoing it on the rare occasions that someone had put dishcloth to dishes. Another was unalarmed by the mould on the muck-caked saucepans but intensely exercised by dirt on the kitchen floor. For my own part, I was thoroughly sluttish until the mess passed a certain threshold, at which point I would clear it all up or try to bully someone else into doing so.

What struck me at the time was the profoundity of each person's conviction that theirs was the right and the only acceptable attitude. I suspected that the beliefs came from our particular family experiences, whether we were imitating our parents' convictions or reacting against them, and recently a series of important experiments have proved that all of us do indeed constantly transfer our childhood experiences, often lock, stock and barrel, on to our adult circumstances. Participants in one experiment were asked to describe their parents. Some time later, in a different context, the researchers fed back to the participants a description of a person who shared most of the characteristics that they had attributed to their parents. For example, a woman who had told of a father who is 'interested in politics, athletic and not very happy' would subsequently be told about another man who was political and athletic, but with no mention of the 'not very happy'. Asked to recall this man a little later, the woman would be liable to say he was political, athletic and not very happy, imposing on the new person *all* the characteristics of her father. Triggered by information linking someone new to someone from the past, like these characteristics, the participants were liable to transfer additional traits from the old to the new. Other triggers might be a person's looks or their name.

Extending the studies, it was shown that we tend to feel more favourably towards strangers who resemble loved parents or siblings. We expect strangers who remind us of past figures to react to us in the same way as our relatives once did, and these expectations in turn feed back into how we feel. For example, if a person resembles our brother or sister, beside whom we always felt inadequate, being in this person's company may make us feel that way. We project our sibling's traits on to the newcomer, then reinvent ourselves as inadequate by comparison. Most dramatic of all, we are liable to manipulate people to behave in ways that actually resemble the originals – so powerful is our need to reconstruct the past in the present. For example, most of us are aware of repeating patterns in our friendships or love life, and up until now you may have attributed this to a tendency to fall for the same sort of lover or to seek out certain kinds of people as friends. This certainly happens, but in addition it seems we actually manipulate those who are close to us to behave in the ways that we were used to as a child. So if we found a parent domineering, we may actually cause a new person who resembles them to behave like that too, by baiting them or encouraging them to control us. The implication is that the sort of people we get close to are the ones whom we can persuade – all done unconsciously – to play the role we demand of them from our past. But even that may not be enough for intimacy to happen, because it works both ways: on top of what we need from them there is the matter of who they need us to be, based on their childhood. Friendship and love, it seems, go beyond two people finding compatibility based on their pasts; to achieve this kind of relationship, both parties must feel at home with being fashioned by the other to fit the other's precise childhood prototypes.

Auditing your role in the family script

So, what was your role in the family script? Your gender affected it, as did your place in the family, your looks and the unique combination of projections that each of your parents placed on to you. Remember how these influenced Prince Charles, how the dinner-table jokes at his expense for not knowing arcane equestrian terms and his special role as the recipient of the negativity in his father's inner life undermined his self-confidence. Whether you are a high or a low achiever, your script will have played a major part. If you were seen as special by one of your parents or intimates, or if you were an unwanted ugly duckling, that will still be affecting you.

First, look back through this chapter and list the points which struck home, the moments when you found yourself on familiar ground. Then proceed more systematically, using your memory to define how you believe you were influenced by the various factors:

- Your gender
- Your position in the family
- Your parents' aspirations for you at school and in your career

- The degree to which you were favoured or had an ugly duck-ling status
- How all this can be seen as the past in your adult present.

Make brief notes under the following headings: gender; birth order; achievement; ugly duckling; past in present.

To fill in gaps, if you are still close to a family member you might want to show them your account when you have finished it. If they are willing, you might even ask them to write down who they thought you were, using the same headings. If you are no longer close to anyone who knew you in your original family, there may be people you have long forgotten who could help, such as uncles and aunts – in fact, anyone who witnessed what went on back then.

There is no need to be definite about your ideas – you can put question marks after any uncertain statements you have made about yourself. Use this exercise as an opportunity to speculate about what was going on, including what your parents' motives might have been or what makes your siblings tick. Your audit might be as brief as the following one for a woman who had an older and a younger brother:

GENDER: Being the only girl meant Dad gave me more attention. Mum's hopes for me because I was a girl: to be the sexbomb she was?
BIRTH ORDER: Being second meant older brother resented me. Younger brother not much competition (because he's not bright or good-looking?).
ACHIEVEMENT: Mum dumped all her insecurities on me – she would have been threatened if I'd done well at school. Older brother got Dad's aspirations for achievement. Why I lack intellectual confidence?
UGLY DUCKLING: Younger brother had 'unwanted' status.
PAST IN PRESENT: Authority problem at work. Anger at bosses = anger at Mum for not taking me seriously?

There is no 'right' way to do this audit. Everyone's will be different. Another example, for a man who had a single older brother, might be:

GENDER: Being a boy a problem: Mum didn't really like men much. Dad leaving and Mum not keen on me being keen on rugger meant only my brother to share manly pursuits? In fact, I'm very masculine – why, given my mum's antipathy to it? Being butch a compensation? Do I like women?

BIRTH ORDER: Brother coming second lucky for me. He's more feminine – Mum's frustrated need for a daughter? Being oldest meant I had to be the man when Dad left.

ACHIEVEMENT: I don't know that Mum ever cared much how well I did at school. She seemed quite taken aback that I got good GCSEs. Dad was pleased. But why haven't I done better in my career, given that I did well at school?

UGLY DUCKLING: I'd say neither of us were much favoured by Mum. But I was lucky Dad favoured me. Brother such a workaholic because nobody really loved him?

PAST IN PRESENT: Mum always being positive, however bad things were, makes me unrealistic – idealizing women, over-optimistic about work. Always looking for my Dad in bosses, tend to be too much of a people-pleaser. Trouble with women – expect Mum's apathy towards me from potential lovers. Angry towards women because of Mum?

Having completed the audit, put it to one side to be joined later by those you will encounter at the end of each succeeding chapter.

Most of what I have described in this chapter is familiar. We can clearly recall the impact of having been the oldest or youngest, the only girl or boy. The experiences analysed in Chapter 3, from the ages of three to six, are less accessible, although most of us have some distinct memories from that time. It is time to return to those half-remembered years, to explore the origins of our conscience.

Chapter 3

SCRIPTING OUR CONSCIENCE, AGED THREE TO SIX

... But they were fucked up in their turn
By fools in old-style hats and coats,
Who half the time were soppy-stern
And half at one another's throats ...

Were you badly behaved at school ... or perhaps you were a goody-goody? Do you have a hatful of parking offences to your name, or are you one of those people who hates to break even minor traffic regulations?

Do you tend to fall out with bosses? Or are you a model employee who welcomes the hierarchical embrace of large corporations?

Do you find that sex is best when it's naughty? Are you prone to promiscuity or infidelity? Does familiarity in a partner quickly become a turn-off? Do you hate dirt and untidiness, or perhaps they don't bother you?

Do unwanted thoughts get immovably stuck in your mind? Are you more fastidious than most? Are you prone to damning judgements of others, ablaze with righteous indignation?

Your answers to these questions are a clue to which of the three kinds of conscience you have: punitive, weak or benign.

People with a punitive conscience tend to be conformist, taking the side of authority. They are often markedly similar to one or both parents in many respects, perhaps parroting their opinions almost word for word. They are liable to be sexually inhibited and prone to obsessional cleanliness or tidiness. They find themselves ranting against the wickedness or depravity of others, furious at signs of lax morality. They are diligent at school, college and work, if unoriginal.

By contrast, if you have a weak conscience you tend to be more of a rebel, someone who found teachers and rules objectionable and now finds bosses difficult to obey. Your sex life is erratic, you tend not to stick with partners and promiscuity may be a problem. You were not scared to steal as a child and, if caught, tended not to learn from the punishment. You are not very close to your parents and can be very aggressive. You tended not to do well at school, as at work now, and are liable to be unpopular. You like to drink and smoke, and when younger were not averse to illegal substances.

Unlike the preceding two types, those with benign consciences are not troubled by authority and its manifestations, although they are not slaves to it either. They rub along pretty well at school and work, being thought of as a 'decent sort'. If things go wrong, they try to understand rather than to blame. They enjoy sex, mostly as part of a stable relationship. Dirt or mess don't bother them, but there are limits to what they can stand. The benign type can take or leave alcohol and cigarettes, and when young only dabbled in drugs: they know when to stop, when to say no.

Very few people, if any, conform exclusively to a single type, although overall you should be more like one of the three. Whilst your conscience may be benign in most respects, for instance, it may be punitive in others and weak in still others. This patchwork effect comes about because both parents vary in their reactions to you with regard to the key issues that concern conscience: sex, obedience to rules and self-indulgence. How did your parents go about channelling your instincts, especially in the years between three and six, in regard to each of these? If they struck a good balance between restraining your youthful exuberance on the one hand and encouraging it on the other, you will tend to be benign. If they reacted with fierce prohibitions, it will have made you punitive towards those desires. If they were too permissive and inconsistent in punishing you, you are liable to be the weak variety. There is probably no one to whom both parents behaved consistently in only one of these ways with regard to all issues. Whilst one parent may have been liberal about letting you buy sweets, making you benign in your attitude to eating, the other may have been forceful about homework, making you punitive towards yourself in your career. They will have shown different attitudes to your siblings as well, so

94

the consciences of children with the same parents vary considerably, both from issue to issue and overall.

In my case, for example, because I suffered from carsickness as a child, when my father and I were in the car alone he would sometimes allow me to steer. In doing so he was being kind, but he was also encouraging a cavalier attitude to traffic regulations. Perhaps that has something to do with my difficulty in obeying them, my weakness of conscience in this regard. Yet in other respects I have a punitive conscience, as in my attitude to lying, and in others still, such as my feelings about food, I am benign. Whilst, overall, I should say I am benign, I am weak and punitive in some areas. This multiplicity is typical.

Parental scripting of different types of conscience

The starting point for any explanation of the different types is the theory of Sigmund Freud. According to him, a continual struggle is being waged within us to reconcile a number of competing demands. Instincts are constantly frustrated because the external world can rarely meet their urgent demands immediately. You cannot just reach out and grab anyone whom you find desirable, or stuff the most delectable morsels on the dinner table into your mouth as soon as they come into view. Enter the conscience, a set of beliefs about right and wrong, which polices the psyche through the guilt born of morality and the shame triggered by breaches of social convention. Its job is to censor greedy instincts and force them to delay gratification. But if instincts and conscience were the sum total of our mental apparatus we would simply lurch between extremes of abstinence and self-gratification; so a third agency, which has been steadily developing since our infancy, comes into play. This is the ego, a less moralistic, highly pragmatic, diplomatic 'fixer' who negotiates deals between slavering instinct, po-faced conscience and the practical feasibilities of the external world.

The work of the conscience becomes increasingly complex as we grow older, and the more developed the society the greater the demands it makes on us to curb our animal selves. The history of a typical university-educated person is an example. At every stage we must tell ourselves to be patient: 'Work hard for your GCSEs so you can take 'A' levels; work hard at them so you can go to university; get a good degree so you can get a good job; stick at your over-demanding, underpaid first job so that you can get a better one . . .' and so on, to the grave. To achieve all this we must constantly put off present pleasure in the name of future reward – write the essay rather than watch the TV programme, get an early night rather than stay for another drink. Our conscience is in charge of seeing that this and much else happens. It comes into play in all situations where repression of instinct is demanded, from whether to have a second slice of chocolate cake to regulating our sexual desires on a first date.

The sex instinct prompts a good deal of fantasizing before you meet the potential sexual partner. In the pub or over a meal the ego negotiates some real sexual expression through flirtation, but the conscience is already raising objections. These may be reasonable, like 'She may be attractive but, admit it, you don't really like her' or 'How will he look in the morning without the benefit of six vodkas inside you?' Or they may be irrational, like 'Sex is wicked' or 'I'm useless at sex.' Asked in for coffee at the end of the date, the cacophony from the conflicting psychic agencies can become deafening as each argues its case. This may continue even as you leap into bed, and it may be only for a few brief seconds during orgasm that instinct is supreme. Given half a chance, conscience is liable to pour immediate cold water on you in your post-coital state, sneering that you have made a terrible mistake or performed badly.

To what extent conscience plays a starring role in this or more everyday dramas that demand restraint, like whether to have a cigarette or buy an unneeded garment, depends on how punitive, weak or benign it is. The punitive is the psychic equivalent of a police state; if you are too punitive you are too inhibited. At the other extreme a weak conscience may instigate too much fun for your own good, resulting in indiscriminate, promiscuous pleasure-seeking. In between is the benign variant, better able to maintain the delicate balance between future penalties of self-gratification, like a hangover or getting pregnant, and the need to fulfil oneself in the present.

As we all know, Freud put sex at the centre of his explanation of most things. In the case of conscience, he traces the reasons for their differences back to earliest infancy, beginning with the uncomfortable assertion that small children are sexual beings whose sexuality has to be repressed. According to him, mouths are the first site of sensual pleasure because the infant's life revolves around sucking the nipple. As soon as they are able, they make sense of a new object by putting it in their mouth. If their oral cravings are not satisfied by feeding on demand and a tolerant attitude to mouthing of objects, the infant is supposed to become orally fixated in later life. There is some evidence to support this. Alcoholics or chocolate addicts are more likely to have have been weaned at earlier ages and, as infants, to have been denied opportunities for sucking. Numerous studies of smokers have found them to be more preoccupied by oral craving themes, like perceiving mouths when asked to react to pictures of random shapes, than non-smokers. (Incidentally, smoking is twice as common amongst depressed people and almost all schizophrenics who have ever been studied smoke. This fact may be linked to the several studies which have found associations between infantile feeding deprivation – which may be the norm for many schizophrenics – and smoking.)

In most cases the parental mission to civilize 'the beast in the nursery' does not begin in earnest until the second year. At ten months, 90 per cent of what the average mother does with her infant is affectionate, playful and devoted to meeting his needs; rarely does she seek to scold or thwart. But very soon comes a big change. She is transformed from unconditional meeter of needs to agent of control, so that in the child's second year much of her energy is devoted to inhibiting many of the activities that he most enjoys, like exploring, or bladder and bowel functions. Tantrums at being thwarted become commonplace and mothers issue a command or disapproval at least every two minutes at this age.

Freud called this the 'anal' stage, because toddlers are particularly excited by the process of excretion and enjoy the results of this activity. How the mother goes about curbing the child's messy pleasures affects his later attitude to his instincts. If her response is rigid, condemnatory and angry, the child develops an 'anal personality', comprising obsessive orderliness (from being made fearful of mess), obstinacy (still angry at being forced to excrete on demand) and parsimony, especially about money (it becomes equated with faeces and, in later life, being tight about money may symbolize holding faeces in). In experiments, adults who had previously been assessed as anal were found to be more liable to a host of uptight behaviours. These ranged from carefully arranging magazines that were in disarray to compulsive stamp collecting, greater obstinacy in argument immediately after being asked to play with a faeces-like substance (play-dough), and dislike of praise as a reward compared with money. The anal personality is most common in the offspring of anal mothers – anality breeds anality.

During the second or third year, increasing discouragement of the child's joy in mess culminates in toilet training, and the sensual focus shifts from the anal to the genital region. You don't need to be a wide-eyed convert to Freud's doctrines to observe that sex play amongst three- to six-year-olds is the norm. Be it Doctors and Nurses, Mummies and Daddies or 'I'll show you mine, if you show me yours', most of us can remember something of the sort. Even in the far more sexually inhibited atmosphere of 1948, when Alfred Kinsey published his famous study of sexual behaviour, more than half of his male respondents could recall mutual masturbation in childhood with other boys and one third remembered touching girls' genitals. More recent surveys, based on mothers' accounts of their children as well as on adults' memories, also confirm that it is normal for both boys and girls to be sexually active between the ages of two and six. Boys have been averaging three erections a night for two hours at a time whilst in the womb, and postnatally both sexes are capable of orgasm from early on. Although small boys cannot ejaculate, by five months they can achieve a striking duplicate of adult orgasm. Girls are capable of real orgasms from three years of age, if not before.

Two-thirds of young children are reported by their mother to touch their own sex parts and they are very liable to be voyeuristic and exhibitionistic, sometimes engaging in outrageous flirtation and adopting a seductive manner, often impersonating their parents, older siblings or television characters. Boys favour public masturbation more than girls, who are partial to exhibitionism. According to their mothers, three-quarters of two- to six-year-old girls walk around in their underwear, and two-thirds do so in the nude (versus half of boys) or sit with their crotch exposed (versus one-third of boys).

But these days of sybaritic living are numbered, and as primary school age approaches there is a marked decrease in overt sexuality. Like Adam and Eve, children begin to feel uncomfortable if others see their naked bodies. They demand bathroom privacy and become self-conscious when changing to go swimming at public baths. Eight-year-olds are four times less likely than four-year-olds to be reported by their mothers as touching their sex parts or showing them to other children or adults. Female modesty replaces exhibition, so that older girls are five times less likely than younger ones to walk around in the nude or their underwear.

Freud explained this new self-repression as being the result of the 'Oedipus complex'. Using as his template the myth of Oedipus, a man who killed his father and had sex with his mother, Freud claimed that this is a universal, genetically inherited script and that it is the foundation of conscience. He believed that, as small children, all of us long to possess our opposite-sexed parent and to kill the same-sexed one. As time goes by, we realize that our same-sexed parent is much stronger than us, and to avoid posing a threat to them we repress our sexual attraction and identify with their conscience – adopt their moral precepts as our own. In short, that conscience is the result of fear and based on that of our same-sexed parent.

Whilst Freud was right about children having a sexuality, it now appears that he exaggerated the importance of the Oedipus complex in forming conscience. Far from it deriving from fear of parents, the evidence suggests that being loved by them is much more important.

In my case, for example, as described in Chapter 2, there were two occasions when my father made it clear that he passionately wanted me to buckle down and do well academically, as he had done. After my 'O' levels, when he presented me with three alternatives for my future, he made it unmistakable that it really mattered to him that I get my act together. Many parents have tried to cajole, bribe and otherwise persuade their sons to work harder, but without success. My father succeeded because he had real purchase with me, having been intimately involved in my childcare since birth. He had been unstintingly enthusiastic and encouraging throughout, a very tactile person who offered hugs and affection liberally. Although he occasionally lost his temper completely, for the most part he was not a frightening presence. I believe I complied with his aspirations for me not out of fear but out of a sense of obligation and love for a kind man. Also, I identified with his scholarly ways. A doctor and psychoanalyst, he was a voracious consumer of nonfiction which provided a model of how to be disciplined. In the final analysis, I did not want to hurt him.

This is consistent with the scientific evidence. If a parent is friendly and loving we are more likely to identify with them than if they are scary and frightening: boys who are beaten up by their fathers, or girls whose mothers are savage, tend not to identify with them. Paul, whom I interviewed for a TV documentary, is an example. Along with his three brothers, sister and mother, he had been horribly abused by his father, a cruel and brutal man with whom he did not identify.

Now, if I care to think about all the unpleasant things that happened in our childhood, I could actually write a book. My father was a very violent man indeed. The worst case I can think of is one particular time I'd gone to the Baptist church and come back home on a Sunday.

I'd be about ten. My mum was crying her eyes out in the back kitchen and I happened to look on the wall and there was a handprint of blood. So I turned round to my mum and said, 'Oh Christ, what the hell's been going on?' And she said, 'Your dad hit Jane [his sister] with such a force she went headfirst against the wall, and there was a bang just like a bag had bust and she split her skull wide open from here right the way back, and she's had twenty-two stitches in her head.' And my mother was rather hysterical about it, as you can imagine. My father had put one hand at both sides of the skull to press the gash together, and he told my mother she was making a fuss over nothing.

Jane eventually went to the hospital, and when she came back she told my mum that the hospital couldn't say there and then if she had a fracture of the skull but if she had trouble seeing things or if she was sick she was to come back. Well, she wasn't in the house ten minutes when she turned round to my mum and said, 'Mum, Mum, I can't see! I can't see!' So she was rushed back into the hospital and she was under – what's that thing where they say that a person is touch and go for a week? – and my mother prayed like she'd never prayed before and my father never showed no concern whatsoever.

My father didn't only do certain atrocities towards the kids, he also did things to my mum quite a lot. I can recall one particular incident. She'd just come out of hospital and that particular time, the doctor told her that she'd have to spend some time in bed because she had bronchitis and she called my dad upstairs. 'Look,' she says, 'I can't go round to the Post Office to get my Family Allowance. Would you kindly go round to pick it up?' He came back about half an hour later and told her that he'd lost it and she said that she didn't believe a word, called him some names, and he got hold of my mum and he punched her quite a few times. Now, bear in mind, she'd just come out of hospital.

Paul was a sensitive fellow, implacably opposed to everything his father stood for and strongly identified with his mother. He had his mother's religious beliefs, her warm personality and her concern to put others' needs before his own.

I'm very close to her indeed. She'd share certain things with me. I'd a lot of compassion and a lot of feeling for my mother, and I used to think to myself it would be a nice thing for my mother to be growing old gracefully. But, instead, I'd sit back and I'd find my mother getting old before her time. She was doing everything at home, vacuuming, washing and ironing, out of routine, right? And she was taking pills like they were going out of fashion – sleeping pills – and she was becoming nothing more than a zombie. I found it very hard to put up with. And there was my father – he was in his own little world. Nothing penetrated it.

Scripting your sexual conscience

That Paul identified with his mother's, rather than his father's, conscience is typical: he wanted to be like the parent who loved him, not like the one he feared. It is also interesting that Paul was homosexual, because our sexual conscience – what seems right for us sexually – is tremendously shaped by our family scripts. Of course, it is partly fashioned by the wider *mores* of our society: for instance, there have been societies like ancient Egypt, in which even incest was permissible, and others where sex was so repressed that, as in the case of Victorian Britain, it was unacceptable for a woman to show even a bare ankle in public. But within these broad social influences, it is the mixture of attitudes to sexuality directed towards us by our other family members that generates our specific 'lovemap' – the set of strong preferences for sights, sounds, smells and personality characteristics in a potential mate. Genetics have been shown to play little or no part in these. According to John Money, the inventor of the notion of lovemaps, their topography is drawn by the end of childhood, and he asserts that 'once a lovemap has been formed it is, like native language, extremely resistant to change'.

Based on our childhood experiences, we develop notions of an ideal partner, a lover with whom we enact fantasy scenarios that we may hope to carry across into reality, linking love and lust. If we are deprived of childhood sexplay or are sexually exploited, our lovemap is vandalized. Lust may be wholly inhibited, so that only love is possible, not sex. Alternatively, lust may become disconnected from love and we may turn into a sex addict, constantly craving genital sex but without forming a relationship. Finally, lust and love may be connected but in pursuit of self-destructive relationships or lonely sexual outlets, like fetishes.

Since most people are heterosexual, they tend to base at least some of the features of their lovemap on opposite-sexed family members. They are the prototype for subsequent relationships. Although there is considerable variation in the answers when couples are asked about the way their parents related to them as children, they are more likely to have chosen a partner who is like their opposite-sexed parent: what their same-sexed parent was like does not predict their partner's attributes as much. When teenage girls describe their fathers and separately describe elements in men that they find sexually desirable, the two often overlap, whereas their accounts of their mothers and of desirable men do not; similar findings exist for women. A very ingenious study tested the same theory for both sexes, but far more scientifically.

The group studied consisted of 980 Hawaiians, all of whom had one parent of one race and one of another, and all of whom had married twice. If the theory that we are more attracted by attributes of our opposite-sexed parent were true, the Hawaiians should have been more liable to marry partners with the same ethnic background as their opposite-sexed parent. This turned out to be the case not only for first marriages but for second ones as well: if you were a woman and your father was white and your mother black, you were more likely to marry a white than a black. Overall, two-thirds of marriages were to a partner sharing the ethnicity of the opposite- rather than the same-sexed parent.

At the next wedding you attend, take an unvarnished look at the bridegroom and the bride's father. Sometimes the physical resemblance is uncanny, and the similarities in personality can be equally striking. Likewise, when you look at second or third marriages ask yourself if the new spouse is not remarkably similar to the previous ones. Alas, when it comes to love most of us find it hard to learn from our mistakes. All too often people go through horrendous relationships only to make exactly the same choice again. Doubtless this is part of the reason why divorcees who remarry are even more likely to divorce than first-time couples.

Our lovemaps are so tenacious because childhood sexuality is repressed. When feelings are buried they don't disappear altogether – only from conscious awareness. Repressed desires continue to seek expression, which probably explains why most people are turned on to some extent by the idea of 'naughty', forbidden sexual partners or practices. It is not only opposite-sexed parents who were sexually out of bounds in our early childhood; so were same-sexed ones, siblings, friends, older children, all adults – in fact, everyone. Whilst childhood sexuality does find expression, it is often accompanied by guilt. Fifteen per cent of us can recall sexual activities with siblings, and far more than that percentage probably felt desires towards them which had to be repressed. Although I know of no research which examines the extent to which our choice of partners in later life is based on what our desired opposite-sexed siblings were like, I am sure it would demonstrate that this is at least as influential as parental attributes. Nor does sexual repression end at the door to home. At school and elsewhere, if children display overt sexuality it is stamped upon, usually with a ferocity that quickly indicates to them just how wrong it is perceived to be. Repressed desires for schoolfriends and teachers may also be a major influence on adult desires.

But repression is not the only topographer of our lovemap. In choosing partners we are also influenced by a plain and simple affection for loved parents and siblings. If we got on well with one of them, this is likely to make us favourably disposed to people who remind us of them. The configuration in the homosexual Paul's family home may be a case in point.

Despite the huge publicity accorded to them, studies purporting to prove that male homosexuality is caused by genes or brain abnormalities have turned out to be largely groundless. Very recently published studies have found little or no genetic effect and there is a large body of evidence, of which Paul is in many respects typical, to suggest that homosexuality has a significant environmental component.

Gay men are twice as likely as heterosexuals to come from a distinctive family constellation, half of them painting the following picture of their boyhood: the mother likes to be the centre of his attention and they are unusually close; she regards the rough-and-tumble play enjoyed by most boys as dangerous, and is excessively anxious about his health and safety; this inhibits his aggression, and he is clinging and highly anxious at being separated from her; in early adolescence she may be flirtatious with him and she is a dominant, powerful woman who is uncomfortable with masculinity – emasculating even; she takes the family decisions more than the father, and is the stronger personality; the father is rejecting or withdrawn or weak or absent – emotionally, literally or a combination of these – and the marital relationship is disharmonious.

The studies on which this picture is based show that gay men are much more likely to report having been 'feminine' boys, enjoying activities such as playing with dolls and girls, compared with heterosexuals. About two-thirds of feminine boys become gay men, and the more effeminate the boy the greater the likelihood. Whilst effeminacy may be partly genetic in cause it can be caused environmentally, by the combination of a dominant mother and a hostile or remote father, as the stories of Elton John and John Gielgud in Chapter 2 suggested. Gay men tend to have had negative relationships with their fathers, half of them (compared with a quarter of heterosexuals) feeling anger, resentment and fear towards fathers whom they deem cold, hostile, detached or submissive. They do not see their fathers as a role model. Over 70 per cent feel dissimilar to them while growing up (compared with a third of heterosexuals) and more similar to their mother. About half feel their mother did not want them to be like their father, and more than two-thirds feel that their mother dominated their father, who had less of a say in family decision-making than she did. As in Paul's case, the intense, loving relationship with the mother leads to an identification with her rather than with the distant or hostile father. Although it remains very possible that future studies will show that at least a quarter of gay men do have a strong genetic tendency to desire their own sex, it seems that the remainder are largely nurtured (interestingly, male homosexuality has turned out to be far less common than the 10 per cent of men originally proposed by Kinsey's 1948 study. At most, only 3 per cent are exclusively homosexual, with perhaps 2 per cent more having been bisexual at any time since puberty. However, if it is true that sons of absent fathers with dominant mothers are more likely to become homosexual, then in theory the increased divorce rate combined with the increasing assertiveness of women and greater tolerance of homosexuality could result in more homosexual men in the future.).

Gay men are often asked: 'When did you first realize you were homosexual?', but the same question could equally validly be asked of heterosexuals regarding their penchant for the opposite sex as partners. As the evidence regarding childhood sexuality demonstrates, many of our initial experiences, at the ages of four or five, are with our own sex. At least partly, we learn to be heterosexual by subsequent social moulding; and our very specific lovemaps originate in early family relationships.

The sexual is only one of several domains in which the conscience holds sway. Attitudes to authority, criminality and conscientiousness at work are some of the others. In the following pages I describe the basic causes of the three different kinds of conscience, but please remember that there is no one who completely corresponds to the types I am describing because the pattern of care from each of our parents is liable to vary from one issue to another.

Scripting the benign conscience

The kind of parenting that creates a benign conscience is called 'authoritative'. It is both responsive to our needs and demanding of us. On the one hand, our parents demand that we participate in the family, closely supervise our behaviour, discipline us using reason and persuasion, and confront disruptive behaviour before it becomes established. On the other, by being responsive to our special needs they foster our individuality and self-assertiveness. They devise situations in which we feel effective and offer praise for effort and achievement (but, interestingly, not for intelligence; praise for being clever runs the risk of making us feel we are never living up to our potential or of giving us a sense of automatic superiority regardless of the reality). Only occasionally do they use 'power-assertive' methods, like shouting or threats, when wanting to stress the importance of something crucial in the face of defiance. Early in our childhood they point to direct consequences, like 'If you keep on pushing him, he'll fall over.' Later, there is explanation of motives, like 'Don't yell at him. He was only trying to help', becoming increasingly sophisticated with age, such as 'He feels bad because he was proud of his sandcastle and you knocked it down.' Using these techniques, internal control replaces external, prompted by guilt at distressing our loved parent after misbehaving or pride for conforming to the wishes of an admired one. Eventually, moral behaviour, being 'good', is rewarded by ourselves.

The resultant benign conscience meant that as a child we were competent for our age, were self-motivated rather than requiring constant supervision, tended to succeed at school and got along well with our peers, neither suppressing feelings nor disposing of them in others. In later life we were less likely to be tempted by self-destructive habits like drug abuse, or by delinquent activities such as disruptive classroom behaviour or stealing and violence. From a young age we weighed up the competing demands of empathic concern against hedonism. In other words, taking the needs of others into account, as well as self-gratification, regulated our conduct.

Being totally dependent on parents for survival, all children naturally want to please them and display a preoccupation with their wellbeing from a very early stage. By the age of two they already display more concern if their mother shows distress than if they witness it in a stranger. Children of authoritative, empathic parents are themselves more empathic, making them better able to take on board another person's point of view, whereas lack of parental empathy generally results in self-centredness. The deprived child is far more preoccupied with getting his own way than with worrying about how others are doing, although under certain conditions deprivation creates a compulsive, excessive and inappropriate concern with the welfare of others. Toddlers with depressed, and therefore often unempathic, mothers, may suppress their frustration and distress more when mishaps occur, to avoid upsetting their mothers. This is particularly true for daughters who are used as comfort objects, where the mother often combines physical affection with sadness, creating worry and misplaced feelings of culpability, transmitted down generations from mother to daughter. Later these children are particularly prone to feeling guilt, to taking responsibility when things go wrong and to having excessive concern for others, compared with children of undepressed mothers. In later life, this can result in a complete incapacity to care for themselves and an obsessive seeking out of the 'walking wounded' to care for.

Of course, both our parents were not authoritative all the time, in all respects. There were areas in which each fell short of the ideal, so that one may have been authoritative towards us regarding homework but less so with bedtimes, whilst the other may have been inconsistent about pocket money. Nor do parents behave in the same way in their dealings with all their children, despite their natural desire to claim that they do. For example, my father showed great interest in my school reports and took considerable trouble to help with my homework, yet one of my sisters recalls that he did not even read her reports and once told her that if she wanted to leave school at 16, that would be fine by him.

Such parental differences also affect the development of a benign sexual conscience and behaviour. We have already seen that relationships between mothers and sons can affect whether a boy is homosexual; those between father and daughter can also be highly influential. For example, if daughters were close to their fathers as children they are less likely to begin having sexual intercourse at an early age, and if they felt able to talk openly and congenially with him they are more likely to report a satisfying adult sex life – to have a benign sexual conscience. But a father may have been close to, and able to talk well with, one daughter – and not so with another. He may have been comfortable being naked in the presence of one but made anxious about this, perhaps because of her particular beauty, in the case of another. Outsiders observing these two daughters in their teens may say, 'Look how different they are with regard to sex. One is relaxed about it – the other gets embarrassed or standoffish as soon as it is mentioned. They had the same parents and upbringing, even went to the same schools – so it must be genes, not nurture, that made them so different.' But, effectively, they did not have the same father. It is his differing relationship to them, not genes, that explains the difference.

Such variations from both parents, manifested again and again over many years, are crucial in creating the differences not just between our own and our siblings' sexuality, but in personality, abilities and mental health. They also affect what aspects of our different parents each of us identifies with. But if identification with parents and a benign conscience are largely the result of being loved consistently, love alone is not enough.

It almost goes without saying that authoritative parents need to be teaching the right rules – children of parents who teach them that stealing is right are, not surprisingly, more likely to steal. Quite simply, the wrong rules are being offered. Conversely, it is helpful when good parental models are presented – it is easier for us to enjoy academic work, for instance, if we have witnessed a parent doing so.

On top of this, we tend to expect to be treated by others the way our parents treated us. We get used to things being a particular way and, having learnt this, bring it to bear on what happens outside our family and in later life. For instance, if our father was fierce we are more likely to see bosses at work that way. A study of junior doctors found that their feelings about managers directly reflected the degree to which their fathers had been unsupportive, strict and domineering. With fathers like these, doctors were more likely to perceive their manager the same way.

But benign conscience is not merely taught. Towards the end of his life Paul Feyerabend, the modern German thinker, summed this up rather well. 'I conclude that a moral character cannot be created by argument, "education" or an act of will . . . Like true love, it is a gift, not an achievement. It depends on accidents such as parental affection, some kind of stability, friendship and – following therefrom – on a delicate balance between self-confidence and a concern for others.'

It is through relationships and out of emotion, not just via the transmission of knowledge, that benign morality develops. If all that were required to create good citizens was to teach them right and wrong at school, like teaching the alphabet or biology, it would make the job of governments far easier. But whilst the analogy between humans and computers is useful for some ideas, when it comes to morality it is not sufficient merely to think of a child as hardware in need of the right software programming. When a four-year-old girl grabs her sister's present from her on her birthday, just telling her that this is wrong will not work. There needs to be enough love for the parent who tells her not to do it that she feels guilt or distress at having let that parent down. She needs to feel an element of shame at having acted in a way that her parent disapproves. Love creates those crucial emotions far more effectively than fear.

Scripting the punitive conscience

In contrast to the benign, the punitive conscience, if that is what we have, crushes individuality and allows little space for instincts. We fear authority. We are likely to be fastidious and excessively concerned with cleanliness and tidiness. We may find it hard to sort our thoughts out, because we are liable to be obsessive, as well as prone to other symptoms of neurosis such as hysteria (most common in women), phobias and panic attacks. We are easily and frequently afflicted by a guilt which may amount to depression. Damning judgmentalism is never far from our mind, whether it be ourselves or others whom we are censuring. We are liable to think, 'I'm fat' or 'I'm stupid' or 'I'm lazy' when we are not, or to condemn others who we deem to have these problems. Our righteous indignation may be attached to a passionately felt ideology, often political or religious, determined to coerce or manipulate everyone else to behave in the 'right way'.

We may have managed to convert these aspects of our personality into great achievement. Perfectionism can be a virtue in some sciences or in administrative posts; judgmentalism can become a crusading zeal that creates a forceful leader. But in most cases, punitiveness makes us underachieve. Our educational performance may be rigorous but unimaginative – intellectual acuity dulled by obsessiveness. In adult life our moralizing is offputting in the workplace and we are attracted to jobs which are secure and predictable, especially regimented professions such as law enforcement and the military. Law-abiding and fiercely critical of people who are freer in their self-expression, we are secretly envious of them and greatly enjoy being the agent of its curtailment. It would be interesting to know if those traffic wardens or car clampers who love their work are more than averagely likely to have punitive consciences.

All these generalizations may be overthrown, however, because the punitive 'psychic police state' is as vulnerable to revolution as any other dictatorship, even if it takes time. The politician who was a good boy at school, clean-cut at university and married young to a nice sensible woman just like his mother, to give himself the right image when seeking selection during his twenties or thirties, may run wild in middle age. He may burst out of the chrysalis of his pleasant but dull, people-pleasing personality. The wife is replaced by a pretty young researcher and his name may appear in the tabloids: it is the adolescence he never had. People with punitive consciences dare not openly rebel against their parents in their teenage years.

If we are punitive we are liable to associate sex with forbidden fruit, feeling it to be wrong and becoming inhibited as a result. Our partners are particularly likely to be someone who closely resembles, or has a similar personality to, opposite-sexed parents or siblings, because repression of our childhood sexuality was fierce. We choose them quite unconsciously. But familiarity subsequently breeds contempt, at least partly because the chosen mate's actual similarity to forbidden figures from the past becomes more conscious and undeniable when we get to know them well, reactivating the fiercely imposed incest taboo. Now we find ourselves strangely diffident about, or even disgusted by, the same body and person whom, not long ago, we desired passionately. We may find 'naughty', illicit sex especially exciting, whereas love is a turn-off because it is too close to home (our original, family one). Forbidden sexual targets may be irresistible: uniformed nurses or schoolgirls for men, for instance and for women unsuitable charmers or rotters.

Of course, whilst this is particularly marked in people with punitive consciences, familiarity reduces sexual activity in most couples to some extent. The definitive British sexual survey of twenty thousand people, published in the 1990s, revealed that the amount of sex we have in general is not nearly as much as might be thought. Fully 52 per cent of women and 38 per cent of men had had only two or fewer partners in their lifetime. There is a steady decrease in the amount of sex between couples over time, so that, compared with the first year of being together, the graph of sexual frequency has fallen considerably by the third. Ageing certainly accentuates all this, so that the average frequency of sex among 16–44-year-olds in relationships is five times a month, whereas it is only twice a month in the 45–59-year-olds. These trends are accentuated among people with punitive consciences.

A patient whom I treated for two years illustrates what causes the punitive conscience. Mrs B was referred to me for once-weekly psychotherapy because she was suffering from frequent panic attacks. A thirty-two-year-old full-time mother of three young children, she had moved to London from Glasgow six months earlier so that her husband could take up a new job. Tidily, unostentatiously dressed, with modestly styled shoulder-length blonde hair, she was slim and of average height. She had a polite, pleasant manner and seemed as eager to be helpful as to be helped.

In the early sessions she described some terrifying experiences. On one recent occasion she had had to travel across London, to her a huge and unfamiliar city. As she found herself further and further away from her home she became convinced she would never be able to find it again. She drove literally and figuratively in circles, from one roundabout to the next. From the vantage point of her car her sitting room seemed like a safe haven but when, eventually, she did make her way back home, the agoraphobia she had just experienced reinvented itself as claustrophobia. It was intolerably oppressive. The walls seemed to close in on her and she felt a desperate need for fresh air, yet the thought of leaving the house was accompanied by an equal dread. Whatever she did she would still be at risk, in a state of nameless fear.

Mrs B's conscience was extremely punitive towards her instincts, and her panic attacks were partly caused by this. She was passive and fearful of expressing aggression, strictly law-abiding and slavishly compliant to authority-figures such as bosses and teachers. Despite having done well at school and in her career, she lacked self-confidence and was liable to denigrate her achievements. She was fearful of upsetting others in social situations, and never allowed herself to let go.

As she became more aware of the way that she was restricting herself, she realized that there was a prohibitive 'little voice' in her mind which belonged to her mother, not to her. She did not drink at the start of the treatment, but began to do so after a Damascene moment one evening when her husband, Jim, offered her one. Before answering 'No thanks' she suddenly thought, 'Why shouldn't I?' Acknowledging that it would have been her teetotaller mother rather than herself who was refusing it, after that she was able to drink moderately.

Mrs B's 'little voice' made her very similar to her mother in a number of respects, sharing the same phobias and panicky feelings as well as personal style and feelings. At the start of the treatment her greatest fear was that she would never be different from her, but as it progressed she became sophisticated at identifying and over-riding the 'little voice'. She reported becoming very angry with her son when he sat on her lap and kissed her, but was able to realize that this was just what her mother would have done, with her active dislike of displays of affection and physical embraces. Mrs B's son developed mild school phobia, but it disappeared after she became aware of how she was inducing it in him, just as her mother had induced phobias in her.

Typical of those with a punitive conscience, she reported that she had enjoyed sex less since getting married, familiarity having bred disinterest even if not contempt. Her husband was keener, liable to complain because they only did it once or twice a week. Also typical, sex felt 'dirty' and she was mildly phobic about this: she would take a shower beforehand because either he or she was 'smelly'.

Mrs B's story illustrates the role of the psychological mechanism known as 'projection' in causing punitive conscience. Parental projections on to us, forcing us to feel the way they do, can instil a strong fear of being a bad person. In Mrs B's case, her panic attacks were partly caused by feeling that she was wicked and about to be punished for prohibited thoughts and feelings.

Parents are constantly attributing their thoughts and feelings to us, mistaking what is theirs for what is ours. If we deviate from these scripts, they may work vigorously to fulfil the prophecy. Some of these projections do actually benefit us. For example, my mother always took great delight in my attempts to dribble a football aggressively, projecting on to me this persona, which in later life stood me in good stead on the pitch (although my team mates may disagree – one of my friends used to chant 'Selfish James Never Scores, Selfish James Never Scores' in a vain attempt to persuade me to part with the ball). In making these projections my mother was hoping that I would be able to act in a way that she would like to have done, with pleasing results for both of us. However, many parental projections are of unpleasant feelings – ones that they cannot bear alone.

Parents routinely use their children as recipients of their own intolerable feelings, as well as vehicles for desirable aspirations. If our mother was angry and shouted at us for being dirty out of her own horror of filth, it may have made us feel angry too. Having connected these emotions with this issue, we then became like our mother in this respect. Perhaps in turn we shout at our own offspring in the same way, for the same reason, and so on down the generations. In simple terms, as several studies have shown if you feel bad, making someone else feel bad is quite a good short-term method of coping. Now it's their problem, not yours. You may then deal with it in the same way, projecting it on to siblings and friends in the playground and, in later life, on to subordinates at work and on to partners and children at home.

Although everyone does this to some degree, parents vary considerably in terms of what and how much they project on to their children. Punitive consciences are caused by a stream of negative projections, creating pathological guilt and fear of wrongdoing. In Mrs B's case, this was illustrated by a chain of thought that began with a dream. She was five, and her mother was talking to a neighbour over the garden wall. Mrs B began to urinate until the flow became a flood covering her head and she feared she would drown. After recounting this dream, Mrs B recalled that she had been toilet-trained at fourteen months, very abruptly and fiercely. Throughout childhood she continued to wet herself from time to time and, at the age of ten, did so in a car park, to her mother's rage and embarrassment. This single incident led to her mother sending her for tests to identify a medical condition, but the doctor diagnosed a psychological one. The true problem lay in the mother, not the daughter. Having made Mrs B neurotic about bladder control by the way she had been toilet-trained, her mother's anal personality, with its fear of loss of control, was transferred into her daughter.

Mrs B experienced her mother's injunctions as a voice separated off from the rest of her experience. Once its real author had been located she was able to distinguish herself from it. It had been forced into her, rather than incorporated as part of herself. In fact she identified more with her father, who was loving and supportive. With the help of therapy, she was able to ignore the little voice's dictates with rapidity and to great effect. Indeed, she went on to provide 'authoritative' parenting to her own children, rather than simply reproducing her mother's cycle of punitiveness.

At the heart of parenting which creates punitive consciences is the conviction that the child's will must be 'broken', as the psychoanalyst Alice Miller has so graphically described. She labelled the notion that parents must be cruel to be kind as 'Poisonous Pedagogy'. Such parents see even the newborn infant as wilful and capable of deliberate disregard for parental authority, which is perceived as insulting to the adult and dangerous to the child's later character if not ruthlessly quelled. He or she must be taught unquestioning obedience as early as possible, preferably while still too young to understand what is being done. The most strenuous and manipulative methods must be employed, up to and including physical abuse.

An example of how commonplace these ideas are comes from a discussion of the validity of corporal punishment amongst a group of young women. An advocate of it asked, 'What about infants? Even if they're not wet or hungry or sick, babies cry.' Offered non-violent alternatives for dealing with these problems, she replied, 'All they want is attention. If you let them get away with it, before you know it they'll control you completely.' Similar reasoning is often used to justify leaving toddlers in the care of strangers or with large groups of other toddlers: 'They have got to learn to be independent, so it's best if they learn it young. They may be upset for a few minutes, but it's for their own good.'

The most extreme kind of punitive conscience is known technically as the Authoritarian Personality (not to be confused with the similar-sounding 'authoritative' parenting). The type was identified after the Second World War in an investigation of the causes of fascism. If we have this personality, we are hostile to 'legitimate' social targets – minorities such as Jews, blacks or homosexuals. We are submissive to strong leaders, glorifying them. We avoid introspection, preferring toughness and cynicism, and regard others with suspicion, attributing ulterior motives to the most innocent behaviour. We are superstitious, and tend towards repetitive and ritualistic behaviour. We are heavily preoccupied with others' supposed sexual improprieties, so that a sexually arousing scene in a film or a provocatively clad passer-by may trigger an outburst of condemnation because we are disposing of feelings that were impermissible in our childhood on to others, in particular despised groups. Having projected all our 'bad' traits on to women or Jews or blacks, it makes complete sense to punish these 'despicable' people for their 'oversexed' or 'sneaky' or 'aggressive' behaviour – ways of being that cannot be acknowledged as our own secret wish.

The Authoritarian Personality results from rigid and punitive parenting, in which sexual and aggressive childhood impulses were not tolerated. If we grew up in this background we had no alternative but to ban such wishes from our consciousness, to become unaware that we had them and anxious if they threatened to erupt, requiring heavy defences. Because our parents were so frightening, using beatings or ridicule to obliterate the slightest sign of sexuality or aggression, we dared not acknowledge that we felt rage as well as fear at such maltreatment. Instead, we put our parents on a pedestal, idealized them as wonderful, and all our rage became directed against those social groups regarded as despicable. Raging against the objects of our parents' prejudices, neatly helps to conceal the true target of our fury – the very parents whose cause is now championed.

Authoritarian Personalities are attracted to illiberal political leaders and ideas. One group that has been closely studied is the American Christian fundamentalists, comprising fully 9 per cent of the US population. They believe the Bible to be the literal truth and are in thrall to a hierarchical God whom they perceive to participate actively in their life. They are convinced that we are living at the 'end time' of war and famine that precedes Armageddon, the end of the world. They are aggressively hostile to various social subgroups like homosexuals, women seeking abortion and abortionists, some of whom they have actually murdered. Compared with ordinary Christians, children of strict fundamentalists are more sexually inhibited. Their parents support the use of corporal punishment. The more that fundamentalist parents believe in the literal truth of the Bible, the more likely they are to use severe physical chastisement and to believe that it is the duty of children to care for and support their parents, rather than the other way around; also, the less empathic they are in the care they provide. Fundamentalist parents are actually as likely to nurture in these ways as parents who have been convicted of child abuse.

Aspects of fundamentalism are widespread in America. Half of all Americans agree with the statement that 'the Bible is the actual word of God and is to be taken literally, word for word' and nearly one third describe themselves as 'born again'. Given the role of corporal punishment this is not surprising, since 84 per cent of American parents regularly employ it. Of course, adherence to such beliefs is not caused purely by strict parenting, but also results from social and political trends. Fundamentalism has created for itself a central role in US political life through the Christian Coalition, a militant group that has successfully infiltrated the Republican party machine in many southern American states, including Texas, where the President at the time of writing, George W. Bush, was governor. The Coalition was an important factor in his election, as it was in that of his father in 1994.

Bush Junior provides a remarkably fine example of the Authoritarian Personality. When still an alcoholic, approaching his fortieth birthday in 1986, he had achieved nothing he could call his own. He was all too aware that none of his educational and professional accomplishments would have occurred without his father. He felt so low that he did not care if he lived or died. Taking a friend out for a flight in a Cessna airplane, it only became apparent he had not flown one before when they nearly crashed on take-off. Narrowly avoiding stalling a few times, they crash-landed and the friend breathed a sigh of relief – only for Bush to rev up the engine and take off again.

Not long afterwards, staring at his vomit-spattered face in the mirror, this dangerously self-destructive man fell to his knees and implored God to help him and became a teetotalling, Fundamentalist Christian. David Frum, his speechwriter, described the change: 'Sigmund Freud imported the Latin pronoun "id" to describe the impulsive, carnal, unruly elements of the human personality. [In his youth] Bush's id seems to have been every bit as powerful and destructive as Clinton's id. But sometime in Bush's middle years, his id was captured, shackled and manacled, and locked away.'

One of the jailors was his father. His grandfather, uncles and many cousins attended both his secondary school, Andover, and his university, Yale, but the longest shadow was cast by his father's exceptional careers at them. On the wall of his school house at Andover there was a large black and white photograph of his father in full sporting regalia. He had been one of the most successful student athletes in the school's 100-year history and was similarly

117

remembered at Yale, where his grandfather was a trustee. His younger brother, Jeb, Governor of the State of Florida, summed the problem up when he said, 'A lot of people who have fathers like this feel a sense that they have failed.' Such a titanic figure created mixed feelings.

On the one hand, Bush worshipped and aspired to emulate. Peter Neumann, an Andover roommate, recalls that 'he idolized his father, he was going to be just like his dad.' At Yale, a friend remembered a 'deep respect' for his father and when he later set up in the oil business, another friend said, 'he was focussed to prove himself to his dad.' On the other hand, deep down, Bush had a profound loathing for this perfect model of American citizenship whose very success made the son feel a failure. Rebelliousness was an unconscious attack on him and a desperate attempt to carve out something of his own. Far from paternal emulation, Bush described his goal at school as 'to instil a sense of frivolity'. Contemporaries at Yale say he was like John Belushi's character in the film *Animal House*, a drink-fuelled funseeker. He was aggressively anti-intellectual and hostile to East Coast Preppy types like his father, sometimes cruelly so. On one occasion he walked up to a matronly woman at a smart cocktail party and asked, 'So, what's sex like after fifty, anyway?' A direct and loutish challenge to his father's posh sensibility came aged 25, after he had drunkenly crashed a car. 'I hear you're looking for me,' he sneered at his father, 'do you want to go mano a mano, right here?' As he grew older, the fury towards his father was increasingly directed against himself in depressive drinking. But it was not all his father's fault. There was also his insensitive and domineering mother.

Barbara Bush is described by her closest intimates as prone to 'withering stares' and 'sharply crystalline' retorts. She was also extremely tough. When he was aged 7, Bush's younger sister Robin died of Leukaemia and several independent witnesses say he was very upset by this loss. Barbara claims its effect was exaggerated but nobody could accuse her of overreacting: the day after the funeral, she and her husband were on the golf course. She was the main authority-figure in the home. Jeb describes it as having been 'a kind of matriarchy . . . when we were growing up Dad wasn't at home. Mom was the one to hand out the goodies and the discipline.' A childhood friend recalls that 'She was the one who instilled fear' whilst Bush put it like this: 'Every mother has her own style. Mine was a little like an army drill sergeant's . . . my mother's always been a very outspoken person who vents very well – she'll just let rip if she's got something on her mind.' According to his uncle, the 'letting rip' often included slaps and hits.

Boys with such mothers are at much higher risk of becoming wild, alcoholic or antisocial. On top of that, she added substantially to the pressure from his father to be a high achiever by creating a highly competitive family culture. All the children's games, be they tiddly-winks or baseball, were intensely rivalrous – an actual 'family league table' was kept of performance in various pursuits. At least this prepared him for life at Andover, where emotional literacy was definitely not part of the curriculum. Soon after arriving he was asked to write an essay on a soul-stirring experience in his life to date and he chose the death of his sister. His mother had drilled it into him that it was wrong when writing to repeat words already used. Having employed 'tears' once in the essay, he sought a substitute from a Thesaurus she had given him and wrote, 'the lacerates ran down my cheeks.' The essay received a fail grade, accompanied by derogatory comments like 'disgraceful'. This incident may be an insight into Bush's strange tendency to find the wrong words in making public pronouncements. 'Is our children learning?' he once famously asked. On responding to critics of his intellect he claimed that they had 'misunderestimated' him. Perhaps these verbal faux pas are a barely conscious way of winding up his bullying mother and waving two fingers at his cultured father's sensibility.

The outcome of this childhood was an Authoritarian Personality. They impose the strictest possible discipline on themselves and others, the sort of regime found in today's White House, where prayers precede daily business, appointments are scheduled in five minute blocks, women's skirts must be below the knee and Bush rises at 5.45 am, invariably fitting in a 21-minute, 3-mile jog before lunch.

As noted above, authoritarians are organized around rabid hostility to 'legitimate' targets, often ones nominated by their parents' prejudices. Intensely moralistic, they direct it towards despised social groups. As people, they avoid introspection or loving displays, preferring toughness and cynicism. They regard others with suspicion, attributing ulterior motives to the most innocent behaviour. They are liable to be superstitious. All these traits have been described in Bush many times, by friends or colleagues.

His moralism is all-encompassing and as passionate as can be. He plans to replace State welfare provision with faith-based charitable organizations that would impose Christian family values. The commonest targets of authoritarians have been Jews, blacks and homosexuals. Bush is anti-abortion and his fundamentalist interpretation of the Bible would mean that gay practices are evil. But perhaps the group he reserves his strongest contempt for are those who have adopted the values of the Sixties. He says he loathes 'people who felt guilty about their lot in life because others were suffering'. He has always rejected any kind of introspection. Everyone who knows him well says how hard he is to get to know, that he lives behind what one friend calls a 'facile, personable' facade. David Frum comments that 'he is relentlessly disciplined and very slow to trust. Even when his mouth seems to be smiling at you, you can feel his eyes watching you.' His deepest beliefs amount to superstition. 'Life takes its own turns,' he says, 'writes its own story and along the way we start to realize that we are not the author.' God's will, not his own, explains his life.

According to Frum, when Bush talks of an 'axis of evil' he is identifying his enemies as literally satanic, possessed by the Devil. Whether he specifically sees the battle with Iraq and other 'evil' nations as being part of the end-time, the apocalypse preceding the day of judgement, is not known (nor is it known whether Tony Blair shares these particular religious ideas). What is certain is that, however much George W. Bush may sometimes seem like a buffoon, he is also powered by massive, suppressed anger towards anyone who challenges the extreme, fanatical beliefs shared by him and a significant slice of his citizens. Bush's deep hatred, as well as love, for both his parents explains how he became a reckless rebel with a death wish. He hated his father for putting his whole life in the shade and for emotionally blackmailing him. He hated his mother for physically and mentally badgering him to fulfill her wishes.

But the hatred also explains his radical transformation into an authoritarian fundamentalist. By totally identifying with an extreme version of their strict, religion-fuelled beliefs, he jailed his rebellious self. From then on, his unconscious hatred for them was channelled into a fanatical moral crusade to rid the world of evil. As David Frum put it, 'Id-control is the basis of Bush's presidency but Bush is a man of fierce anger.' That anger now rules the world.

But it is not only right-wingers and fundamentalists who are liable to fascist punitiveness. It is often found in people with passionately held beliefs that entail virulent judgmentalism about the way other people conduct their lives. These may include socialists, ecologists, animal rights activists and other opponents of the status quo who are determined to force others to realize how 'wrong' their behaviour is. Instead of becoming right-wing authoritarians who attack their parents indirectly by persecuting oppressed minorities, they choose targets which represent a more obvious attack on their parents, like the military or big business, and are just as totalitarian in pursuing them.

Despite the rise of fundamentalism and the continued pervasiveness of Poisonous Pedagogy, surveys show that the amount of 'parental fascism' has declined because, although corporal punishment remains the norm, it is less severe and frequent, and overall, childrearing practices have become more permissive. A particularly striking example is Germany. Germans were more authoritarian and less committed to democratic values in 1945 than their American counterparts, but by 1978 this had reversed. An important part of the explanation was a change in upbringing. In 1950, 86 per cent of young Germans reported having been punished with severe beatings; only 9 per cent did so in 1985. Whereas two-thirds of Germany's older generation today recall having had strict or very strict parents, this is true of only one third of 15–24-year-old Germans. In both the USA and Germany today, parents play with their children six times more than they used to; and the opinions of children are similarly more likely to be sought and respected. Much the same can be said of British and other parents from most developed nations compared with 1950 – part of the general shift towards the child-centred care advocated by Benjamin Spock.

This generational shift is well illustrated by the difference between the way Mrs B cared for her children as a result of her therapy and the way she had been raised. Her mother was so anxious about physical contact that when Mrs B and her siblings used to get into their parents' bed in the morning, her mother would immediately get up. By contrast, Mrs B was able to allow her children into the parental bed without anxiety. She was furious with her husband when he was fiercely strict and used corporal punishment in dealing with her son, for instance when he came downstairs feeling frightened at night (an experience with which Mrs B was all too acquainted). The change in attitudes to sex was illustrated when her son asked her where babies came from. At first she ducked the question and her face flushed as she talked to me about it, but towards the end of the therapy she was able to tell him without embarrassment. This was in marked contrast to the sex education of her own childhood. On one occasion, her relatively liberal father had begun to explain the facts of life to her brothers when her sisters, mother and maternal grandfather were all present. The grandfather stormed out of the house, furious that such matters were being discussed in front of women.

These changes have probably resulted in a substantial reduction overall in the number of adults with punitive consciences. The family patterns that were normal in Freud's day have largely disappeared. The unquestioned authority of stern fathers in particular, and of men in general, has ended. The burgeoning divorce rate means that many children do not even have a biological father in their home. Victorian parents, teachers, police and bosses had a positively totalitarian power compared with today; commentators in the press and other media directly relate this change to the rise in crime, drug use, loutishness and sexual promiscuity, and the decline in religious observance. But whilst fewer of us have punitive consciences and more of us have weak or weaker ones, the cause is not the decline in the number of fierce, chastising parents. Rather, it results from an increase in the amount of chaotic, erratic childcare.

Scripting the weak conscience

If we have a weak conscience we are liable to random promiscuity and unstable relationships. A symphony of antisocial behaviour, in which the instincts have all the best tunes, gets us into trouble from primary school onwards. Whether or not we end up in prison, we fight a lifelong battle with authority. Our career is likely to be erratic and unsuccessful, but in a few cases the freedom from constraint gets translated into an exceptional ability, especially in careers that demand only short bursts of discipline and actively reward spontaneity – rock music, say, or certain sports.

Weak consciences can be the result of a lack of identification with either parent, or identification with loving parents who themselves possess weak or defective ones. But at least as important as identification is the method by which the child is disciplined. One of the most sophisticated accounts is provided by Jerry Patterson, an American psychologist. He points out that all families have 'coercive processes', behaviours like whining, teasing, disapproval, yelling, humiliation, negative commands, noncompliance with demands and hitting. 'Taken individually,' he writes, 'most of these aversive events are trivial, a psychological mote. Rather than cataclysmic episodes . . . coercive family processes change with glacial slowness, a process that is composed of events that are inherently banal.' In families which create a weak conscience, such coercion is much more common. The parents are bad models for how to behave, creating aggressive, instinct-dominated children. Gradually the aggression escalates, until the family seems to be permanently on a war footing. Whereas the average five-year-old child does something its parents regard as naughty every three minutes, children from these families do so twice as often and are twice as likely to persist in naughtiness after being admonished.

Both parents in coercive families, but especially the mother, report feeling depressed and irritable. They use punishment more to express anger than as an instrument for altering the child's behaviour in accord with consistently applied principles, so it is very erratic. They are also prone to paranoia, attributing malevolent intentions to the child with no real basis for doing so. Fully two-thirds of attacks in such families are not the result of any discernible provocation from the victim, demonstrating that they are frequently responses to parental mood rather than attempts to inculcate discipline. If the child becomes a paranoid and violent adult who commits assaults on strangers, it is not surprising. Having been raised in a chaotic home, there is every reason for them to expect physical attacks and humiliation from nowhere; that is why the most common remark preceding city-centre assaults is 'What are you looking at?' In most cases, the object of this question was not 'looking at' anything. The assault was the wholly unprovoked act of a paranoid person, just as it was in the abuse-dressed-up-as-punishment which they suffered as a child.

The fights in the homes from which they came take place against a background of what Patterson calls parental 'nattering', a continual scolding of the child at the smallest sign of trouble. The constant anger and negativity amplify otherwise harmless acts into the cause of major rows. Threats and other non-physical methods cease to work, and in order to make a point the parent is forced to go to the extremities of physical coercion to get his or her way. The violence inflates as the currency of parental discipline becomes devalued and bank vaults of punitive coinage become neccessary to buy even a modicum of purchase.

Children from coercive families suffer from arrested social development, with the relatively uncontrolled behaviour of a four-year-old found in eleven-year-olds. Given the random nature of their upbringing they are liable to become pragmatic and manipulative adults, short-term in their thinking, impulsive and opportunistic. Proof that it is the specific patterns of coercion which are critical is provided by the fact that changing them changes the child. Patterson is just one of numerous clinicians to have shown that, when therapy helps parents to be more authoritative, the child's behaviour becomes less antisocial. He states that 'to change the problem child requires that one change the parents'.

There are many signs that more of today's young adults have experienced this kind of coercive care, or less extreme versions, with inconsistency and lack of supervision. As a result there are more with weak, or weaker, consciences, and increases in crime, teenage pregnancy, drinking, drug use and incivility are direct evidence that the instincts have the upper hand in more young people. Of course there are also many other causes for the rise in these problems, including greater affluence and the pressures from the advertisers and other cheerleaders of rampant consumerism in our culture of material indulgence. But all of us are exposed to these pressures and only some of us veer out of control. The fundamental cause is chaotic parental permissiveness, as I know only too well.

My parents were in the vanguard of the liberalizing of childcare when raising me in the 1950s. At one point they found themselves caring for four children under the age of five, which put particular pressure on my mother. There was therefore a good deal of mayhem, making for erratic punishment. Combined with their permissive childrearing values, considerable confusion reigned about what was right and wrong. That was why, in later life, I had a weak conscience in regard to some issues, making me largely delinquent until my mid-teens.

My family, however, was not as out of control as the extreme that Patterson describes. Children from these families have not internalized a rule-governed moral code and, unlike me, may have received very little love from either parent. Such care results in an amoral, impoverished capacity to empathize with other people which, *in extremis*, amounts to psychopathy. I have interviewed over 150 convicted violent men, many of whom would have qualified for this diagnosis – although I doubt whether there is anyone in existence who is totally without conscience and feeling for others. The nearest to it that I have come was Joseph, an American who claimed to have murdered 161 victims. He was the only violent interviewee who I felt might actually pose a risk to me personally, despite the presence in the room of an armed guard. He had been sexually and physically abused by his mother, a prostitute whom he described as 'the whore'. His victims were mostly tramps, but the first was a sixteen-year-old boy.

It was the day before Christmas and the whore threw me out of the house. I was only twenty-something myself when I met this kid and we got rapping.

He said, 'Come on, I'll take you home.'

I said, 'No, son, I can't go home, I got no place to go', but he meant to his home.

I said, 'Before we go, I'm gonna go get myself another bottle of booze.'

And he said, 'No, you can't do that.'

I said, 'Now wait a minute, you're not telling me what I can do', and we got into it. I picked up a rock and I crushed his skull.

Q: Did you mean to kill him?

A: Yeah, yeah, because I thought he was against me.

Q: How did you feel after you'd killed him?

A: I didn't feel anything, he was just another dead body.

Q: You didn't even feel, 'Shit, I've killed someone'?

A: No.

Q: You weren't worried about being caught?

A: No.

Q: You didn't feel any kind of pity for him? This guy was trying to help you.

A: No, I thought he was wrong in telling me I couldn't do something that I wanted to do.

Q: And looking back on it now, do you think you over-reacted?

A: Absolutely not. I think I was right.

Q: But you've since been told it wasn't right . . .

A: And it's beginning to get to me and it's making me think, 'Was I really right? Or was I really wrong?' Well, I was right. I have great arguments about this.

Q: But it's one thing to be in the right, it's another thing to actually kill someone else for being in the wrong, if they're not physically threatening you.

A: I killed a lot of people and I never had any remorse for one of them and I never will, I don't think.

Q: Was there any difference between, say, stealing from someone and killing them?

A: No, they had something I wanted. I took it.

Q: Do you think of yourself as being a person without any morals?

A: No, but at that time I didn't have any morals at all.

Q: What do you you mean by this?

A: I just didn't care about human life or anything else. I figured if they caught me, they caught me. What could they do to me? They could only kill me once.

Try as I might, I could find no evidence that he felt pity for his victims.

Q: There must have been situations when your victims were pleading with you, 'No, please don't kill me . . .'?
A: I'd stab them again, put another bullet in them.
Q: So it wouldn't make you feel something?
A: It just had to be done.

It 'had to be done' as a direct result of the kind of care he had received from his mother. She neglected him, was cruel and violent, yet his greatest anger was at the way that he felt she betrayed his father by prostitution. The connection between his murders and his anger towards her could hardly have been clearer, because on some occasions he actually hallucinated her face on his victims.

Q: You really saw your mother's face?
A: Absolutely.
Q: You felt this enormous resentment towards your mother. Why didn't you just kill her rather than strangers?
A: Because of my father. If I did that it might have killed him, and I loved my father more than anything in the world. If he'd been dead, I'd have killed her. I'd have chopped her up in pieces. The best I could do was to piss on her grave, and if I could have defecated I would have done that.
Q: Do you think that, if you had been able to kill her, you wouldn't have killed anyone else?
A: It's a good question, but I don't know. I really don't know.

If there was no pity, I did find some evidence that he had developed a vestigial sense of remorse. Contrary to what the psychiatric profile of the psychopath would predict, Joseph did feel depressed at times. He felt strongly that he should have been given the death penalty after his first murder, because that would have saved the lives of his subsequent victims. Since depression is nearly always provoked at least in part by guilt, this suggests that he was not wholly without conscience. He was so depressed sometimes that it was hard to understand why he had not made a suicide attempt.

Q: Why do you think you didn't kill yourself?
A: Oh, I tried. In fact they even had my obituary in the *Trenton*

Times – you can look that up. They thought I was dead. Oh, I tried. I did a good job. They gave me handfuls of pills and I took over a hundred, and I slashed myself all up. They had to go inside of me – I wasn't faking or anything. They had to go up here and sew me up inside.

Q: You wanted to die at that point?

A: Oh yeah, yeah.

Q: Did you ever get depressed?

A: A lot of times.

Q: Did you get depressed as a child ever?

A: In fact, a lot of times – a whole lot of times as a child. I tried to kill myself when I got out of the Marines.

Q: Are you glad that there's no capital punishment?

A: I just got through telling you I asked for the death penalty. If they'd have killed me the first time, you'd have had . . . what? We could say a couple of thousand people living today, if you include the ones that would have had babies and everything like that. I wouldn't have got out to kill again, would I?

Although these last comments may have been intended to curry favour, if they contained even a grain of sincerity they were hardly the words of a man without any conscience.

Given that strict, physical chastisement is the cause of the punitive conscience, you may be wondering why Joseph had a weak one. Surely the severe childhood abuse he suffered was no less traumatic than that of people who develop fascist personalities, who have punitive, not weak, consciences, despite routinely being the object of beating and slapping?

A crucial difference lies in the consistency with which the violence is administered. Whereas Joseph's mother was only predictable in her unpredictability, Christian fundamentalist parents, for example, are strict according to reliable principles. Although it amounts to abuse the chastisement is accompanied by an explanation, so that if the child avoids unChristian behaviour he or she can be sure of escaping beatings; whereas for Joseph there was no safe pattern with which he could comply. Also, unlike Joseph's mother fundamentalist parents would genuinely believe that they were trying to do the best for the child by breaking his will, perhaps even showing love alongside the violence. On top of this, the person with the punitive conscience has been supervised closely from birth, whereas there is a chaotic lack of supervision in the history of people with weak ones.

In early childhood, those of us with weak consciences were liable to be left to our own devices for long periods, and as a teenager allowed to run riot. If our parents were not usually at home after school and made no arrangements for us to be supervised, it put us at greater risk of developing a weak conscience. If we were a 'latchkey kid' we did less well academically, truanted more, did less homework and were at greater risk of committing crimes. We were twice as likely as children whose parents were at home after school to take drugs, smoke cigarettes and consume alcohol, starting at younger ages. This was partly due to the fact that our parents were not authoritative even when they were there. But it also resulted from the simple absence of a parent to supervise our activities, creating much greater vulnerability to peer pressure. We were at greater risk of 'getting in with a bad lot', or, indeed, of being that 'bad lot' ourselves to other children.

This persistent neglect also fosters a weak sexual conscience. Children who have had extreme neglect, for instance those raised in an institution, are liable to be indiscriminate in their choice of sexual partners, to have sex with more partners and to start at younger ages. But neglect also affects sexuality in children raised in a family environment. For example, the absence of a father has a big influence on a daughter's sexual conscience. On average, a girl whose father divorces or separates from her mother and leaves the home before she is aged ten, comes into puberty six months earlier than a girl from an intact family; her body is physically changed by his absence. This pubertal precocity is not very helpful, because girls who arrive at puberty young are at greater risk of a host of later problems, including sexual promiscuity and teenage pregnancy. Fatherless girls are also liable to have negative attitudes to men and to declare themselves less interested in long-term, stable relationships. Nor is the impact of fathers limited to whether they are physically present. In intact families, girls reach puberty later if they have a positive rather than a negative relationship with their father, and the more he is involved in her care the later it is. If the father is absent through illness or work rather than as a result of divorce or separation, the girl's pubertal age is unaffected. What is more, the quality of a girl's relationship with her mother does not predict when she comes into puberty.

Of course, neglect is not the only determinant of weak sexual conscience. Sexual abuse or inappropriate sexual interest from parent to child also have a profound impact. In summary, the host of studies show that the amount of harm caused by sexual abuse is greater the earlier it begins; the longer it goes on; the more it entails penetrative sexual acts; and the more that force, or the threat of it, are employed. It is also more damaging if the perpetrator is our biological parent rather than a step-, foster- or adoptive parent.

Whilst actual sexual abuse is relatively rare, inappropriate sexual interest by parents may be commoner than we would like to think. Covert flirtation between parent and child happens in many families, as does voyeuristic interest in teenage offspring's bodies, with fathers unable to conceal their desire towards bikini-clad daughters and their friends, or mothers taking an excessive interest in the details of their children's sex lives. I know of no studies on these matters, but I suspect that mild, purely playful flirtation, or unashamed appreciation of the attractiveness of teenagers to their parents, is beneficial to their self-esteem. The problem arises when parents express lustful desire. Certainly, differences in the way these things are handled create differences in the way we turn out sexually.

One of the most extraordinary features of abuse is the tendency for abused to become abuser. Common sense might lead us to suppose that if we have been horribly maltreated we would not want to inflict it on someone else, especially not our own children, yet the great majority of physical abusers were themselves abused, as were at least half of sexual abusers. What seems to happen is that, unable to get over it, the abused replay the experience in their minds with the repetition containing a hope that this time it will have a different outcome. The rerun can take the form of putting themselves in the position of the victim again, so that abused women are particularly prone to pairing up with abusive partners, or even to becoming prostitutes, in the secret hope that this time they will be able to change the abuser – a 'happy ending'. For abused men it is more common to turn the tables by becoming an abuser.

This expresses the profound rage that they feel, whilst also having a different outcome from the childhood one because it is not they who are left filled with rage, despair and humiliation. They are able to project the badness out of themselves and into someone else.

I encountered this idea at its most literal in a rapist who suffered delusions and had difficulty in distinguishing between symbols and the things they symbolized. As a child he had been anally penetrated many times by an abuser, and years later felt that the abuser's semen was still inside him. He pictured it as a poison, and in committing his rapes he believed he was ejaculating it out of himself, cleansing himself of the toxin by transferring it into his victims. This is as graphic a metaphor for the projection of unwanted feelings as one could hope to find. It also helps to explain why a single rape is often enough to destroy the mental health of a previously well-balanced and sexually active woman. Although many people find it hard to believe, sexual gratification is only a very small part of the rapist's motivation. Far more important is to put the fear of God into the victim or to make them feel rage, and to do so in a way that is incredibly hard to be rid of. For the rapist's part, the relief he feels afterwards is not sexual but like having exorcized a demon.

Neil, a rapist whom I interviewed, provides a telling example of how repetition of abuse works. In order for this to be clear, to understand why he acted as he did towards his victims, it will be necessary to describe some explicit details. In his early teens he was repeatedly abused by a man he met in a wood near his home, creating a very specific lovemap.

> This bloke come to the side of me and started talking about fishing and birds and stuff, and then he said, 'Would you come into the bush with me and I'll give you ten pounds?' I didn't think anything wrong with it. I just thought I was going to get ten pounds, whatever he did, don't matter, so I went into the bush so nobody could see and undone my trousers and pants and he started sucking me off.
>
> At first I tried to back away because I was a bit frightened. After a while I got used to it and let him do it. And when he finished, he gave me ten pounds. He said he'd like to see me again and he'd give me ten pounds again. Why not?

So I met him, done mostly the same things, he give me ten quid and we met again, went on for so many months. Then he started coming in his lorry.

He'd pick me up from school and take me back to the forest over where I lived and we'd go to the back of his truck. This time he wanted to take all my clothes off. This time he got me to suck him off. I didn't want to, but he said that he wouldn't give me the money unless I did it. When he first came in my mouth I was so scared I just wanted to throw up, but after the weeks gradually got on I got so used to it that it didn't really bother me. Then he wanted to push himself up me and that hurt. But I still went after he done that.

Q: So it wasn't completely unenjoyable?

A: It wasn't that . . . it was like . . . he was there to talk to and he was giving me money.

Q: He was quite a nice guy, then?

A: Yeah. 'Cos mum was only giving 20p, 30p a week pocket money. Never had no other money coming in otherwise . . . It went on to where he started hitting me and hurting me more, shoving things up my backside, different things. Glass bottles, sticks, bits of brick, mud . . . but I still went with him to get the money.

Q: Even though he was doing those awful things?

A: And hurting me more.

Q: Was he still nice to you when he wasn't doing that?

A: Yeah. So I could talk to him. Had no trouble. See, there's a loss of communication with my mum and step-dad.

Q: Was he the person you were closest to in the world at that time?

A: He was the only person I was close to, mostly – only friend that I could talk to.

Neil had such an appalling relationship with his mother that he was happy to be emotionally engaged with almost anyone, at almost any price. But there were consequences. A few years later he carried out a series of violent assaults in the same woods where he had been abused. First, he attempted to snatch an old lady's bag, then he tried to rape a nurse, then a schoolgirl. His accounts of these incidents had an indeterminate quality to them. He would feel the urge to do

132

something destructive, but to begin with he was not quite sure what. For example, the old lady threw her bag at him but up to that point he had not thought of trying to steal it. The assault on the nurse was equally goalless in terms of sex or money. 'I had the knife out in my hand and told her to come with me into the bushes, but at first I didn't know what I was going to do. But then I knew I was going to rape her and I just grabbed hold of her arm and started pulling her, but she kept pulling back. Then she started screaming and I got scared, so I brought up my knife and cut her across the face.'

By the time it came to the schoolgirl he was clear from the outset that he wanted a sexual experience. He had seen her on previous occasions and was using her as part of his lovemap for masturbation fantasies.

At night-time, before I got to sleep, I masturbated in thinking of this girl that I saw just came out of this school. [In the fantasy] I've grabbed hold of her, just ripped all her clothes off, and starting hitting her and slapping her around, and just have intercourse with her.

Q: How?
A: Every way you can think of, forcing her to do things.
Q: Like what?
A: Sucking me off, playing with me.
Q: And anal intercourse with her?
A: Uh-huh.
Q: Not forcing things up her?
A: No.

It seemed that in this fantasy he was putting the girl through the same experience that he had suffered at the hands of the man who abused him, but this time he was the one in control. At the same time, in hurting her he was expressing the rage he still felt towards the man.

Q: And how is she reacting to these things in your fantasy?
A: Shouting and screaming, and saying not to do it, begging me not to do it. But I still go out and do it.
Q: How is she feeling?

133

A: Frightened probably, scared.

Q: Is that how you used to feel when you were abused in that way, when the old man hit you and did things?

A: At first, but then I got so used to the pain and hurt.

Q: At the simplest level you seemed to want to do similar things to the schoolgirl that were done to you. Is that right?

A: Yes.

Q: How do you account for that, if you have those ghastly things done to you that you then want to do them to someone else?

A: It's a feeling, a feeling you can't really express in words: that you've had it done to yourself and you've got to do it to someone else, so they know what it feels like.

Q: Had you ever masturbated about the idea of doing that to boys?

A: Never.

A boy was abused by a man in a wood. A few years later he fantasized about doing the same to a schoolgirl because he wanted the girl to have the feelings that he had. Now he graduated from fantasizing, via an ill-defined need to externalize by enacting his anger, to a specific wish to inflict sexual violence. Just how specific his lovemap was only became apparent when he saw two boys in the woods. In the attacks on the nurse and the schoolgirl he had been half-hearted and easily scared off, but, having been an indecisive, incompetent rapist up to this point, when he saw the boys he suddenly became highly effective.

It was in the same forest where me and the bloke done our stuff together. His face was there in my mind. I saw the boys, one was ten, the other twelve. I had started to carry a knife in my pocket and I got it out. I pretended I was in the army, told the boys that they were my prisoners and they had to come with me. I took them for a run further into the forest until I was in the bushes and waited for them. When they came, I told them to pull their trousers and pants down. One of them said they didn't know they did this in the army, and I said to him, 'Just get 'em down, or I'll kill you.' I told them to lay on their backs and went down and played with them, sucked 'em. And I told them both to lie on their stomachs and I undone my trousers, pulled them down, wanked myself a bit so I got a hard on and I got on top of one of the boys and forced myself up him into his bum. He cried out, and the other boy was crying as well, seeing his friend getting hurt. I had intercourse with him, came inside of him.

When I came it was nice, but there's nothing there that really was significant. They were there to get my anger out of me, do what I want to 'em – they got no say in the matter.

Q: So it wasn't a sex thing? Was it power as much as anger?

A: It was power and anger mixed together, power to dominate 'em . . . After I came, I got off and pulled my trousers and pants up, told them they could get up and get dressed. Then I lost my knife and told them to help me look for it or they'd get killed. They found it and I said they could go.

Q: Was the face of the man who abused you . . . ?

A: Gone by the time they'd left.

Q: But there beforehand?

A: Was there beforehand.

Q: At what point did it come to your mind, the face?

A: When I was in that forest.

Q: The man who abused you – would you kill him?

A: Them feelings are still there, but I've had to keep them to myself. But if I saw him when I do get out [of prison] I probably would do it even if I get caught. It wouldn't bother me, knowing I'd get my own back on him. It would give me great relief in all aspects of everything that happened between me and him.

Typical of the weak conscience, as a child Neil had been little supervised for much of the time and treated erratically and violently when his mother was present. But above all, he was unable to express his rage towards her.

I was a loner, I never had no friends, always going out on my own. Could never talk to my mum – my mum and dad divorced when I was two and we moved up to London, and from the age of five upwards I never had no friends, always on my own, doing anything I wanted to, always getting into fights and arguments. Mum always hit me – she got into trouble for it. I had bruises on my legs and the

135

back of my legs, and at the school – it's hard to sit down on your legs – I told my teacher all about it and it went from there to the Welfare and they talked to my mother and she got into trouble.

See, if I'd done something wrong she'd hit me and she never really found out what the argument was completely about. If it was me and my sister she'd take one side of the story, hear mine but disregard some of it, and the first person she'd get hold of usually was me and hit me. My mother couldn't be bothered with us. She brought us into the world, she started bringing us up and then she couldn't be bothered any more. That's the way I felt.

Q: And hasn't that made you angry?

A: It makes me angry, but . . .

Q: Before you were sexually abused you were having fantasies about killing people. Do you think one of those people you'd like to kill would be your mum?

A: No, definitely not. Not even in my mind, I wouldn't even hit my mother.

Q: Were you able to express your annoyance at her at all?

A: Not at her – I can't even talk to her about it.

Q: Don't you think that may be why you get so angry with other people?

A: It is, 'cos I can't express myself to my mum, 'cos I don't know what she'd say about it. Or she might just blow her top about it, so I just keep it to myself.

Q: You keep it to yourself and you have fantasies of attacking other people, but don't you think it's her you'd really like to be expressing it to?

A: There's always the anger there that I'd like to get up and hit her sometimes. 'Cos she always used to shout – but that was no fault of hers, which I found out later. But I never could get up and hit her. I was powerless to say what I wanted and mean what I say. I could get back at other people, take it out on someone else except for my mother.

Q: Are you still frightened that she might somehow get back at you, or she might tell you to shut up?

A: Well, to tell me to shut up, yes. But not to get back at me. See, if she told me to shut up I'd have to take the defensive on someone else.

136

Q: Offensive, d'you mean?
A: Yes, on someone else.

The inhibition of Neil's aggression, caused by his mother, delayed his discovery of the central purpose behind his attacks in the wood. He had developed the habit of denying his aggressive wishes, unable to acknowledge them to himself. This was the temporary barrier to his trip across the bridge from fantasy into reality that is essential in every rape; it was an obstruction that he demolished gradually, starting with his attack on the old woman.

His slip of the tongue in saying 'defensive' instead of 'offensive' perfectly illustrated his fundamental dilemma as a child, a common one in weak consciences. For much of the time he felt depressed, turning his anger against himself, but at others he fantasized about getting his own back or took it out in fights. He had a choice between hitting a strict, violent, depriving mother and saying and doing the opposite. In his dealings with her he felt offended and offensive, but acted (and said) defensive instead.

Almost invariably, if a convicted violent man is asked about his childhood he will describe an astonishing catalogue of abuse. If he is then asked to give an account of the crimes he has committed against strangers, the displacement of his rage towards his parents will be demonstrated in the detail of what he did to his victims, with precise re-enactment of some of the things that were done to him. There is a close relationship between depression and aggression, homicide and suicide. Neil had made several serious suicide attempts since going to prison, and told me that he would succeed in taking his own life soon. Visible on his wrists were dreadful scars which testified to the credibility of this claim. He felt, with some reason, that he had nothing to look forward to after his release, being socially backward, personally inconsequential, physically unattractive and lacking any professional skills. He could see no hope of his constant depressive mood relenting. Convicted violent men are liable to be depressed, and six times more likely to have self-inflicted scars than men convicted of non-violent crimes. A study of British murderers found that an astonishing one third subsequently committed suicide. I do not know if Neil has committed suicide since I interviewed him, but it would be very understandable if he has.

Just like Joseph, the serial killer, Neil saw the face of his abuser when acting violently. Mrs B described her mother's dictates within her as a 'little voice'. Like ghosts, the sounds and sights of their childhoods haunted these people's adult reality. But it would be a great mistake to dismiss them simply as mentally ill people who have something wrong with their brains and bear no connection to what we ourselves are like.

We, too, hear little voices; we see the past in our present. Joseph, Neil and Mrs B's ghosts were largely damaging, but not all are like that. They are every bit as much the basis of a good sense of humour, a skill or just a likeable tendency to get on with people. Identification with parents and their projections on to you make you who you uniquely are in so many ways and, often against your conscious wishes, make you like them. They are the main reason that, when you walk into a party, you are attracted by the appearance of some guests and not others, that when you get talking to them you have certain favourite subjects and particular styles of conversing. As the night wears on you are coached by, and react to, your parental ghosts as you decide whether to have another drink or puff of marijuana or sniff of cocaine. They influence how you construe the people you meet, the projections you make on to strangers based on your childhood prototypes. In constructing your particular bubble of positive illusions, only hearing what you want to hear and seeing what you want to see, you are governed by these internalizations and projections. Move to another town or another continent, but these ghosts will remain at your shoulder.

Genes and conscience

Of course, in theory genes could partly or wholly explain Neil's weak conscience, just as they could explain Mrs B's punitive one. Indeed, as Freud writes, 'Experience shows, however, that the severity of the super-ego [his word for conscience] which a child develops in no way corresponds to the severity of treatment which he has himself met with . . . a child who has been very leniently brought up can acquire a very strict conscience.' Freud goes on to accord a role for nurture as well (in a classic piece of 'bit of both' bet-hedging), but he certainly believed that strength of conscience reflected inherited tendencies, such as a strong sex drive. Some studies of identical twins do find a high degree of heritability for antisocial personality disorder, impulsiveness and a hyperactivity which is closely linked with antisocial behaviour. It could be that, whatever kind of care Neil (or for that matter, Mrs B) had had, his conscience would have been weak.

However, twin studies are flatly contradicted by the more reliable findings of studies of adopted children. These estimate such traits to be no more than half as heritable as twin studies do – only a minor influence on behaviour. If Neil had been adopted at birth, he would have tended to be like his adoptive, rather than his biological, parents. Adopted children with biological parents who have been convicted of crimes are only half as likely to be convicted of crimes themselves. Despite often having come from poor homes with parents who are prone to criminal behaviour, adoptees' crime rates change to reflect the middle-class homes into which they are often adopted. If adopted children are raised in authoritative and loving environments, where the mother is five times less likely to be psychopathic than the biological mother, any genetic propensity to commit crime in adoptees is greatly reduced by such parenting. On top of all this, the twin studies showing high heritability for antisocial behaviour are not what they seem. A recent reanalysis of them showed that they actually demonstrate much less genetic influence than their authors claimed.

Nor does the bigger picture support much role for genes in causing antisocial behaviour. There are huge fluctuations in the amount of crime over relatively short periods of time. These fluctuations could not possibly be explained by changes in the gene pool, which take millennia. The number of murders in America tripled between 1964 and 1993 (from 7,890 to 23,271) yet between 1993 and 2001 it almost halved. That has nothing to do with genes. The same is shown by British statistics. In 1950, in England and Wales, the police recorded 6000 crimes of violence; in 1998 the number was 258,000. There were similar rates of increase for most other crimes. This 45-fold difference proves the over-riding importance of nurture and social change in causing crime. Even more specifically, there has been a huge increase in violence in England and Wales since 1987, so that (from the viewpoint of 2002) the rate at which the number of crimes increased has, on average, trebled for the last thirteen years. In a scientific monograph I explained this as the direct result of the increase in the proportion of boys being raised in low-income families, the effect of government policies in the early 1980s. Since parents on a low income are far more likely to care for children in the coercive ways described by Jerry Patterson, it should be no surprise that the proportion of boys who become violent men has increased, with a concomitant increase in the violence statistics. When my theories were printed in a newspaper, one reader wrote a letter making the simple point that, if genes explained crime rates, Australia should have much higher rates than other developed nations given that most of its population has the blood of deported British convicts in their veins.

There is also a major contradiction between twin studies of anti-social behaviour in general, and violence in particular. Such studies demonstrate that general antisocial tendencies are highly heritable, but violence consistently emerges as being only slightly heritable or not at all; the same is true in studies of adoptees. That parental care is far more important than genes in creating the difference between the exclusively violent and other kinds of antisocial people, like those who steal, has been demonstrated by Jerry Patterson. Whilst half of his sample of antisocial boys both stole and were violent, one quarter were purely given to stealing and the remaining quarter were solely violent.

Compared with those who were solely violent, the boys who only stole had parents who were more distant and cold, much less involved in the child. These parents misclassified dishonest behaviour, often failing to grasp that, by stealing, their son had engaged in something that was 'wrong', and consequently they did not punish him. Their ideas about ownership of property and what constituted stealing were poorly defined. When they did recognize a violation they 'nattered' rather than imposing an adequate punishment. By contrast, parents of violent boys were 'enmeshed' with their children, constantly punishing them and highly irritable. Their fathers were significantly more likely to attack the child physically, imposing highly inflexible yet incon-sistent systems of rules. If the rules of parents of boys who stole were vague, those of parents of violent boys were the opposite – pathologically so. But it was the violent boys' mothers who made the biggest difference.

These mothers were exceptionally irritable and scolding, threatening at the slightest provocation; yet when the child was aggressive they did not confront him. The behaviour of the mothers was found to be the key difference between the two groups. Maternal depression makes for paranoid hostility. On days when mothers report feeling down, they are more coercive and their sons are correspondingly more aggressive on that day.

Attempts to locate the origins of delinquency in children from birth also founder on the rock of studies that have followed large samples of boys while they grow up. There is little or no connection between what a baby is like and whether he will turn out criminal, but a lot of evidence that the parental care is critical. In theory, the fact that mothers of aggressive boys are more coercive could be a response to supposedly genetic traits in their sons, like hyperactivity or irritability. But when these mothers are asked to care for unaggressive boys whom they have never met, they are still much more liable to be coercive. All these problems (childhood aggression, maternal depression, coercive care) are far more common in homes where the parents have a low income. Class differences in criminality have much more to do with inequalities of wealth than with genes. The more unequal the society, the more violent, and the same applies to regions within a nation. The states of the USA which have the least welfare support for low-income citizens are the most violent.

If genes do play a part in the formation of conscience, it is largely through the 'goodness of fit' between the parental discipline style and the child's temperament, insofar as that is genetically inherited. Some parents rub along best with outgoing and brash children, others prefer the quiet and demure. A mismatch can increase the likelihood of a punitive or weak conscience.

Some toddlers are fearless and unconcerned by danger or novel challenges, whilst others are more nervous. These differences could be caused by genes, and they affect conscience development by varying the extent to which the child is apprehensive about being in the wrong and being punished. In general, fearful toddlers are

141

made more anxious by being punished for being naughty, so they are more likely to try to be good. On average, by the age of five children who were fearful toddlers have stronger consciences and internalize parental dictates more. By contrast, being shouted at or punished holds less sway with fearless children because it does not frighten them as much. Given these differences, it would stand to reason that a mismatch between parental discipline style and the child's fearfulness would affect conscience, and indeed it does.

Neither group flourishes if the parent is coercive, but the fearless children particularly need a responsive parent. They are more likely to have benign consciences if they feel certain that their mother will be responsive to their needs. Because fearless children are difficult, parents may be tempted to become increasingly coercive, even to the point of violence, but when a cooperative approach is used it over-rides the fearlessness and results in compliance with rules. Such children need closer supervision and without it they are liable to become aggressive, with a weak conscience, in middle childhood. More compliant, fearful toddlers need less supervision to develop a benign conscience. Gentleness is what helps them, and coerciveness only makes them punitive.

So the fit between parental style and child temperament does affect the development of conscience, but even this does not necessarily prove a major role for genes. That depends on the extent to which temperament is genetic, and as will be seen in Chapters 4 and 5, care received during the first three years is far more crucial than genes in determining temperament.

Auditing your conscience

What is the patchwork that is your conscience, often varying in degrees of punitiveness according to context? Unlike the audit at the end of Chapter 2, here you will have difficulty recalling much detail from the time period relating to conscience formation (when you were aged three to six). So unfortunately this time there is a little more work to be done, in three stages.

STAGE 1

You will recall that there are three different types of conscience, benign, punitive and weak, and that you can have more than one type depending on the particular aspect of your psychology. The three main aspects are summarized under the headings of attitude to authority, sex and conscientiousness.

For each of the three aspects you will need to answer a few questions to identify if you are punitive or weak. If you agree with two of the statements, that is probably your conscience type for this aspect. If you agree with three or more, it is definitely this type. If you come out as neither punitive nor weak, you are probably benign for that aspect.

Attitude to authority

PUNITIVE

Do you agree with two or more of these statements?

- I never, or almost never, break traffic regulations and I find it very annoying when other people do.
- Hierarchies at work are vital and I do my best to do what my boss tells me.
- I should obey laws and rules even if they sometimes seem rather silly.
- Strict punishment of children, including corporal punishment, is sometimes vital.
- I am very similar to one or both of my parents in many of my views or interests.

WEAK

Do you agree with two or more of these statements?

- I have three or more convictions for traffic offences. I am not

143

scared to break the speed limit or to park illegally, if I think I can get away with it.

- I always seemed to rub teachers up the wrong way at school and I don't like having to obey bosses.
- All politicians are hypocrites who are only out for themselves.
- One or both of my parents were oppressive of my individuality.
- I will never let The System get the better of me.

Attitude to sex

PUNITIVE

Do you agree with two or more of these statements?

- I was quite an inhibited teenager compared with the people in my class at school, and I started my sex life later than most.
- I find it hard not to be condemnatory if friends or acquaintances have affairs.
- I often use adjectives like 'dirty', 'naughty' and 'wicked' when talking about sex.
- I am easily embarrassed or annoyed by public displays of sexuality, such as drunken friends behaving in a sexually overt way in my presence, or witnessing very explicit scenes on TV or in films when I am in the presence of other people.
- I sometimes wish I could loosen up more in my sex life – be a bit more experimental. Or, at least, one or more of my sexual partners have sometimes seemed to suggest I should do so.

WEAK

Do you agree with two or more of these statements?

- I started having full sexual intercourse soon after coming into puberty, and had already had a lot of sex with different partners by the age of seventeen.
- I was one of the most sexually active people in my class at school.

- My relationships, at least before the age of thirty, tended to be short-term compared with my peers.
- I found it more difficult than most of my friends to remain sexually faithful, my eye being prone to wander, and I tend not to feel guilty if I stray.
- I see sex, or used to do so when young, as a pleasurable game in which I am trying to lure others into bed, if necessary using deception. The getting there – the game of seduction – is often more fun than the arriving.

Attitude to conscientiousness

PUNITIVE

Do you agree with two or more of these statements?

- I hate to leave a task uncompleted, like not finishing reading a novel.
- I have very high standards for myself or others, so that even if I have done my best (or they have) it is often still not enough.
- I sometimes get terribly bogged down in trying to explain myself to others, or find myself behaving very repetitiously.
- I get extremely angry about the moral or professional laxity of others.
- I believe it is important for people to curb their emotions, and guilt is the right response if they fail to do so.

LACK OF CONSCIENTIOUSNESS (WEAK)

Do you agree with two or more of these statements?

- You should please yourself in life, live for the present.
- I dislike judgmentalism and don't go around judging the goodness or badness of others.
- Artificial conventions, like dress codes or 'good manners', only create unhappiness by curbing our natural selves.
- Life is too short for feeling guilty, and I rarely do so.
- I have my own values which are made up for me, by me.

Having completed this task, you should have an idea of your conscience type in relation to these three aspects. If you have come out the same on two or three of the aspects, for example, being punitive for sex and authority but benign for conscientiousness, this is your overall conscience type.

Now we can move on to the question of what caused your type.

STAGE 2

Despite the fact that you will have few real memories from the age when the foundations of your conscience were laid down, you can still probably recall something. To help you focus on the specifics relevant to conscience formation, write down what you can remember under the following five abbreviated headings:

- childhood sexplay
- parents/carers curbing instincts
- parents punishing
- pupil at school
- closer to father or mother

These are the kind of things you should trawl your memory for:

- *Childhood sexplay, in other words anything you recall regarding sex in your early childhood*: parents' or siblings' reactions to you flashing your knickers or willy-waggling, playing about with your peers, envy of other children's lovey-dovey friendships, any dubious advances from adult strangers or family members. Were your parents quite relaxed about you seeing them naked when you were small? Did they let you run around without clothes on the beach or if friends came to visit? If you cannot remember anything at all, make a guess as to what your parents' attitude might have been to any childish signs of sexuality.
- *How parents/main carers curbed your other instincts when small*: their attitude to dirtiness, tidiness, time-keeping, aggression or loudness, fantasy play (did they share in it with you, enjoying your exuberance?), showing them up in public. Was it a major

146

aim of one or both your parents to coerce you into being an especially good child?

- *How your parents punished you*: can you recall a specific instance when one or both used physical punishment, like a slap? Did this happen regularly? If so, what was the feeling you associated with it – did they do it to satisfy a need in them, like to express anger or offload depression? Would you say your parents were often inconsistent, punishing you for a particular naughtiness, like fighting, one day and not doing so the next? Did your parents contradict each other either directly or indirectly, as to what they punished? Were they sometimes very frightening, making alarming threats or looking scary? Did they use icy-cold stares or humiliate and mock you?

- *What sort of pupil you were at nursery and primary school, especially in the way you reacted to school rules and teachers' authority*: did you buckle down to learning the three Rs or were you pretty reluctant? Did you obey instructions from teachers? Did you resent having to go to school?

- *Were you closer to your father or your mother*: you probably take after both of them, but which do you take after the most? Can you recall a shared activity, a special joke, a particular family setting in which you and one parent would exchange a secret look, that reflected an understanding between just the two of you.

STAGE 3

You may not have the time or inclination to do this, but try to do it if you can. Run through the list in Stage 2 with someone old enough to remember those days, like a parent, an older sibling, the parent of a friend from that time or a relative who was adult when you were small. Add what they have to say to your existing observations in Stage 2.

APPLYING WHAT YOU HAVE LEARNT

If you have completed the audit so far, you should now have

accorded yourself a type of attitude to authority, sexual conscience and conscientiousness. Then you will have made some notes on what you recall from your childhood that are relevant to this, and, if possible, asked someone else for their memories. The final step is to link this up to what you have learnt about the causes with the three different aspects of your adult conscience.

Under three headings:

- attitude to authority
- sexual conscience
- conscientiousness

apply what you have discovered about your past to this aspect of your present. From your past, the information about your childhood sexplay obviously relates directly to adult sexual conscience. The other formative factors (how you were curbed and punished, how you were at nursery school and how close you were to your parents) can explain your attitude to authority and conscientiousness. In making your analysis, focus on how you were treated differently from your siblings when you were small. Remember also that, in pondering the connections, you tend to identify with aspects of parents if they were loving to you and that other similarities can result from them having projected their feelings on to you.

Before attempting the application, reread the summary in the first few pages of the section in the chapter which refers to your overall type.

Here is an example for a man in his early thirties with two brothers:

Attitude to authority

Punitive (agree with three statements: believe in hierarchies, corporal punishment, have my dad's views). Closest to Dad, who was strictest with me. If something went wrong he was liable to blame me, e.g., aged four to five, when some toys were mixed up with clothes in the washing machine, breaking it – not my fault, but he belted me when he got home from work. If he was in a bad mood

I was the one that got it (projection of badness on to me?). But he was loving too – why we agree about politics and I took up his career? Mum was fairly strict about tidiness about the house and very tough on homework. Brothers seemed to get away with more with her – why they are less punitive?

Sexual conscience

Benign overall but with one punitive feature (agree with: started sex life late). Parents were both pretty relaxed about nudity. Recall exciting sex game aged five to six, in neighbour's house with two girls and another boy of same age. Don't recall it being seen as 'bad', but we were secretive so must have been scared of being caught. Don't recall mother being very cuddly or lovey-dovey. Speculation: they probably would not have reacted well if I was particularly sexy as child. Oddity that neither of my parents explained facts of life to me. Possible reason for why started sex life late: both seemed to completely ignore that I came into puberty. Not the same for my brothers, so for some reason I was sexually *persona non grata*. Never been very confident chatting up women, maybe a bit limited sex repertoire – didn't practise enough when teenager?

Conscientiousness

Punitive (agree with: don't like tasks uncompleted, have high standards and guilt for uncurbed emotions). Dad and I often shared looks when Mum was drivelling on about need for us to enjoy art and 'creativity'. Earliest memory of her trying to get me to paint picture and being frustrated because it didn't look like anything. Dad very unhappy if I did things badly, same with himself. I can't abide leaving a job half done – Dad's motto 'If a job's worth doing it's worth doing well.' Dad = source of my perfectionism. Perfectionism = why I'm so hard on myself and tend to feel down?

Put this audit with the one for your family script.

By the age of six, all of us have a more or less benign, punitive or weak conscience in regard to our instincts. Which kind you have will affect the role accorded to you in the family drama, interacting with the many influences described in Chapter 2. If you have a punitive conscience, for example, it may be accentuated by being the eldest child. If you have a weak one and you are also unfavoured, it could make you even more of a black sheep in the family. The permutations are innumerable. But now it is time to return to an even earlier period in your life, one that is largely not available to your conscious mind – the period before you were three years old. As you travel further and further back into the past, your experience is ever more obscured from your firsthand knowledge; and yet the way your parents cared for you then also becomes ever more influential on who you are now. In particular, the way you were cared for during your first three years is still profoundly influencing the kind of relationships that you make today.

Chapter 4

SCRIPTING OUR RELATIONSHIP PATTERNS IN OUR FIRST THREE YEARS

... They fill you with the faults they had
And add some extra just for you ...

During the 1960s and 1970s John Bowlby, an English doctor and psychoanalyst, developed what he called his Attachment Theory to explain insecurities in relationships. These will be discussed in detail later in this chapter. Meanwhile, which of the following, which psychologists call 'patterns of attachment', most closely resembles how you are in relationships?

PATTERN 1

'I am comfortable without close emotional relationships. It is very important to me to feel independent and self-sufficient and I prefer not to depend on others or have others depend on me.'

PATTERN 2

'I want to be completely emotionally intimate with others, but I often find that others are reluctant to get as close as I would like. I am uncomfortable being without close relationships, but I sometimes worry that others don't value me as much as I value them.'

PATTERN 3

'I am somewhat uncomfortable getting close to others. I want emotionally close relationships, but I find it difficult to trust others completely or to depend on them. I sometimes worry that I will be hurt if I allow myself to become too close to others.'

PATTERN 4

'It is relatively easy for me to become emotionally close to others. I am comfortable depending on others and having others depend on me. I don't worry about being alone or having others not accept me.'

If you chose the fourth pattern, you are among the 50 per cent of adults who are confident that they will be loved, who don't rock the boat when things are going well at work or play, and are secure in relationships. You have what is called a Secure pattern of attachment. But the other half of us are not like that. If you chose one of the first three patterns, or feel that none precisely fits you (in which case, you are probably pattern 3), your pattern of attachment is insecure. You shy away from intimacy altogether, or become clingy and frightened if a relationshp gets intimate, or just never feel sure about your partners. These first three are called respectively the Avoidant, Clinger and Wobbler patterns of attachment, and we shall be looking at all four in greater detail later in this chapter.

Our pattern of attachment is profoundly affected by the kind of childcare we received from the age of six months to three years, a particularly sensitive period for developing fundamental expectations about others. Largely powerless to do much to control your destiny, with minimal language and social status, you are at the mercy of whoever is looking after you. Your experience with them generalizes out to form a bedrock of assumptions which you bring to all your relationships with others: Can they be trusted? Are they going to like you or do you expect rejection or indifference? Can they be relied upon to meet your emotional, sexual and other needs? If you are repeatedly let down during early childhood, either because the people you most rely on keep physically disappearing or because they are emotionally unresponsive when they are there, then this is what you tend to expect of people you depend on in later life, at work and in love.

Although not a mental illness in itself, an insecure pattern of attachment predicts a greater likelihood of most of the nastiest psychological problems that can beset a person. Differences between insecure and secure people run deep, are far-reaching and evident early on. Toddlers with eating disorders, like self-starvation and vomiting, are far more likely to be insecure towards their mothers than merely picky or healthy eaters. The insecure child has difficulties in making friends and a tendency to be bullied or bullying. They are at greater risk of being aggressive, depressive and antisocial. Their brains and bodies differ from the secure, with different electrical patterns in the right side of the brain, and differing heart rates and levels of the stress hormone, cortisol.

As adults, the insecure are far more likely to suffer from mental illness. When asked in detail about their childhood and adulthood relationships, virtually all schizophrenics emerge as insecure. The depressed, neurotic, substance abusing and anorexic are all more likely to be, so are violent men and intriguingly, three quarters of fascists. The insecure are more likely to separate or divorce and as parents, to act in frightening, rejecting or abandoning fashion.

The idea that this collection of problems could all be connected with the way we were cared for in early childhood may seem odd. But there is now a mountain of evidence that the distant past hugely influences our present. The next section provides a backdrop to the specific details of those early experiences that caused our relationship patterns.

The distant past and our present

At any one moment there are patterns of electrical waves fizzing through our brains and of chemicals sloshing up against each other which, in total, are our thoughts and feelings. The particular configuration of electro-chemistry in your brain is highly responsive to what is happening in the here and now around you. If a pedestrian walks out in front of our car, our brain registers the fact and instructs our body to change the vehicle's direction. If we are watching a sad movie it causes changes in our brain, producing a weepy mood. But at least as important as what happens in the present are the experiences we have during our first few years. They establish the electro-chemical patterns with which we subsequently intepret life. Advances in the measurement of brainwaves and chemicals in the last two decades of the twentieth century have proved that there is a web of pre-existing neural connections out of which we make sense of the here and now. This web is established by the kind of care we received during childhood: the earlier the experience, the more enduring the pattern.

In both animals and humans, if a particular part of the brain is stimulated in early childhood it gets bigger and develops more connections between the neurones (the cells of which the brain is made up). The greater the frequency and intensity of stimulation, the more sophisticated that area of brain becomes. Also, if frequently repeated, the stimulation creates a pattern which becomes stable and not easily susceptible to change. The sum of these patterns from our childhood becomes the physical basis of our adult personality, mental health and intellect.

A major reason for the greater impact of early experience is that this is the time when the brain is growing most rapidly. About a quarter of a million connections are forged between brain cells in rats during every second of the first month of their life. During the first three years human brains grow with a similarly explosive vigour, never to be repeated, so that the brain of a two-year-old actually has twice as many synapses (connections between neurones) as that of his mother. Because so much is happening so young in the laying down of our mental 'wiring', the effects are more lasting and more important.

Rats that were deprived of maternal care in early infancy still have raised levels of the stress hormone cortisol months later. By contrast, rats raised in early environments that have been enriched by extra stimulation have superior performance and brain chemistry. Where they are placed in a large cage, with the chance to explore a maze and a toy-filled section, they do far better, developing more neurones and connections in key brain areas, than rats raised in pairs or alone with no other stimulation than food and water. Enriching the rat's social environment like this has far less effect when done in adulthood. Where adult rats are divided into the same three different types of living conditions as described above, the brains of those in the enriched ones do not benefit nearly as much as when this is done in childhood.

Similar findings come from experiments with monkeys. The kind of early care a monkey receives precisely predicts both its brain chemistry and the kind of adult it will become. Rhesus monkeys separated from their mothers at birth and reared together until the age of six months are more easily scared of strangers and unfamiliar experiences than mother-reared ones, and slide to the bottom of the hierarchies of status that develop in groups of monkeys. The more secure, socially assured, mother-reared monkeys are at the top. When threatened by social separation or isolation in later life, those monkeys separated from their mothers at birth have different brain and body chemistry. When they become mothers themselves, the females of this group are significantly more neglectful or abusive of their offspring than those that were mother-reared, repeating the cycle of deprivation.

Less extreme variations in early care also emerge as having powerful and enduring effects on monkeys. If a group of infants are only briefly and occasionally separated from their mothers during the first fourteen weeks of life, they are fully as insecure as young monkeys reared solely away from their mothers; and when tested at four years old they still have depleted brain chemicals.

Patterns of mothering are passed down from mother to daughter through the specific amount of care given and received. When the daughter of a monkey becomes a mother herself, the amount of contact she had with her mother precisely predicts the amount that she bestows on her own daughter.

Of course, the similarity in mothering across generations could have been simply a genetic inheritance; but this has been disproved. The amount of contact with the particular daughter has been compared with the mother's average for all her daughters. The daughter's subsequent mothering reflects her particular experience rather than the average for all her sisters. It is the unique care that she received which determines her subsequent pattern of mothering, rather than any genetic tendency inherited from the mother.

Another theory is that inherited differences in the child could cause the mother's pattern of caring: a difficult baby could make the mother uncaring. This was contradicted by a study of what are called Highly Reactive infant monkeys – ones that are very difficult to care for because they over-react to the slightest sound or movement. They were fostered out to either average mothers or exceptionally nurturing ones. The exceptionally nurtured young monkeys grew up even more socially well adjusted than normal infants fostered by average mothers. Nurture was so influential, in other words, that it could even turn a difficult infant into a superior adult. Furthermore, when the generation of offspring in the study grew up and themselves had infants, their parenting style, whether exceptionally nurturing or average, precisely mirrored the kind of care they had received as infants. This was regardless of whether their original infant personality had been Highly Reactive or not.

What is true of monkeys is not necessarily true of humans, but here something very similar does seem to be the case. In general, the earlier a negative pattern of experience occurs in our childhood, be it sexual abuse, physical abuse, neglect, lack of adequate nutrition during the first (compared with the second) half of the first year, parental divorce or separation, parental mental illness or parental financial misfortune, the greater the likelihood of it still affecting us in adulthood. The effects of experience in middle childhood (aged five to ten years) are less enduring and extreme than those in infancy, but more so than experiences in the late teens or afterwards. In the case of sexual abuse, for example, the earlier it happens the more sub-personalities the person is liable to have in adulthood and the greater the damage to the Hippocampal region of brain. In the case of physical abuse, in a sample of 578 children assessed in kindergarten through into adolescence, those who were physically abused in the first five years of life were significantly more maladjusted in adolescence than those who were first abused after age 5 or not at all. In the case of divorce, if parents do so when their children are in their twenties it does still have an effect, but much less than on five-year-olds. Stress in infancy sets a pattern of abnormally low or high cortisol levels whereas later stress does not affect one's baseline to the same degree.

Whatever electro-chemical way of being was established in our childhood, we bring it to bear in choosing and relating to people in adulthood, and in the conduct of our lives. If as children we constantly secrete large quantities of cortisol to make us ready for

156

'fight-or-flight' in the face of parental maltreatment, it can cause permanent brain damage making us hyper-responsive to seemingly innocuous events: panic or anxiety, or a collapse of self-esteem, are very easily triggered. Infants of depressed mothers, and those who have been abused or neglected, have heightened levels of stress hormones such as cortisol. Brain scans reveal abnormal patterns of brainwaves in the frontal lobes and right side of the brain, which is associated with depression in adults. As we shall see, these are caused by the sort of care received, far more than by genes. The precise kind of disturbed behaviour manifested at the age of three predicts which mental illness will be suffered in adulthood. Such adults have abnormal brain chemicals, such as low levels of serotonin (the neurotransmitter that is increased by Prozac and other antidepressant drugs), and patterns of brainwaves. In short, patterns of abusive or frustrating or rejecting care that recur in childhood become established in the brain as a set of expectations which are difficult to change in later life.

Our earliest experiences have greater impact not only because the brain is rapidly expanding, but also because we have been programmed by evolution to be more responsive to certain cues at particular stages as we grow. These are known as Critical and Sensitive periods for development. Humans are not alone in this. In some bird species, males must hear the entirety of their mating song, sung precisely right, before the eightieth day of their life and are incapable of learning it later – it is a Critical period.

Compared with other species, humans are less Critical than Sensitive in their learning periods. With Sensitive periods some learning can still occur after a certain age, but with curtailed success. A good example of both kinds in humans is the acquisition of first and second languages. If we have had no exposure whatever to human speech before the age of six we cannot learn it at all, making the first six years a Critical period for acquiring language. By contrast, there is a Sensitive period up to puberty for learning any second language once a first has been acquired. Thus, among a group of immigrants who had moved to America ten years earlier, those who started learning the new language after puberty were substantially less proficient than those who had begun in childhood.

Sensitive periods are also found in relationships and patterns of emotion. Rats that have been distressed by early separation from their mother have raised levels of the stress hormone cortisol; the longer the separation, the higher the level. In older rats maternal separation does not have this effect – it only happens if they were separated from their mother for their first two weeks. In the early separated rats, the stress levels are jammed at a level that is still measurably higher in adulthood than that of unseparated rats. Something similar has been proven in humans. Children who were insecure in their relationships when small still have high cortisol levels and abnormal brain patterns in middle childhood.

The crucial role of early experience has been further proven by recent studies of children who were placed in institutions and later adopted. Romanian children who lived in orphanages where they were severely deprived, and were subsequently adopted by middle-class Western parents, still bore the electro-chemical scars when tested six years after being adopted. If they had spent more than eight months in the orphanage they had elevated cortisol stress levels, whereas if they had only spent four months there the levels were more normal. As we shall see, the period between six months and three years of age in humans is a Sensitive one for developing Secure attachments, and the pattern a child will have at the age of eighteen months can be predicted with a high degree of accuracy by measuring his mother's pattern before he was born. Just like monkeys, we pass these patterns down the generations, establishing them physically in the brain by the way we care for our child in the early years.

Being cared for by a constantly changing array of carers and as a result, receiving less personal care, has dramatic effects on the physical growth of a child. For every three months spent in an institution she loses one month of growth, because lack of care reduces the amount of growth hormones produced. Mental abilities and language development suffer similarly. These deficits can sometimes be corrected if the baby is adopted and given adequate care but the same is less true of the child's emotional development. Many of these children suffer long-term damage – aggression, delinquency, hyperactivity, emotional insecurity, signs of autism, indiscriminate friendliness – even if they are adopted by loving parents. The amount of damage depends almost completely on the precise care received at specific ages.

Children who, soon after birth or having been in an institution in early infancy, are adopted into responsive, loving and financially secure homes actually do just as well as children raised by their biological parents in similar homes. Damage only results if any of a number of different adversities, or a combination of them, are encountered. First, if the child is subjected to maltreatment from his biological parents before being taken into an institution, he is at greater risk. If the mother is depressed or violent, the child suffers deprivation and abuse during this Sensitive period for developing a 'happy' pattern of electro-chemistry and the damage is often still evident years later.

Secondly, it is generally the case that the longer the child spends in an institution, the more at risk he or she is. As noted above, children who spend more than eight months do worse than those who have spent less than four. However, thirdly, the quality of care he receives in the institution, and the continuity of that care, can protect him from damage. Children who are in institutions that provide only generalized, group care with minimal stimulation and a lack of one-to-one individual relationships with staff, or who suffer frequent moves between institutions or foster-parents, are more damaged several years after adoption than children placed in institutions with more personal care regimes. Furthermore, if the child is lucky enough to become the particular favourite of one of the carers, he may be less damaged. Fourthly, the later that the child is adopted, the greater the likelihood of problems. Children who are not adopted until after the age of two are at greater risk than those adopted earlier. This is strong proof that early adversity does more harm than later. Finally, the quality of care provided by the adopting parents affects the outcome. Children adopted into homes that are less loving and less consistent do worse. If two or more children are adopted, as in the case of Mia Farrow's large adoptive family referred to in the Introduction, they tend not to do as well.

Taken together, this body of evidence is strong proof that the earlier in life we suffer adversity the more enduring and severe is the damage done to our emotional development, and that the role of genes in causing problems is not nearly as great as the particular care we receive. The same conclusion is drawn from studies of the effects of our mother suffering depression at different times in our childhood.

In general, mothers have a greater effect on how children turn out than fathers – not surprising, since in the vast majority of cases it is the mother who spends far more time with the child. Infants whose mothers are depressed tend to be withdrawn, inactive and less cheerful. Their attention span is shorter and they seem to be less confident that they can master their environment. The development of their language and mental abilities is delayed. Subsequently as toddlers (aged one to two) they are less communicative and co-ordinated, tending to be hostile and aggressive rather than friendly, and in later life they are at greater risk of both depression and aggression. The reason is clear, and not surprising. Such mothers did not respond much to the child's attempts to communicate, were unstimulating and tended to put a negative gloss on events. Their voices had a negative tone and their utterances were filled with lugubrious, 'glass half empty' observations. That it was the mother's behaviour that was causing the child's problems is suggested by a simple comparison. Eight per cent of children whose mothers are not depressed, but who are disadvantaged because she was suffering from a physical illness during their early years, are liable to be emotionally disturbed. By contrast, this is true of four times as many children of depressed mothers: the depression is causing the mothers to care for their children in a disturbing fashion.

But does maternal depression when the child is very young have a greater impact than if she is depressed later on? The studies suggest that it does. The longer and more profoundly depressed the mother, the less sensitive the care she provides and the greater the likelihood of later insecurity and depression in the child. If our mother was depressed when we were a toddler but she subsequently recovered, we remain more likely to experience problems several years later – having a depressed mother has been shown to affect our electro-chemistry not only at the time, but subsequently. Infants and toddlers with depressed mothers have atypical brainwave patterns, and the degree of abnormality is in direct proportion to the severity of the mother's depression. Even more detailed study shows the precision of the impact of the mother's behaviour on the infant's brainwaves. When she was being negative or rejecting, the infant's brain responded accordingly, so that the brainwaves changed at precisely those times when she was insensitive to his attempts to engage with her. The infant took his brain abnormality with him, wherever he went. When measured away from the company of the mother the abnormal patterns are still present, showing that they have generalized to encompass all situations – the infant is bringing this way of experiencing to bear, whoever he is with. Above all, it would seem that the younger we were when our mother was depressed, the greater the impact.

Scripting our pattern of relationships

This new evidence of the greater impact of early over later experience would have come as no surprise to John Bowlby, author of the Attachment Theory mentioned at the beginning of this chapter. According to him, our species evolved an instinct to form intense and specific attachments from around six months of age because it would have been useful back in primordial times. There were many predators and, just as it benefits monkeys living in the wild today, it would have paid to do our level best to stay close to mother once we were capable of crawling. If separated, we would let her know where we were by crying out and seek to attach ourselves to her physically by clinging. But if there is an instinct in toddlers to become attached that does not explain why about 40 per cent of children are insecure, rising to 50 per cent of adults. What accounts for the difference between secure and insecure?

According to Bowlby, above all things it is the kind of care we received between the ages of six months and three years. Emotional unresponsiveness or the physical absence of the carer at this age, or the combination of the two, create a state of fear that the carer will be emotionally unavailable (throughout this chapter I shall be using the words 'carer' and 'mother' interchangeably because, in the vast majority of cases, mothers are the main carers). This anxiety continues into adulthood and is triggered by subsequent intimate attachments. Not having our attachment needs met at an early stage in our lives leaves us permanently stuck in attach mode, unable to relax and to feel confident that everything will be all right. Insecure adults fear being abandoned or rejected by their intimates, because that is what happened in early childhood. Having once learnt to expect intimates to behave in a certain way, they assume it will always be like this.

Of Bowlby's two main explanations for insecurity, the absence of the mother is the more controversial. The notion that leaving small children with substitute carers is liable to be harmful could hardly have chosen a worse moment to make its debut than the 1960s, the decade in which large numbers of mothers began to leave the home for the workplace. The subject is still hotly debated today, the evidence picked over like carrion, but a consensus is gradually emerging. Few experts now dispute that leaving a toddler with complete strangers on a regular basis will promote insecurity. The real issue has become how to organize substitute care that does not have this effect – for this is now also proven to be possible.

Insecurity in toddlers with working mothers

Bowlby's theory predicts that if the main carer is physically absent on a regular basis when the child is still under three, and there is no adequate substitute, the child becomes anxious because his attachment needs will not be met. To avoid insecurity, the substitute carer needs to be as much like a responsive mother as possible. The ratio of children to carers should be low, the younger the child, the lower the ratio. Each child should be assigned to a specific carer – the same one every day. Turnover of staff must be very low, and days when the carer is absent carefully handled. The carer should be well trained, affectionate and nurturing. The transition from maternal to substitute care should not be abrupt; the child must get to know the substitute carer before being left in their care for long periods.

If these conditions are met the child should be secure, even if his mother works full-time. However, if they are not met a host of problems will arise. A series of films made in the 1950s and 1960s by Jimmy Robertson, a British psychoanalyst, and his wife Joyce, vividly documented what happens when a small child is abruptly separated from his parents and left full-time with strangers for days or weeks. John, for example, aged seventeen months and left by his father in a group setting for nine days because his mother is undergoing an operation, protests when first abandoned. Over the succeeding days he becomes totally withdrawn, suffering what looks frighteningly like a nervous breakdown, falling apart to reveal rage, sadness and blank despair. It is a heart-rending film, ending with John's unforgettable look of reproach when he sees his mother on their reunion.

Few would dispute that doing this to a small child is damaging; doing so repeatedly runs a serious risk of permanent injury. But this kind of extreme care, twenty-four hours a day for days on end, is not what psychologists argue about. There is a more subtle range of substitute care conditions to choose from when mothers return to work.

Studies in the 1970s simply compared the security of children who were reared entirely at home by their mother with those having day care from substitutes. This left out critical factors, like the quality of the substitute care and the length of it. However, with the 1980s came a host of more sophisticated studies concluding that, overall, children who had experienced twenty or more hours of day care per week during their first year of life were at a significantly greater risk of insecurity. Only 26 per cent of the home-reared children turned out insecure compared with 43 per cent of day care children. The adverse effects continued into later childhood. Whether measured at four years or ten, children who had experienced day care when young were much more likely to display poor emotional wellbeing, bad work habits, disharmonious relations with other children, aggression and antisocial behaviour. Although since that time there has been one important American study which did not find that day care alone was problematical, it did find that the combination of insensitive mothering with day care was. The conclusion today must be that this large body of evidence proves that day care can promote insecurity. But, as with all research, there are a great many ifs and buts.

In either type of care 26 out of 100 children will be insecure, but there is something about day care that adds a further 17. At its simplest, this is because a good deal of the substitute care on offer does not match up to what an average mother can provide; there are too few carers, and in any case they keep changing and so there is lack of continuity. But the working mother equation is

163

more complex than this. It is not necessarily a simple matter of the inadequacy of substitute care alone causing insecurity – it can also do so by influencing the mother's relationship to the infant. For example, the longer and earlier the day care, the more insensitive a mother is likely to be and the less engaged with her child. She may find it hard to bond if she has not seen much of him in the earliest months and years. At the same time, upset by lengthy, perhaps unresponsive early day care, the child may be difficult to handle. This turns into a vicious circle of unresponsive mothering and increasingly unmanageable child behaviour.

Another factor is the kind of woman the working mother is and the way in which she perceives the importance of early childcare. Those women who are more likely to leave their child in day care for long periods from an early age are also more likely to perceive children in general as resilient creatures who do not need intense one-to-one care in order to flourish. They are not upset or concerned by leaving their small child and going to work and, by implication, believe that, given basic nurture, genes will take their course and determine the child's capacities and temperament. Working mothers who believe that it is actually good for their children's development to have early substitute care may be less responsive with their children and less engaged with them at play. They also tend to put their children in poorer-quality care arrangements, at earlier ages and for more hours per week than other mothers. Their unresponsiveness and their unconcern about poor-quality care make sense, since they do not believe that the way the child is cared for has much influence on its wellbeing. Not surprisingly, given all this, the children of such mothers are more likely to be insecure.

It is also worth noting that security of attachment is not the only characteristic affected by day care. It is now clear that, overall, more than ten hours a week of day care in early infancy predicts a significantly greater likelihood that the child will be aggressive at the age of four, as reported by teachers as well as mothers. This is not just a matter of being more assertive: the child fights more, is more bullying and gets angrier when frustrated. There may be other more subtle consequences, some positive, some less so, and only some of which have so far been measured. One that has been is indiscriminate friendliness as a way of coping, which has been

164

identified in children who have multiple, changing substitute carers from a young age. Although attempts have been made to put a positive gloss on this, as showing greater sociability, in later life these children run the risk of being unable to form deep, stable, lasting friendships and loves, and of early sexual promiscuity, teenage pregnancy and juvenile crime. As adults they may be more likely to have traits such as superficial charm, people-pleasing and manipulativeness.

The findings regarding the effect of day care on intellectual development are more equivocal. Some studies suggest that it can actually promote faster development, others that it reduces academic performance in the long term. It is possible that some of the accelerated development occurs because the child has to grow up younger, learning to talk and walk in order to survive, and in families where being 'clever' is the only way to attract love the child quickly uses the acquisition of skills as a way of parent-pleasing. The sort of mother who works when the child is small is sometimes also the sort who does not enjoy babyishness and who values adult skills and competitiveness – although, of course, this is also true of plenty of full-time mothers. Such mothers may find it difficult to take pleasure in the many aspects of childhood which have no obvious goals, like fantasy play or babbling wordplay, that have no discernible connection with being a high achiever.

One third of the minority of mothers in Britain with small children who work full-time have relatively high-powered jobs. The same is true in America. Like male high achievers, such women are often the kind of people who like to be organized and in control of their life. This is simply not possible when caring for a small baby, because its rhythms and mental life do not accord with those of adults. A baby eats and sleeps at variable times, and is living in a very different mental space. To do a high-powered job during the day and then to adjust to this utterly different way of being is not easy. No wonder that so many of these mothers reach for books that advocate strict adherence to sleeping and eating routines which get the baby into a pattern that suits the mother.

Amazingly, remarkably little scientific research has been done on the effects of forcing the baby to fit in with the mother from birth. Alas, academic psychologists have shown little interest in the issues that concern parents of infants. There is one study showing that infants sleep and feed better if the care is child-centred, but I know of none that have examined the long-term consequences of either kind of regime. The most widely used book, entitled *The Contented Little Baby Book*, should possibly be retitled *The Contented Little Parent Book*. But then, it could be that, in the end, it is better for the baby if its parents get their sleep because they are then able to meet its needs more calmly and empathically.

A particular problem for high-powered mothers is to allow the women who substitute for them to do a good job. A successful manager, for example, is used to ordering her subordinates to do her exact bidding and, if they do not, to getting rid of them. I know of several female executives who have sacked their nannies or childminders for not doing precisely as they were told – as if they were just another employee. These executives were totally unaware that, in doing so, they were seriously destabilizing their toddler, creating separation anxieties and insecurity. In some cases the executives were actually jealous of the minder's relationship with their child, fearing that it was 'too strong', and failing to grasp that this was precisely what their child needed if it was to be secure.

Overall, there is little doubt that being left in inadequate substitute care does create an increased risk of insecurity in toddlers. This is not quite as large a problem as one might expect in America and Britain. A surprisingly small 55 per cent of American mothers of a child under one do any paid work. In Britain (contrary to what one might imagine from reading the newspapers) astonishingly few under-three-year-olds have a working mother. Only 11 per cent have one who works full-time, and part-time work is done by a mere 27 per cent of mothers of under-threes: to put it the other way around, 66 per cent of toddlers have a mother who has no paid employment. Even where the mother does work, the majority place their children with relatives or friends who often provide excellent substitute care. The main problem is for that tiny minority of British toddlers who are left in day care with professionals, be they childminders or creches. The quality of at least half these set-ups is poor, making a good many children insecure. But there is another side to this coin.

For many women, staying at home with a small baby is deeply depressing. This creates just as much of a risk of insecurity – in fact, a greater one. If we have the best interests of the child at heart, it is better for him or her to be in good substitute care than for the mother to stay at home and get depressed.

In Chapter 2, I described how the family script of Prince Charles made him the recipient of unwanted emotions, especially those of his father. His childhood also provides a fine illustration of the potential advantages of not being cared for by your mother. There is reason to suppose that Princess Elizabeth, as she then was, would not have been very responsive; instead, he was fortunate enough to have had more than adequate substitutes. The experience of separation from his working mother for most of his childhood during the daytime, and when she was away on affairs of state, probably did him no harm at all. It might actually have been worse for his wellbeing if she *had* been his full-time mother.

Writing to a friend soon after his birth on 14 November 1948, Elizabeth commented that 'I still find it difficult to believe I have a firstborn of my own.' Her subsequent behaviour suggests this incredulity continued because she spent very little time mothering him, doubtless encouraged by a husband who was not present at six of the first eight of Charles's birthdays. Elizabeth only saw her son for half an hour at 10a.m., and for bathing and bedtime in the evening. It should come as no surprise to learn that his first word was Nana – his name for his nanny, Helen Lightbody.

Even these brief daily encounters were curtailed by frequent absences. In November 1950, a week after Charles's second birthday, Elizabeth flew to Malta to join her husband where he was stationed as a naval officer. She was there until late December, during which time Charles stayed with his grandparents. Her perception of Charles's importance in her life at this time is powerfully illustrated by what happened when she returned. Her biographer, Sarah Bradford, states that the separation from her son had not caused 'any obvious consternation' and therefore she did not 'find it necessary to rush back to him'. Accordingly, she spent time at their home in London catching up on her correspondence and other administration. She also went with her mother to watch one of her horses compete in a race. Only after four days had passed did she finally visit her two-year-old son. Although upper class mothers at that time often played a minimal role in caring for their children, few mothers would be so cool about seeing their toddler after a two-month gap, suggesting that she was not emotionally engaged, not longing to find out how he was developing or herself eager to enjoy the love that comes from a small child.

Although both his parents were in Britain on Charles's fifth birthday, neither made the journey to celebrate it with him at Windsor, where he was staying with his grandparents. This suggests how out of touch both parents must have been with what matters to a small child, and soon after his birthday they departed for a six-month tour of the Commonwealth. At its end their son was flown out to North Africa to be reunited with them on the deck of their yacht. Instead of approaching his mother, he joined a line of dignitaries who were waiting to shake her hand.

In terms of a crude reading of Bowlby's Attachment Theory, this sounds like a prescription for insecurity. But had Elizabeth been present more of the time there is good reason to believe that she would not have made a very responsive mother. As a small girl she had symptoms of Obsessive Compulsive Disorder (OCD). Marion Crawford ('Crawfie'), her governess from the age of five, described a night-time ritual involving thirty toy horses, one foot high, on wheels at the end of her bed. Crawfie wrote that 'Stable routine was strictly observed. Each horse had its saddle removed nightly and was duly fed and watered.' The preoccupation went far beyond a normal child's craze, lasting the whole of her childhood. There were many other signs of OCD. At night she placed her shoes exactly under her chair at a particular angle with her clothes carefully folded on it, leaping out of bed to check the alignment. She had an obsessive way of lining up the brown coffee sugar granules given to her as a treat after meals.

Elizabeth only saw her parents for brief daily snatches and these obsessions were probably the result of the regime of her nanny, Alah Knight (who had also nurtured the Queen Mother). Named after the Islamic God, Alah was harsh. Crawfie said of her reign that 'the nursery was a State within a State. The Head of the State was Nanny.' She saw her task as to suppress and channel the unruly instincts of the child. Crying babies were not to be indulged, hungry babies must learn to adapt to regular, externally defined feeding times. Potties must be introduced as soon as possible and regularly, always after breakfast. Given this obsessive regime, it is hardly surprising the princess had obsessions.

Her most personal relationships were with animals. She told Crawfie, 'If I am ever Queen I shall make a law that there must be no riding on Sundays. Horses should have a rest too.' She may have sought animal company because she was lonely. Crawfie recalls she 'was a very neat, serious, perhaps unusually good child. Until I came, she had never been allowed to get dirty . . . other children always had an enormous fascination, like mystic beings from a different world, and the little girls [Elizabeth and Margaret] used to smile shyly at those they liked the look of. They would have loved to speak to them and make friends but this was never encouraged. I often have thought it a pity.' Repression of emotion was the norm within the family. Lord Harewood, her uncle, said, 'It was a tradition not to discuss anything awkward', and the Queen Mother 'always swept the awkward things of life under the carpet'.

Just as Elizabeth was to leave her son frequently when he was small, so she had been left. When she was nine months old her parents disappeared on a state visit to the Antipodes. When they returned six months later, she did not recognize them. This need not have mattered had the substitute care been equivalent to that of a mother, but all she had was the tyranny of Alah. That her son had a very different experience when she, in turn, kept disappearing was totally due to the warmth and responsiveness of his substitute carers.

His nanny, Helen Lightbody, was in her early thirties when Charles passed into her care. She was passionate about him, and as expressively maternal as his mother was repressed. Despite all the disturbing experiences which followed, as described in Chapter 2, his infancy was probably a good one. One strong piece of evidence that Lightbody was nurturant and loving is that Prince Philip was against her. When it came to childcare, whatever he was against was almost certainly very good news for the mental health of a child. Jonathan Dimbleby, Charles's official biographer, sums up Philip's attitude to Lightbody as follows: 'From his well-meaning but unimaginative perspective, the Duke detected in Helen Lightbody, the Prince's nurse, an impediment to his son's proper development, an inclination on her part to favour his son over his daughter, indulging the boy's "softness".' Luckily for Charles, Philip did not arrange to get rid of Lightbody until he was old enough to withstand the loss, after the age of three. By then he had developed a relationship with an adequate substitute, Mabel Anderson, who had been Lightbody's junior since Charles's birth. After Lightbody's departure, Anderson became 'a haven of security, the great haven . . . a surrogate mother', according to Dimbleby. Charles remained very close to Anderson in adulthood.

Thus, if a physically absent mother may be one potential cause of insecurity, a loving, consistent substitute can not only compensate for the absence but may actually be better for the child than the original. Indeed, one study showed that children who had been made insecure by disturbing mothers were more likely to prosper if they were subsequently cared for during the day by substitutes. Another study found that infants of depressed mothers were less likely to be insecure if their mothers were out at work than those cared for full-time by a depressed mother. This is in accord with Bowlby's second prediction regarding the causes of insecurity, unresponsiveness. Being at home with a depressed or distracted mother is every bit as bad for a child as being cared for by a substitute, and may even be worse.

Insecurity in toddlers with full-time mothers

There are many reasons why a mother may not be very well attuned to the world of infants and toddlers. She may find it hard to communicate without using words, and may be easily bored by the lack of intellectual activity or competition. Perhaps her own childhood has left her with only limited capacity to empathize with people of all ages, as well as babies. Drugs, alcohol, neurosis, depression or a disordered personality may make her very preoccupied with herself and unable to see life from another's standpoint, whatever the age, let alone being able to tune into an infantile world dominated by sensation rather than thought.

The unresponsive carer is liable to be intrusive, imposing meanings on the infant which he does not feel, telling him when he is hungry or sleepy or in need of a cuddle, instead of offering these in response to his signals; at other times the carer may simply ignore communications altogether, or not recognize them for what they are. Such parents are liable to equate being responsive with 'giving in' to infant demands, believing that such 'indulgence' only serves to make them even more demanding (the doctrine of Poisonous Pedagogy, described in Chapter 3), an equation emphatically rejected by Bowlby. According to him, it is the infant whose needs are met who will be the undemanding one – not needy because she is satisfied. The clingy, angry, greedy child is like that not because she has been over-indulged but because she has been deprived and always left wanting more. Bowlby argues that you cannot spoil an infant and that it is only as she becomes older, after the early months, that the long hard slog of disciplining and denial of the child's instincts should begin – the demanding, as well as responsive, authoritative parenting in later childhood, which makes for a benign conscience.

There is a rhythm between responsive carers and their infants. They knit together their emotions with each other's behaviour, like ballroom dancers in time with each other and with the music. Unresponsive carers cannot manage the reciprocity, the open-ended give-and-take involved. But if a dancer is ignored by his partner, no lasting harm is done; the same is not true of an unresponsive carer and an infant.

Babies whose full-time mothers are unresponsive in early infancy are more likely to be insecure at twelve or eighteen months. Some sixty studies show that, on average, 62 per cent of infants with unresponsive mothers are subsequently insecure. Even more signifi-cant, the impact of early unresponsiveness on security is still present decades later. One study measured responsiveness of care at one, eight and twenty-four months for infants who were then followed up into adulthood. The degree of earliest responsiveness predicted whether that infant would be an insecure adult fully eighteen years later, despite a huge number of intervening events.

Studies like this suggest that about two-thirds of people's patterns stay the same between infancy and early adulthood, and if the pattern changes there is usually a clear reason. In a sample of people followed from infancy to twenty years, 78 per cent of those who had had no major problems in their lives, like divorcing or drunken parents, did not change pattern. In the people whose pattern did change, there had been major adversities such as the loss of a parent or abuse. In another very stable sample whose families suffered minimal divorce or deaths, 77 per cent retained the same pattern of attachment. Without exceptional life events, the pattern stayed the same.

The kinds of experience likely to change attachment patterns are as Bowlby's theory would predict. For example, in firstborns the arrival of a sibling can result in a shift from secure to insecure if the parents become less responsive. In adulthood, a love affair collapsing or a bereavement increase the risk of insecurity. Conversely, therapy can decrease it. In one study of insecure adult patients who had undergone therapy, 40 per cent had changed to a secure pattern by the end. But on the whole, people's attachment patterns tend to stay the same. Three-quarters of eighteen-month-olds still have the same pattern at the age of three, and the same proportion of adults do not change in any five-year period.

The vicissitudes of children of depressed mothers, half of whom are insecure, show how early and firmly the attachment pattern is laid down. If the mother ceases to be depressed, the child will generally remain insecure despite the fact that she has become more responsive. This suggests that the child has developed an unresponsive prototype of what the world has to offer, which is hard to shift. Likewise, children who have been appallingly maltreated early in life tend to remain insecure even if they are subsequently given responsive care. For example, 60–70 per cent of Romanian orphans who experienced severely deprived early infancies are still insecure several years later, despite having been adopted by generally responsive Canadian parents.

The best evidence of the enduring effect of early care on security comes from a study that has followed children from birth to nineteen years. Some of the children measured as secure at eighteen months did, none the less, become disturbed during middle childhood (five to ten years) when faced by adversity. But compared with children who had been insecure when small, they were more likely to become stable again in adolescence because early security had an inoculating effect against later adversity. Thus, if children who had been secure at eighteen months had difficult adolescences they were

172

more likely to recover by early adulthood than children who had been insecure. Overall, whether it was performance at school, at work or in relationships, or whether it was rates of delinquency or of mental illnesses such as depression, having had a secure pattern of attachment at eighteen months predicted better outcomes at nineteen years. Given that the children's degree of security at eighteen months was caused to a large extent by the responsiveness of early care, this study seems to prove that early responsiveness has a profound effect on the way our lives turn out in most important respects.

All this goes to show that there is no simple equation of a 'working mothers bad, full-time mothers good' variety. Many women do not particularly enjoy little babies and are at risk of being depressed by caring for them full-time. It is far better for the child that the mother goes back to work, if she finds this keeps her on an even keel and the substitute care is good. However, mothers are just as likely to be made depressed by working full-time if they would prefer to be at home minding their baby. In general, part-time mothers are less at risk of depression than full-time ones, and part-time work seems to suit about half of all mothers very well. The key is whether or not the mother, and her partner, are happy with the arrangement. Where mothers find themselves in the wrong social space for them, working when they would prefer to be at home or vice versa, they are most at risk of getting depressed, making them unresponsive. If their partner is unhappy with the set-up, it increases that likelihood.

The most common task performed by therapists is wrestling with the effects of unresponsive early care in their patients. It demonstrates itself in all their relationships, including the way they respond to the therapist, and if they are made aware of its origins their symptoms often disappear. A good example is Mrs B, the patient described in Chapter 3, referred to me because she was suffering from frequent panic attacks.

During the treatment it emerged that great insecurities existed in all her intimate relationships. Fears of abandonment or rejection dominated her emotions towards her mother, husband and children, and at the same time their feelings towards her. These fears proved as central to the panic attacks as the factors already described, the guilt and terror of being bad that her mother had projected into her creating the self-critical 'little voice'.

Mrs B could not recall sharing a single positive experience with her mother, a woman who was herself extremely anxious and prone to depression. Mrs B was the second oldest of eight children in a working-class family and her mother had been overwhelmed by their demands. There was little care that was personal or loving, and she could not remember ever having been praised. Her mother's approach to childcare was summed up by a memory. Mrs B recalled being with her mother at the zoo when her three-year-old younger brother became separated from them. They could see him some way off searching for them and crying desperately. Her mother had said, 'Leave him be – it will teach him a lesson.'

Mrs B was filled with fears about being separated from her mother. Partly, she wanted her mother to need her, but as the therapy proceeded was able to see that the deeper desire was for her mother to be responsive and caring. Frequently during her childhood her mother had threatened to abandon her. Alongside the fear of this occurring – for a toddler, a fear as strong as that of death itself – Mrs B was left with a tremendous, murderous anger towards her mother for threatening her in this way. So intense was this fantasy that it prevented her from being either assertive or angry when her mother made demands upon her as an adult. Instead she would become withdrawn or protective of her mother because, in her mind, to begin to express her anger towards her would risk becoming actually violent, as her mother had been towards her. Indeed, Mrs B was terrified of 'finishing off' or 'shrivelling up' her mother if she dared to stand up for herself.

The link between this anger towards her mother and her panic attacks was shown on an occasion when she felt panicky after taking her son to hospital with a broken finger. It prompted a memory of damaging her own fingers when small, and of her mother's hostile, unresponsive reaction. The visit to the hospital with her son triggered the murderous anger, but instead of feeling that emotion she was panic-stricken. In her mind, she was killing her mother – but if she had no mother, who would look after her? Clear connections emerged between feeling a desire to kill her mother for being so unresponsive, and panic at the feeling that she was alone with no one she could depend on. The panic attacks ceased towards the end of the treatment, partly because she was much better able to express anger when she felt it, without expecting to kill the object of her rage and thus rendering herself bereft of intimates.

The connection between her panics and childhood insecurity extended to her other relationships. Several of the attacks had been triggered by the absence of her husband, Jim. On one occasion she had been suffering from flu and her husband had returned home to look after the children. He had told her to go to bed and said he would take them for an outing. Feeling abandoned by him, she was panicky and angry by turns. She recalled how sadness and 'weak' feelings had been jeered at by her mother, and that Jim took a similar attitude; in reality, there were many similarities between them. As the treatment progressed she became better able to express her anger towards him as well as her mother, without imagining that it would kill him. To the extent that she did so, she became less panicky.

Her insecurities were prominent in her relationship with her children. She brought her three-year-old daughter along to the early sessions because the little girl displayed signs of alarm and distress at being left. Mrs B could recall feeling like this herself when her younger brother was born. The fear was exacerbated by her mother repeatedly stressing to her how cataclysmic it would be for Mrs B to get lost. When Mrs B was late to pick up her eldest son from school, he was furious with her. She was all too aware of how he felt. Being let down by her mother in this way was a common experience, by turns enraging and terrifying.

Mrs B became adept at catching herself about to re-enact scenes with her children that she had experienced with her mother. Early in the therapy she reported threatening them with abandonment when she was angry, just as her mother had done, creating blind terror. But later on she became full of insights. She would find herself not wanting to leave them with substitutes, like teachers or childminders, ostensibly because she was concerned for their wellbeing. She was able to realize that it was she herself, not her children, who was frightened to be apart. Instead of alarming them by acting anxiously, she was able to leave them calmly.

The precise ways in which Mrs B's mother related to her had a profound effect on the way she related to me and everyone else. Not only does the kind of early care we receive significantly determine how insecure we are, it also determines the particular form our insecurity takes. In Mrs B's case, if she had read about the four patterns (Avoidant, Clinger, Wobbler and Secure) introduced at the beginning of this chapter she would probably have chosen the second one. Hundreds of scientific studies have demonstrated that each of the four patterns has a specific form of parental care as its cause. I suggest that you reread the summaries on pp. 140–1 before proceeding.

Scripting the Avoidant pattern

The first pattern is the Avoidant. One fifth of us have this allergy to intense involvement with others, wanting self-sufficiency, neither to depend nor to be depended upon. The Avoidant assumes that others will be hostile and rejecting. Anticipating this, we get our retaliation in first by being spiky and stubborn, and if we are none the less forced to become involved we employ a domineering, intrusive style.

Not surprisingly, given our misanthropy, we were the kind of person who did not want to settle down and have children in our early twenties and may still have not done so by our thirties. This disinclination continues so that on average, by the age of forty-three our longest relationship is likely to have been twelve years, compared with eighteen for secure people. If we do overcome our reluctance to tie the knot and marry, half of us are divorced by our mid-forties compared with only one quarter of Secure people. On splitting up we appear not to care, often actually expressing relief and dwelling neither on what has been lost nor on what might have been; likewise if we suffer a bereavement. But this insouciance is a front. Physical measures, such as heart rate or the amount of palm sweating, prove that we are deeply upset on the inside; we simply don't show it.

Sexually, the female Avoidant does not tend to have many partners whereas the male may be more active. He has a penchant for one-night stands, sex without love and with partners who are already spoken for (similar to those with punitive consciences). Both sexes report disliking the lovey-dovey side of sex, such as caressing, embracing, kissing or gazing into a partner's eyes. They favour oral or masturbatory practices – ones that involve less emotional contact.

If we are Avoidants we tend to prefer work to love, maintaining that success in the former creates happiness in the latter. Yet work is a constant source of annoyance because our fellow workers are so incompetent and noncompliant. Highly critical of them (again, we share this with the punitive conscience), we prefer to work alone and to concentrate on solitary processes, like computing, which protect us from the stress of having to deal with infuriating peers. We work long hours that leave little time for a social life and take the minimum of holidays, which we don't enjoy.

Avoidants are more likely than other types to be agnostic, but if pressured to picture a god we imagine him to be a distant figure who is inaccessible and highly controlling. He is rejecting, and we think it unlikely that he cares for us very much. This turns out to be a strong clue to the cause of our Avoidant pattern of attachment. We have, like our view of the supreme deity, an image of our mother as rejecting, controlling and negative – an image which developed because, on the whole, that is what she was actually like.

When observed relating to their babies at three and nine months, mothers of infants who subsequently become Avoidant are far more talkative than those of other types but their talk is not very communicative. Her comments to us were less in response to our acts or sounds than driven by her own emotions or whims. She was controlling and intrusive, determined to interrupt our flow and redirect it in her own desired direction. If we were drowsy, she may have woken us up and forced us to pay attention. If we became fascinated by something in our surroundings, instead of sharing this she was liable to insist noisily that we look back at her, even using physical force to twist our head or realign our body in her direction – although generally these mothers are averse to close bodily contact such as cuddling. The interactions had to be dominated by her, and we soon abandoned attempts to initiate contact because we encountered a steady stream of negativity. Her facial expressions were often angry or condemnatory as she attributed destructive or

177

naughty motives to our actions. Not surprisingly, given that she felt these motives to be present, she often displayed hostility to such a malevolent creature.

This rejecting pattern of care does not end in infancy. Avoidants are liable to have been the object of insensitivity for the remainder of their childhoods, with harsh, critical and unsympathetic reactions to the travails of adolescence. The end result is that as Avoidant adults we expect rejection in relationships and put the boot in first – we reject before being rejected.

Scripting the Clinger pattern

The second pattern is the Clinger, found in about one tenth of us. We want to be completely emotionally intimate, but no one is ever quite as close as we would like. We feel uncomfortable and lonely without intense involvement. Despite the fact that we constantly give of ourselves, the recipients of our emotional largesse don't seem to value us as much as we value them.

Our relationships are prone to highs and lows, to jealousy, conflict and dissatisfaction. We are liable to mother our partners, protecting, feeding, sympathising, smothering. We are looking for unqualified closeness, keen to move in with new partners and to share their life as soon as possible. We seek total commitment and constant affection. We are liable to idealize partners, making it all the more disappointing when the lover turns out to be less than perfect. We are very prone to falling in love, so when a relationship ends we take it badly, unable to accept that it is over; similarly, if we suffer a bereavement we are liable to keep imagining that the dead person is still alive, perhaps thinking we saw them in the street. In general we have a negative view of ourselves, easily losing sight of the positives. We are at greater risk of divorce than those who are secure.

As a child we were liable to be a victim, especially of Avoidants, with whom we may none the less have become friends and in later life choose as partners. At first sight odd, this marriage of opposites has a certain logic. The Clinger's vulnerability and insistence on showing emotion forces the Avoidant out of his shell and engages him with another human. It takes an Avoidant's thick skin to cope with the Clinger's demanding intensity; a more secure person might soon find the strain intolerable. The relationship may not be the most satisfying and joyful, but the mutual difficulties of the two patterns can make for something that works.

Clingers are also the opposite of Avoidants in bed, strongly preferring embraces, caresses and displays of affection to genitalia. Clinger women may get turned on by exhibitionism, voyeurism or bondage, whereas the men tend to be sexually reticent. Both sexes are liable not to have many different partners nor to be very highly sexed.

As a parent we are over-protective, limiting autonomy and exploration, promoting dependence and often causing our child to become too preoccupied with our wellbeing. We are highly responsive to our infants' expressions of fear, but ignore initiative or exuberance. When our infant becomes a toddler we are over-anxious about leaving him with others, so much so that this fear is injected into the child, reducing his capacity for independence. Even when our child is grown up and about to leave home, we still conjure up all manner of terrors about what will happen to them.

The Clinger's work life is fraught with worries. We are terrified of losing our job and find the short-term contracts of the modern job market very unsettling. We prefer working with others to being alone, yet don't feel appreciated by bosses or valued by colleagues, constantly anticipating criticism for underperformance regardless of whether or not it is deserved. We are likely to ruminate about our love life whilst at work and to become so emotionally wound up that it disrupts our performance. Easily distracted, we have trouble completing projects and tend to slack off after receiving praise. On average, we earn one third less than people with other patterns of attachment.

We are likely to believe in the existence of God, who has the distinct advantage over people of not letting us down and not making us feel like an insignificant failure. His love is unconditional and predictable, and it is possible to be worthy of it by following reliable procedures: prayer, ritual, doing good works.

Whereas the parental cause of Avoidance is rejection, with Clinging it is inconsistent, unreliable care. Mothers of babies who are subsequently discovered to be Clingers are confusing. Our mother

179

tried to engage with us by looking or touching, yet she did not respond if we tried to strike up a conversation, simply turning and looking but making no other reaction. For much of the time there was a strong sense that she was not very involved – not filled with passion for us. She initiated little and was a passive, emotionally absent presence. It may be that depression caused a flat, empty mood which even our beguiling infant smiles could not penetrate.

The Clinger's mother was not as averse as the Avoidant's to picking her infant up but she did so with little warmth, unaccompanied by kisses or loving words. She was also poorly coordinated in the way she did it, all fingers and thumbs, making us uncomfortable and mildly distressed. If we were frequently separated from her for extended periods, we were extremely clingy afterwards. The very high rate of the Clinging pattern amongst modern Israelis may exemplify this: many people in that society were separated from their families at young ages and given erratic and unresponsive group care.

At the heart of the Clinger is a fear of abandonment. It causes the clinging, but at the same time we wriggled and resisted attempts to hold us because, like Mrs B, we also felt furious at having been left before and wanted to punish our mother. The pattern was exacerbated if our parents separated or divorced. Of insecure children followed up to the age of eighteen who have suffered family breakdown, 73 per cent are Clingers in adulthood whereas only 20 per cent are Avoidant. Hence, if early childhood has created a potential to be insecure, divorce may direct it towards the Clinger rather than towards other patterns.

The adult Clinger is nervier than adults with other patterns, twice as likely to suffer from anxiety. The inconsistency of our early care has put us at risk of hysteria, the kind of inflated, panic-stricken emotion that is liable to overwhelm the jumpy. We are also more at risk of phobias and obsessions, both of which can be methods for trying to impose order on an otherwise frighteningly unpredictable universe.

Scripting the Wobbler pattern

The third pattern is the Wobbler, found in one fifth of us. Wobblers share features of the other two insecure patterns. They are Clingers to the extent that they want emotionally close relationships, but like Avoidants they find it difficult to trust or depend. We can be a puzzling mixture of both patterns. We may worry that we will be hurt if we allow ourselves to become too close, and then start to wobble when things get serious. Our pattern of relating is perplexingly unpredictable, and sometimes downright weird, from as young as twelve months old. At that age we don't seem to have any coherent, organized strategy for dealing with stress. Perhaps we cry loudly to gain access to our mother's lap, only to fall silent abruptly and become completely still when we achieve our aim. We may approach our mother, but with our head averted. Whilst we seem calm and relaxed, closer observation suggests a trancelike state of mind. A tendency to drift off into our own world is mingled with intense emotional engagement. Our body chemistry differs from that of more secure children, for example we have higher levels of the stress hormone cortisol.

By the age of six we may have developed a very controlling, bossy stance towards our mother, telling her what to do, where to look and what to think – more reminiscent of a parent with a child than the other way around. When playing with other children, we often seem to have difficulty in thinking clearly, acting strangely and aggressively. We might spend some minutes trying to annoy another child by shining a flashlight in her eyes or using a puppet to grab her nose. Yet a few minutes later, with a different child, we may not relate at all but withdraw, to be found lying with our face buried in a pillow.

In adolescence we are prone to dissociation, feeling very 'out of it'. We are likely to say that we sometimes step outside ourselves and that, speaking of ourselves, we feel we are in the presence of someone who is not physically there. To our friends we are something of a mystery. Nothing we could do would be surprising – apart, that is, from behaving in a normal fashion for an extended period.

181

As an adult, we are more likely than other patterns to suffer from mental illnesses and to commit crime. We are more liable to be involved in violent relationships with our partner or children, making us a frightening parent.

The cause of Wobbling is clear-cut: fully 85 per cent of children have this pattern if they suffered severe abuse or neglect. It is the response to extremely disturbing care, with our parents more liable than those of children with other patterns to have been alcoholic or drug abusers (2.3 million Americans take cocaine or crack cocaine, 111,000 are addicted to heroin; a staggering 200,000 British children live with a parent who is addicted to heroin or other hard drugs), violent or mentally ill. Our mother was also more likely to have suffered a recent bereavement or a serious trauma, such as a frightening accident, and our parents' relationship to have been disharmonious. These adversities create scary parents. We are torn between the desire to seek our mother out as a safe haven, and fear of the maltreatment that we have learnt will result from contact with her. Such childcare is paradoxical: our mother is the vital source of comfort from distress, yet her frighteningly abusive and neglectful behaviour is its main cause.

Depending on her particular adversity, at its most extreme our mother may have adopted unusual ways of talking to us in infancy. Sometimes she may have entered trancelike states for minutes at a time during a feed. She may have treated us as if we were an inanimate object or appeared to be strangely fearful of us.

The Wobbler may at times show signs of both Clinging and Avoidance, and at others may actually seem Secure. Uncertain of the best way to cope with our bizarre and disturbing mother, we are liable to feel we are everywhere and nowhere, everyone and no one.

Scripting the Secure pattern

The final, fourth pattern is the Secure, found in half of us. If we are this type it is relatively easy for us to become emotionally close to others. We are comfortable depending upon others and being depended upon by them, and don't worry greatly about being alone or having others not accept us.

As a child we were liable to have equally secure best friends and to steer clear of insecure children, and our adult romantic partners tend to be secure too. When set a problem to solve with their partner, secure men are positive and supportive, trying to help rather than acting as a competitor or getting annoyed. Secure women are likely to seek emotional support from their man and to be happy to receive embraces or other physical expressions of affection and encouragement. Secure couples have the least negative relationships of any combination of patterns – less critical, less conflict-ridden, more warm and friendly. The most common causes of rancour, like the man not spending enough time with the woman or disputes over the division of domestic labour, are less likely to be a problem. If war does break out, they have the best strategies for dealing with it. Instead of focusing on each other's shortcomings ('You never wash up') or endlessly recounting their own heroics ('I always do the cooking'), both are prepared to examine how each other's behaviour impacts on the two of them. They do not tend to attribute motives that only make matters worse, such as connecting malice or envy or hostility with mere neutral behaviour. Not surprisingly, they tend to report that they are satisfied with their romantic or married life. They have greater commitment to relationships and stronger feelings of love for their partners. Followed over time, their relationships last longer and, if they include marriage, are less likely to end in divorce.

The cause of Security is an image of a responsive mother who was there when needed. Although no mother is always like that, this is the basic expectation which has grown from thousands of experiences of it being that way. If our mother was often absent during our early years, a reliable and responsive substitute was there instead. When she was present, she was interested by our communications and early in our lives she allowed us to take the lead. She enjoyed bodily and eye contact, was involved and expressed warm emotion. If things were going wrong in her life, the negatives were less likely to seep into her relationship with us. Mothers of Secure children are anyway less likely than those with other patterns to suffer depression or other mental illnesses, or to be alcoholic or drug abusers.

Re-editing our relationship script

Whichever of these patterns is ours, they hugely influence who we become friends with and how others relate to us. For example, there are telling differences in the way teachers respond to Avoidant and Clinger pupils, fashioned by the child's attachment pattern. Because Avoidants expect to be pushed away and rejected, they act towards teachers as if they are about to behave like that. Consequently, the teachers tend to see Avoidants as unfriendly and not in need of their help. This, in turn, confirms the child's assumption that others are just waiting to give them the cold shoulder. By contrast, Clingers are hugely dependent and vulnerable with teachers, who as a result give them a great deal of attention. However, because the Clinger is often unpleasant as well, the teacher tends not to like her much. The child picks up on this, confirming her expectation of being unvalued. Contrasted with both of these, the positive expectations that Secure children bring to their dealings with other children and teachers make them popular, reinforcing their assumption of a world containing people who are benign and trustworthy unless proved otherwise.

In later life, insecurely attached adults are attracted by the insecure, Avoidants and Clingers being especially likely to team up. The insecure tend not to feel comfortable with a secure partner or friend, whose trustworthiness, supportiveness and warmth seem unnatural and confusing. Instead, they gravitate towards people who are self-fulfilling prophecies of rejection and abandonment. Coupled with their own unreliability in relationships, it makes for a stormy love life and rocky friendships. That the insecure choose each other partly explains why they are twice as likely to divorce and why their relationships are shorter-lived than those of secure people. The cycle of mutual mistrust and rejection confirms the expectation of being let down, leaving them in the state of lonely depletion which they assume to be the inevitable consequence of emotional contact with others. To top it all, they don't find solace in their professional lives. The negative, disruptive assumptions about relationships make colleagues react to them with the negativity that the insecure 'always knew' was bound to happen.

But our pattern of attachment is not set in stone. There is some flexibility, so that our relationship script can be re-edited in later life. Bowlby espoused a 'branching' view of development, analagous to the branches of a tree. We are pointed in the initial direction by the degree of responsiveness of our early care. The longer we receive a particular kind of care, the further we are sent in that direction and the longer it would take us to backtrack. Hence, if care is unresponsive continuously for the first three years, we will be a long way down the branch which assumes this is the norm and it will take considerable time for us to return to the trunk where responsiveness is the norm. The child brings his previous experience to bear on new experiences. If all he knows from his entire childhood is depressed care or maltreatment, even if in adulthood he manages to find a partner or therapist who is highly responsive, his pattern will take a lot of shifting.

Yet children can change for better or worse, and there are clear environmental reasons. A secure child can be driven to insecurity by severe experiences, like abusive or disharmonious parents. Conversely, if an insecure child becomes secure, it is usually the presence of a friendly teacher or caring grandparent that has made the difference. This is illustrated by mothers who manage to break out of a cycle of abuse. Thirty mothers who had been abused as children were examined to identify the difference between those who repeated the cycle and became abusing mothers, and those who did not. All had been severely maltreated as children, for instance being burnt with an iron, thrown against the wall or repeatedly hit with belts. As mothers they vowed that when their own infants were small they would not repeat the past, but eighteen of the thirty none the less did so. These mothers were far more prone to anxiety and depression, with lives that were more chaotic, stressful and poverty-stricken, than the twelve mothers who broke the cycle of abuse. But the key difference was that these latter were significantly more likely to have had an emotionally supportive adult other than their mother available during their early childhood, or to have undergone extensive therapy at some time in their life. This made them more likely to be drawn towards a stable relationship with a partner, which in turn meant that they were better supported, emotionally and financially, in the task of caring for their own children, whom they succeeded in not abusing. In terms of the branching model of development, if a responsive and nurturing alternative to parents was provided in early childhood or if several years of therapy provided one, the course of the journey could be radically changed. Armed with these good experiences, the women were able to seek out more positive environments for themselves and to relate to others in ways that did not simply confirm a negative view of relationships.

Genes and relationship patterns

But such changes in life's route are the exception. Most commonly, we make our own luck based on what happened in our childhood. For example, ten-year-olds who show emotional or behavioural disturbance are twice as likely as those without these problems to go on to suffer negative events, like losing their job or a divorce, when interviewed 20 years later. What they were like at age ten meant that they subsequently made choices and behaved in ways which ended in tears. In the same way, when antisocial children are followed up into adulthood they are more likely to choose partners who are also antisocial, or drug users or prone to relationship problems. This, in turn, rocks their already unsteady boat. Attachment patterns usually remain unchanged and are passed down the generations. In three cases out of four mother and child have the same attachment pattern, and a maternal grandmother's pattern generally predicts that of her granddaughter. Such a strong level of inheritance immediately raises the question of the role of genes. In theory, the similarity in pattern could simply reflect the transmission of the same 'attachment genes' from parent to child. Those who advocate this hypothesis also argue that the 75 per cent similarity of parent and child could be caused by the impact of the child's inherited character on the parent.

Suppose we are born highly irritable and this makes our parents unresponsive and angry. We could become insecure as a result, and in the process our irritability could also change our parents' attachment pattern to our own. We could actually change our parents' pattern from secure to insecure or, for example, from their Clinging to our Avoidance, creating similarity. Far from the popular stereotype that parents cause children's personalities, it could be the other way around.

Intriguing and sophisticated though these theories are, they accord far more power to infants than is truly the case, and, indeed, when tested scientifically turn out to be wrong. If a parent's attachment pattern is measured before the birth and then correlated with the child's a year afterwards, it is as similar to the child's as when the parental pattern is measured after the birth; in other words, whether measured before or after, three-quarters of parents and children have the same pattern. Indeed, the mother's prenatal attachment pattern actually continues to predict the child's concept of the world at the age of five. That proves it is not the child's behaviour which is causing the similarity, because in the case of the parents whose pattern was measured prenatally the child was not there to have an influence. But if the pattern is not transmitted by the impact of child on parent, it could still be a straightforward inheritance of an attachment gene. Again, there is compelling evidence to the contrary.

In the first place, if the mother does not have much contact with the child, his pattern is much less likely to be the same as hers; so environment is critical. The degree of mother–child similarity was compared between two groups of mothers whose infants were cared for by substitutes during the working day. One group returned to their mothers at night; but the other group hardly saw their mothers at all because they were cared for by substitutes all night as well, in communal sleeping arrangements. Only 40 per cent of the sleeping-away infants had the same pattern as their mothers: the normal 75 per cent similarity of pattern does not occur if the infant hardly ever sees his mother. It is, therefore, being cared for and related to by our mother, not genes, that causes similarity of attachment patterns.

An even more convincing proof was a study by a Dutch researcher, Dymphna van den Boom. Interestingly, Boom began her research because, while working as a teacher of troubled children, she had become convinced that innate difficulties in babies were a major cause of later problems. To test the idea she selected one hundred infants whose postnatal personalities should have placed them at very high risk of being insecure. They had been measured as highly irritable immediately after birth, easily upset and annoyed, harder to cope with than smiling, placid babies. If the irritability were a genetic trait it should translate into insecurity a year later regardless of the kind of care received.

When the babies were aged six to nine months, fifty of the mothers received counselling sessions to increase their responsiveness and sensitivity to their difficult charges. Up to this point these mothers tended to have become discouraged by their baby's lack of good cheer, ignoring the baby or blanking him out. Custom-building her help to the particular problems of each individual mother and child, Boom taught techniques for soothing the baby, encouraged play and helped the mothers to connect emotionally.

Meanwhile the other fifty mothers and their irritable babies had received no help at all. When the patterns of attachment of the two groups were tested when the infants were one year old, the contrast was remarkable. In the group who had had no help 72 per cent of the children were insecure, whereas in the assisted group only 32 per cent were. The only difference was the counselling sessions. The study showed beyond much doubt that with the right help most difficult babies can be turned around by their mothers, proving that the problem is not a genetic trait that cannot be changed. Irritability at birth is probably caused by difficulties during pregnancy or delivery, and is reversible.

Boom's study is a decisive indication of the relative unimportance of genes compared with parental care, and there is plenty of corroborative evidence. Studies of the attachment patterns of identical twins show little or no role for genes: despite having much more similar genes, identical twins exhibit security patterns that are no more similar to each other than the patterns of non-identical twins. In accord with this, patterns of attachment can and do change – which they should not if they were genetic; and, as we have already seen, they change in rule-governed ways that clearly reflect changes in the environment, suggesting its importance. Patterns are never fixed in the way that the colour of our eyes is. If they were genetic, they would be the same in our dealings with everyone. In fact, one quarter of children show different patterns towards their different parents, being secure with one and not with the other, or demonstrate differing patterns of insecurity to each; it all depends on the way the particular parent relates to them.

Most telling of all are the many studies which measure the temperaments of babies soon after birth and then check if their degree of difficultness at that age is connected with their degree of insecurity at eighteen months. If the baby is born difficult and goes on to be insecure, its genes could be the reason; a survey of these studies back in 1987 did find a weak connection between Clinging at eighteen months and being easily distressed immediately after birth. More recent surveys disprove this finding. Early difficultness does not predict insecurity; it is predicted by the state of mind and circumstances of the parents and the way they behave towards the child. The fundamental facts of who is and is not secure demonstrate this.

Twenty-six per cent of toddlers who come from stable, middle-class homes and have less than twenty hours of day care a week are insecure. The proportion rises to 43 per cent if they have over twenty hours. Having an unresponsive mother increases the risk even more, so that 50 per cent with depressed mothers are insecure, climbing to 85 per cent among those whose mothers are violent, drug abusing and maltreating. The trend is exactly what Bowlby's environmental explanation would predict: the more emotionally unavailable the care, the greater the likelihood of insecurity. These facts have some interesting implications which justify a brief diversion into the history of childhood.

A short history of childhood relationships

As anybody who has read Charles Dickens knows, it was the norm for children to be maltreated, by modern standards, even as late as Victorian times. Indeed, for great swathes of the population throughout history, the maltreatment which causes Wobbling in 85 per cent of cases was common. Does that mean that for much of its history the world was largely populated by Wobblers? The answer appears to be that this was not only the most common pattern of attachment for much of the last ten thousand years, but, surprisingly

189

enough, that it was actually the healthiest: Wobbling was the best way to be.

By my reckoning, there have been three eras in the history of attachments. In the first, starting when our species came into being perhaps some 3 million years ago, mothers were usually accessible and responsive because they had the time and resources. Societies consisted of small groups of between thirty and a hundred nomadic hunter-gatherers. The men hunted when food was needed and available, whilst the women picked leaves, nuts and berries to supplement this diet. When they had used up the available wildlife and garnered the edible plants they simply moved to a new area, population density being very low.

Under these conditions there is every reason to suppose that childcare was usually responsive, primarily administered by the mother, and therefore that the population mostly had secure attachments. Support for this view comes from observing our closest animal relatives, monkeys and apes, whose nomadic pattern of life is not dissimilar. They form intense relationships with their offspring, and the precise manner in which a monkey is related to in infancy determines its pattern of attachment, passed down generations. In the wild, secure attachments may be the norm.

Human societies which resembled primordial ones and were observed by anthropologists in South America, Africa and Asia during the twentieth century also suggest that security may have been common. A very readable account of one is found in Jean Liedloff's bestselling book *The Continuum Concept*, which describes how the infant is with the mother at all times, whether at night in bed or strapped to her during the day. Older offspring and relatives are available in abundance to help out with other children.

There is no sure scientific method to know whether such a portrait of our distant past is correct, but there is abundant evidence that from about 10,000 BC, in what I consider the second attachment era, life for infants and toddlers took a turn for the worse. Humans began to settle permanently in villages and to cultivate crops and husband livestock. Food surpluses from successful farming created the possibility of barter, of specialized trades (fisherman, basket weaver, blacksmith) and of money. Villages became towns, towns became cities and, as populations grew, resources became more scarce. Social classes and castes arose, with rich and poor, superior and inferior, powerful and powerless. Many mothers did not have the time, health or energy to devote to childcare that affluent members of developed nations have today, or that hunter-gatherers had in the preceding era. This was a prescription for maltreatment.

The interests of children and their parents do not always coincide and, in poorer families, they are brought into conflict by scarcity. It is in a child's interests to make his mother as guilty as possible to get his way, but, if this is not to the mother's overall advantage, it would make sense for her to hit or ignore the child; and, if made to feel guilty, she can be provoked to lose her temper even more. That is why levels of abuse and neglect are highest in overstressed, poor families where parents are unemployed or poorly educated. By physically abusing the child, the parent can avoid expending overstretched resources like time, energy or emotional wellbeing. Physical and sexual abuse, emotional neglect and deprivation were commonplace until very recently.

But fascinatingly, the result of all this maltreatment may actually have been a child who was well adapted to its world. Maltreated children tend to become opportunistic, selfish and amoral adults. These may have been highly advantageous traits for a person living in a world filled with life-threatening adversities. To be insecure and to have amoral tendencies was the best way to survive if, even during peacetime, murder and assault were commonplace. In London the parish records show that the risk of being murdered in the thirteenth century was ten to twenty times greater than at present. Wobbling may still be the best adaptation for people in much of the developing world today, where the likelihood of being assaulted or murdered is twice as great as in developed nations. For much of recent history, Wobbler patterns of insecurity may have been both rife and highly appropriate.

Meanwhile, back in the twenty-first century, a child's pattern of attachment is established as electro-chemical pathways in his brain. This pattern is what he brings to bear in facing a mass of new challenges. How insecure he is, and in what ways, will affect what sort of conscience he develops.

Authoritarian adults with punitive consciences are very likely to be Avoidant in relationships because the sort of parent who is rejecting with toddlers, creating Avoidance, is more liable also to parent three- to six-year-old children in ways that nurture Authoritarian tendencies. Convicted criminals are prone both to being insecure in their attachments and to having weak consciences. Violent criminals are especially likely to be Wobblers, because the sort of frightening parent that causes Wobbling is also likely to parent in the coercive ways that make for weak consciences. What is more, as our family script is being written, the way our attachment patterns and conscience have been moulded will affect what role is accorded to us in the unfolding family drama. An eldest child who is Avoidant, with a weak conscience, is going to attract different scripts from one who is secure and benign.

Whatever the particular permutation in your case, there is one last step backwards that you must take to complete your understanding of how your parents have made you what you are. In Chapter 5 you will revisit the care you received in the first six months of life and examine its impact on the most fundamental aspect of your being, your sense of self. But first (if that is the order in which you have chosen to do so) you need to carry out an audit of your own pattern of attachment.

Auditing your pattern of attachment

The four questions at the start of the chapter (pp. 140–1) should have enabled you to define your pattern, bearing in mind that, if none of them seemed to fit you precisely, you are probably the third, Wobbler, pattern.

To find out precisely what caused your pattern there are two methods you can employ. The goal of both is to establish the extent to which your care was received consistently from carers who were responsive to your unique moods and needs.

METHOD 1: DETECTIVE WORK

Pretend that you are a detective or journalist or biographer trying to discover the attachment history of someone else. Interview your mother or father or a sibling or an adult who was close to the family when you were small. The first question to which you need an answer is:

- *Who looked after me aged six months to three years during the daytime?*

If it was just your mother, you need to ask a further series of detailed questions about her frame of mind and relationship to you in that period. After each question I shall provide a brief summary of what the answer may reveal:

- *What sort of toddler was I? For example, was I amusing, irritating, determined, wilful, happy, angry, lovable, introverted, popular, a quick learner, slow to develop? Was I born that way? Please give examples.*

This usually gets an enthusiastic response. It will tell you a good deal about her and how she envisaged you, as well as about what you were actually like. If she saw you as 'a little devil' or 'a good little girl', this may indeed be what you were like. What she may be unaware of is the extent to which you were like that because of the way she related to you.

If your mother sees you as having had inherited, unchangeable traits, this is a strong clue that she was not very responsive to your needs. Studies of mothers show that those who see their children's behaviour as nothing to do with their care are less consistent or involved, and less child-centred.

If your mother seems to focus heavily on whether you were a quick learner or slow to reach developmental milestones, like walking or talking, it is a strong clue that she was not terribly responsive to your emotional needs – some mothers connect easily with their child's abilities but not to their childish vulnerability or fantasy life.

- *What did you like about me then, and what did you dislike, compared with my siblings?*

This is another way of approaching the last question.

- *Were there any worrying or frustrating events happening in your life at that time, like one of your parents dying, Dad losing his job or not doing very well, moving house to a place that did not suit you, or you wishing you could get back to work?*

With any luck she will be able to open up about this. Although she may still feel sad or angry about the event(s), and be reluctant to go back to them, the advantage of this question is that she will not see it as an attempt by you to blame her. Talking about bad external and unavoidable events is easier than feeling guilty about how she treated you.

- *If there were worrying or frustrating events, did that not affect how you related to me?*

This may be harder for her to think about.

- *Compared with my siblings at that age, were you closer to me or did we get on less well?*

She probably will not answer this, but her reaction may be revealing. If she gets angry or seems distressed by the question, it may suggest she knows she was not able to be very close to you.

- *What was your relationship with Dad like at that time, compared to your relationship with him when my siblings were that age?*

She may feel able to answer if you have a good relationship with her now. She may never have pondered how variations in the quality of their marriage/partnership impacted on her relationship with her offspring at this age. If you are sensitive in handling the question, not seeming intrusive or voyeuristic, it could be very revealing of what was going on.

- *During those years, would you say there were any periods when you were depressed, feeling very negative about the world, hopeless about the future, helpless or desperate?*

194

Since depression in mothers is a major cause of anxious patterns of attachment, it is a vital question. Some mothers who were depressed will simply deny this, having suppressed it. Others may have been depressed but did not realize it at the time and have not acknowledged it since. Mild depression amongst full-time mothers with small children is very common – it occurs in about one third of them. The state of mind of the average mother has been described as dysphoric, the opposite of euphoric. Having to meet the needs of a highly demanding small child for twenty-four hours a day, often without much sleep, tending to feel socially isolated, unvalued by society and disorientated by no longer having the structure that a paid job provides, is enough to make anyone feel low. In evaluating her answers, bear in mind what you know of her since you became an adult – she may not be aware that she has a depressive streak.

- *Do you think you were fairly even-tempered, or did you find yourself quite often feeling irritable or angry or bad-tempered?*

Some mothers may find it easier to recall this than being depressed. If they do, it is a strong clue that they were mildly or severely depressed, because such people tend to be aggressive to their intimates. My mother freely admitted to having been irritable when my sisters and I were small, but only very late in life was prepared to use the words 'mildly depressed' to describe her state then.

- *Would you say that your state of mind during that time was different from how you felt at the same age when looking after my siblings?*

Obviously a tricky one. Most mothers are anxious to see themselves as having been even-handed.

- *How did you cope when I was badly behaved?*

She should be able to tell you this. Most mothers feel their pattern of punishment was justified and the right one.

195

Inevitably, your mother is likely to find some of these questions difficult, so it is important to ask them in a spirit of inquiry, not of accusation. Remember that her illusions are very likely to be threatened by these questions. It is also likely that if there are disturbing truths which she would rather not recall or not tell you about, you will need to ask others the same questions about her care of you. An older sibling, your father or other adults who were around then should be able to help fill in the gaps.

If someone other than your mother or father cared for you for more than a few hours on rare occasions, you need to identify who they were and for how long each one did so. Again, your mother may not be able to recall much, or not want to, and you may need to turn to third party sources.

Having obtained as detailed as possible a picture of the choronology of the substitute carer(s), as well as interviewing your mother about the way she cared for you, you need to ask the questions above in relation to the carer. Best of all will be to track down the carer(s) and ask them, although this may not be easy or even possible.

METHOD 2: PSYCHO-ARCHAEOLOGY

To a surprising extent you can retrospectively infer, from your subsequent and present-day relationships, what your early care was like. First, reread the account of the attachment type that applies to you (pp. 165–72) to remind yourself of the typical patterns and their root causes. I shall now describe what might be the case for the Avoidant and the Clinger. Wobblers will recognize elements of both of them in these accounts.

School relationships

- *Were there any patterns to your relationships with friends and other children at primary or secondary school? Likewise in your feelings about teachers and them about you?*

Remember that the insecure patterns often entail unpopularity or having only a handful of friends. Avoidants tend to team up with each other or with Clingers. Looking back, try to link the kind of early care you have identified with the way you were at school. You probably cannot remember the early care, but you can speculate retrospectively.

An example from an Avoidant man might be: 'My mum was irritable and often rejecting. That made me angry and aggressive, very much like her, e.g. I tended to be selfish and bullying just like her, e.g. I would always tell my friend Jimmy that he was a silly billy, and there was a Jimmy in my life for much of my school career. Mum used to call me silly billy.'

An example from a Clinger woman might be: 'My mum was remote, emotionally. I was terrified of being unpopular and was. At primary school I clung to the teacher, who would get annoyed with the endless whining. At secondary school I had Jill as a friend but she was always leaving me high and dry – not turning up when we'd arranged to meet or being late. Is that what mum did emotionally when I was a toddler? Is that why I had Jill specifically as a friend?'

Sexual relationships

- *Can you see any patterns in the choice of your partners, and the duration and kind of relationships?*

Clingers tend to fear abandonment in love, Avoidants to be rejecting, because they had these experiences in childhood. Searching through the way you have behaved in your sexual relationships, you should be able to identify some trends. To do so, try writing down the names of all the lovers you have had and then, for each one, answer the following questions.

- *Who ended the relationship?*

If a pattern emerges that you usually ended them, you may have been doing to the partner the equivalent of what your mother or father did to you: going silent on you, being cold and distant, hostile, paranoid. If the other party ended them, perhaps you were picking partners of a kind who would throw you out or were behaving in ways that encouraged them to do so. This may have been a reliving of a feeling of abandonment from your toddler days: being left with strangers in a nursery, your mother getting depressed and leaving you feeling bereft, your mother seeming to be caring but actually always having her attention diverted elsewhere.

- *Did you get really close to the person?*

Let us say you have identified a pattern of lack of real intimacy. If you are Avoidant that could be because you would not let the other person into your inner life; or, if you are a Clinger, because you were so demanding that it was all one-way traffic – either you did not find out much about them because you were so desperate to talk about yourself, or the opposite, so you may have been a doormat, never getting to talk about yourself.

The precise pattern of lack of intimacy may reveal with which parent, in what way, you lacked intimacy. If you look at the way that you yourself have prevented intimacy, it may turn out to be the same as the method that one or both parents used.

For example, a Clinger woman might write: 'I was never close to Dad. He tended to say he was going to take us out, e.g. to the Christmas panto, but then cancel because of work. If I tried to hold his hand he would let me to begin with, but at the first opportunity would pull it away. If I told him about my day at school he would only pretend to listen. That sums up my sexual partners, too.'

- *What was their greatest fault? What was yours?*

If you write down the greatest faults, a pattern should quickly emerge. Who has not endlessly made the same mistakes, both in choosing partners and in dealing with them? When you analyze the partner's faults, it could be that you discover you always managed to find someone with the same faults as your parent(s), or that you saw those faults in them even when they were not there, or a little of both. For example, based on what your parent(s) was like, you may have a tendency to pick cruel men or diffident women; but it could also be true that you expect them to be like that even when they are not.

198

Friends and colleagues

Write down the names of your three main friends and of three key figures from work. Ask yourself in each case if there are any patterns in your relationships with them:

- *Do I have frequent bust-ups with one or some or all of them?*
- *Do I fear that they will not like me?*
- *Do I ever bully them?*
- *Am I really close to some or any of them?*
- *Is the relationship usually quite an equal one, in terms of who leans on whom?*

By now you should be expert at seeing that the patterns revealed are potentially traceable back to early childhood. Having identified the pattern, consider if it does not remind you of specific features of your relationships with your parents, or, if there are speculations which come to mind, of how these patterns may reflect ones from when you were a toddler.

Having completed this audit, put with it the others and proceed to the next chapter.

Chapter 5

SCRIPTING OUR SENSE OF SELF IN OUR FIRST SIX MONTHS

... Man hands on misery to man ...

Are you a hungry person, sometimes greedily addictive, prone to bingeing on work or sex or drugs or food? Are you a good mimic, a charmer, a lover of artifice? Are you mistrustful, even of closest intimates? Are you a control freak? Do you tend to extremes of good and bad, to dividing the world into the wonderful few versus the rest? Do you have to be the centre of attention? Do you lurch between grandiose views of yourself and black depressions? Are you prone to antisocial, hostile, impulsive, selfish behaviour? Are you a novelty-seeking risk-taker?

If your answer to any of these questions is yes, you probably have a weak sense of self and suffer from what is known as a personality disorder. You are in good company, for, even if 80 per cent of criminals and about 13 per cent of the general population do so, so do the majority of high achievers, be they in politics, business, the arts or showbusiness. Your weak self can lead to astonishing feats of imagination and dominance, albeit often accompanied by terrible anguish. But your symptoms are not unknown to everyone else. Far more than 13 per cent of people have some of them – all of us do, to some extent, in some situations.

Scripting a weak sense of self

Although no one can directly remember their earliest experiences in infancy, it is back in this forgotten time that the origins of personality disorder and the weak self are found. The psychoanalyst Donald Winnicott claimed that it results from the care you receive in the first few months of life. Early care that lacks empathy creates an immature adult with arrested development, prone to the reckless and amoral acts of a young child, to the 'me, me, me' selfishness and inflated grandiosity found in the fantasy life of the toddler. But the cause of personality disorder is not wholly care in infancy. Subsequent experiences, especially sexual and physical abuse, can also play critical roles. In most cases, genes play little or none.

Winnicott wrote that 'There is no such thing as a baby', meaning that babies are as nothing without the thoughts and feelings attributed to them thousands of times every day by their carers. Indeed, if a baby has solely her physical needs met but receives no personal attention to her emotional and social demands, studies of children in orphanages show that she is more at risk of death and certain not to flourish. Just as we anthropomorphize the behaviour of our cats and dogs, so with babies; but with the important difference that if our mother keeps on attributing human meanings to our babyish gestures, we eventually come to share them. During the first six months our experience is largely physical and primitive – sensations of hunger and satiety, of warmth and cold, of fear and security. Our sense of time is limited to physical changes, like from being fed to not being fed. We cannot label these sensations with words so we cannot represent them as conscious thoughts with which we could predict and ultimately control our world. We have weak boundaries between what is Me (what is inside our body and mind) and what is Not-Me (outside), the two often being blurred. When we suck on the breast or bottle it could be a part of us. When the milk goes into our mouth, we do not know what it is or where it came from. This blurring is still evident in our later vivid fantasy life as a three- or four-year-old, which is filled with such confusions. Only half jokingly we may say, 'I am you' to an admired parent or, when naughty, imagine that our wickedness can be read in our mind long before it could have been detected. By having our needs met reliably we accrue a memory bank of bodily sensations which are the foundation of the sense of where Me begins and Not-Me ends – the sense of self.

If our mother was an empathic carer, she quickly learnt to

adjust her behaviour to fit our natural rhythms of looking towards and turning away. We became partners in a dance which could be frenzied and peaceful by turns, constantly switching types of behaviour. Perhaps we let out three cries of increasing length: 'Eh, Eeh, Eaarg!' Our mother may have responded through a different mode, by waving an arm to the precise beat of the cry, imitating its cadences: small wave, big one, even bigger one, with a slight change in direction to express the 'arg' element of our last utterance. A wave responded to with an equivalent sound, a look with a pat or stroke – such exchanges, known to psychologists as 'cross-modal matching', were going on all the time between us.

In the vast majority of mother–baby couples, it is the mother and not the infant who determines whether this dance takes place at all; and, if it does, how it is patterned. The emotional empathy of Winnicott's 'ordinary, devoted, good-enough mother' is as critical to the infant's wellbeing as food is to his physical health. If our care does not start from where we are, emotionally, then if we look away the unempathic mother is liable to misintepret this action as rejecting or controlling. She will talk at or grab hold of us, acting intrusively and out of keeping with our desires. If we are only fed or picked up when it suits her, she hijacks our capacity to experience our own needs and we grow looking outwards for definition. Winnicott maintained that in this situation we develop a 'false self', feeling unvalued and powerless. In later life that weak self shows up in the symptoms of personality disorder or even, if it is extremely weak, as the mental illness schizophrenia. In a thousand ways, every day and through innumerable tiny interactions during the first few months, the lack of empathy communicates itself. The result is a chronic loss of the capacity for pleasure, for which we compensate in one of two main ways.

The first is to become frenetic. In this hyper mode, as an infant you are locked in a state of vigilant consciousness and wakefulness, regardless of how fatigued we are. Our activities may be speeded up, with exaggerated sensitivity to sounds or other stimuli. We may indulge in repetitive, rhythmic behaviour such as rocking or sucking. There may be muscular tension, stiffness, rigidity of the back and body, and exaggerated intensity and muscular force when we make movements. Breathing may be rapid, with our pulse rate racing and our food gulped in a frenzy. These indicate our desperation at our lack of

202

pleasure and may prefigure a host of severe problems in later child-hood and adulthood, from addictive behaviour to eating disorders.

It may be the precursor of hyperactivity, for example. Although the illness officially known as Attention Deficit and Hyperactivity Disorder or ADHD is widely touted as being of largely genetic origin, and although 3 million American children are being treated with drugs at any one time on the assumption that it is a disease of the brain with a physical cause, there are a lot of good scientific reasons to question this approach. Even during pregnancy, the mother's state of mind significantly affects whether her child is likely to be hyperactive. In one survey of seven thousand cases, twice as many children suffered the problem if their mothers had been anxious in late pregnancy. Raised cortisol levels or stress in the mother during pregnancy predict subsequent behavioural and emotional problems in the child at ages 4, 7 and 9. Postnatally, several studies have shown that when unempathic early care takes the form of intrusiveness it leads to hyper states. Intrusive and overstimulating care measured at six months was strongly associ-ated with hyperactivity a long time afterwards, both at three and at eleven years old. The infant's temperament, measured a few days after birth, did not predict hyperactivity at all, suggesting little role for genes. Another study compared children of maltreating parents who had been placed in group care in an institution with those who had more personal care in a foster home. The ones who had the more depriving group care were much more likely to be hyperactive, again suggesting that genes were not playing much of a part. The importance of care is strongly suggested by the fact that children whose mothers are often overstretched, such as those who are single parents, are three times as likely to be afflicted as children who have both parents at hand. Whilst twin studies show that genes may none the less be significant in as much as half of cases, it seems very likely that the seeds for the illness are often planted by intrusive, unempathic care which fosters the habit of being hyper as a way of dealing with lack of pleasure. (Subsequent progress of the illness is probably the result of erratic or coercive punishment in later childhood, determining whether the infant's potential for hyperactivity is fulfilled and, subsequently, whether this evolves into a weak conscience and criminal delinquency.)

203

The deprived infant's lack of pleasure may equally be compensated for by the opposite of hyperactivity – a depressive turning in upon the self, passive hopelessness. Faced with extremely unempathic care, as infants we fall back on our body as a last-ditch source of gratification: rocking, head-turning or, to create sensation, scratching our skin or masturbating. Finally, still unpleasured, we withdraw from the world altogether, becoming somnolent, lethargic, unreactive to stimuli, floppy and weak; we have shallow, slow breathing and are a reluctant, unenthusiastic eater. Dried tears sit upon a rigid, expressionless, averted face. We evade repetition of the pain of insensitive, unempathic relating from the carer by becoming totally unresponsive.

These hyper and depressive ways of coping with lack of pleasure become established as electrical and chemical patterns in the infant brain: levels of hormones and patterns of brainwave and heart rate differ between babies who have had empathic and unempathic care. This lays down the potential for disorder in later life, like a computer program, although, just as that can be modified by being rewritten, so can the infant's neurology. The right set of subsequent experiences can change for the better the instructions for what to expect from life, because the brain is an organ which is highly sensitive to its environment throughout life.

However, such compensation is the exception rather than the rule, and in most cases the effects of unempathic early care are still visible decades later. It tends to result in an adult with a weak self whose personality is disordered. Our integrity is fragile and in order to stay intact we use 'primitive' mental mechanisms, so-called because they are evident in very small children. Faced with adult reality, we regress to childlike modes. Although we may appear normally adult in most respects, in others that are more or less well concealed we have never grown up.

Personality Disorder

A core stratagem in people who are Personality Disordered is omnipotence, an infantile conviction that we can achieve anything we wish because we are magically all-powerful. Having felt powerless to affect our carers as a baby, we simply reverse the truth and tell ourselves that we can change the world merely by thinking it different. This is visible in normal three-year-olds when they close their eyes and wish something different by casting a spell. So keenly do they feel their fantasy world of play that they believe anything can happen. But in adults, such 'magical' thinking can appear as symptoms of mental illness. At its most extreme, it takes the form of delusions where sounds or sights that conform with the fantasy are hallucinated. The many severely deluded people who are attracted to the belief that they are Jesus Christ may be suffering from omnipotence. Not only did he have the status of Son of God, he was so omnipotent that he could cheat death itself. All of us had grandiose fantasies as children and there can be few adults who have permanently left them behind, particularly at times of severe stress. But the Personality Disordered continue to have them more of the time, into adulthood; sometimes these fantasies are highly creative and unusual, but more usually they just lack realism.

Exactly the same dynamic of reversing the truth fuels narcissism; however, instead of powerlessness being reversed, in Narcissism it is the feeling of being worthless which is compensated for by a grandiose fantasy of inflated status in the eyes of others. We are constantly attention-seeking and only happy when we, or something about which we feel authoritative, are the topic of conversation. The abiding injury we suffered as a baby at being deemed unimportant because our needs were subordinated to those of the carer is temporarily healed by saying to anyone who will hear (or just secretly to yourself), 'I am the most wonderful/beautiful/intelligent/special person' – the precise opposite of how we really feel. We have great difficulty in relating to people realistically and honestly unless we incorporate them as part of ourselves, something easily done because of weak boundaries between Me and Not-Me. We love to create social or professional cliques, elites based on the unconscious dictum: 'I am perfect. You are perfect too, but you are part of me.'

Narcissism and omnipotence are usually present if we suffer from one of the commonest kinds of Personality Disorder, the borderline. We have a pervasive instability and ambivalence in the stream of our everyday life. Emotionally erratic, capricious, impulsive and often explosive, we make a very awkward, manipulative companion. Relatives and acquaintances are put on edge by sullen displays and hurt looks or by obstinate nastiness, eliciting rejection rather than the nurture that is desperately needed. Ever since we learned depressive and hyper modes in earliest infancy, our mood has shifted between extended periods of dejection and apathy and frantic spells of anger, anxiety or excitement. Our despair is genuine but it is also a means of expressing hostility, a covert way of frustrating and retaliating. Angered by the failure of others to nurture us, we use moods and threats to 'get back' and 'teach a lesson'. By exaggerating our plight and acting in a miserable fashion we avoid responsibilities and place added burdens on others. Cold and stubborn silences are punitive blackmail, a threat of trouble to come. Easily offended by trifling matters, we are readily provoked to contrariness if things do not go our way.

We often have rapidly changing and self-contradictory thoughts about ourselves and about odds and ends of passing events. We voice dismay at the sorry state of our life, and articulate our sadness and our resentments: we feel discontented, cheated, unappreciated, misunderstood and disillusioned, and that all our efforts have been for nothing. The obstructiveness, pessimism and immaturity which others attribute to us are dealt with by believing that these are a reflection of our greater sensitivity or of some other special attribute – a narcissistic distortion. Yet, confusingly, at other times we may lurch from this sense of superiority to saying it is our own unworthiness, our failures and bad temper, which are causing misery to ourselves and pain to others. In a state of permanent flux, our thoughts and emotions are precarious.

Not surprisingly, sustaining stable personal relationships is difficult. A study comparing borderline women seen in a marriage guidance clinic with non-borderlines found the borderlines to be filled with self-deceit. They had greater sexual dissatisfaction and depression about their sex lives, yet thought of themselves as highly desirable. They reported more problems in their relationships, greater sexual boredom, difficulties in achieving orgasm and proneness to affairs, yet they invariably identified their partner as the one with the sexual problem. The borderlines also were more likely to report lesbian desires, and half had suffered sexual or physical abuse as children. Borderlines in particular, and Personality Disordered people in general, are more likely to have what is called a ludic love style, in which love is seen as a game – something done to, rather than with, another person. We tend to believe that what our lover does not know about cannot hurt them, and to be unfaithful. Deceiving partners in this way, we gain more pleasure from the playing of roles than from intimacy or sexual intercourse itself.

Intimate contact with others leaves us feeling battered because our omnipotent and narcissistic fantasies are constantly banging against the ceiling of reality. It also leaves us feeling drained and depleted, increasing our loneliness, dependence on others and need for company. We seek contact and set off on the cycle again. We are at grave risk of filling the emptiness with manic addiction to drugs, alcohol, sex, gambling or work, just as we responded with hyperactivity when pleasure-starved by unempathic care in infancy.

A sad example of a borderline personality was the television presenter Paula Yates, with whom I worked on a television series for six months in 1986 and who died from a heroin overdose in 2000. In her autobiography she described herself as 'a mass of contradictions' and she was someone who could change her ideas freely, apparently with no sense of self-contradiction. An ambitious careerist who worked throughout her own children's early years, she said of working mothers, apparently without a hint of awareness that she was referring to what she had done herself, 'If they do not stay at home the emotional well-being of the next generation is in jeopardy . . . if you think you can give birth and skip back to work in a matter of weeks you are being at best irresponsible and at worst, selfish.' On several other occasions she flatly contradicted herself in print, for example writing, 'I never said that mothers should not work.'

Regarding her public image, she stated that, 'My image in the media as a Mata Hari is nothing to do with reality' and complained that she had been misrepresented. Yet those closest to her knew she was sexually promiscuous to an unusual degree, and prone to eroticizing all relationships with men. 'Paula flirts with everyone,' said her friend Sue Godley, and in practice she was eccentrically sexual in her dealings, especially with complete strangers. At our first meeting she walked up to the table and sat on my lap. I saw her do this to other men, too. Part of the intention was to create embarrassment and draw attention to herself. In my case she did it because I was a close friend of someone upon whom she relied for her living; she wanted me to like her in order to advance her career.

As a colleague, Paula was liable to be as egomaniacal as a toddler. Indeed, in many ways she never grew up and this may have been partly because, according to her account, her childhood was traumatic. She described herself as having been a 'whining, whingeing, clinging child' in a home 'permeated with the scent of melodrama every morning'. The man who she thought was her father, Jess Yates, was a manic depressive TV presenter whom she loathed. Shortly before she died it emerged that her biological father was, in fact, another TV presenter, Hughie Green. Her exceptionally selfish and whimsical actress mother was rarely there, but the young Paula still missed her terribly. Her parents separated when she was nine. When her mother subsequently made a rare sign of acknowledging her existence by sending some presents, her deranged father threw them on the fire. Paula recalled that 'Because of incidents such as these I hated being a child. I couldn't wait to be grown up. No one should ever have to feel as powerless as I did as a child.' According to her autobiography, she escaped into fantasy from this intolerable reality: 'I had my own parallel world where the future was everything and the present meant nothing. I withdrew into my own internal world.' The adult Paula still had a tendency to convert fact into fantasy, which on occasion spiralled to such a degree that she had to be admitted for treatment in a mental hospital.

Throughout her marriage to Bob Geldof she gave many interviews stressing how deeply in love they were. Yet those close to her knew that she was having affairs. Paula may partly have been just lying rather than fantasizing. She made extensive use of a public relations consultant whose job was to massage the truth for the media.

Her lack of consistency was typical of the borderline personality. Paula was engaged in a minute-by-minute struggle to make her life seem meaningful, for, instead of a stable, secure self, she had an insecure, shifting void. She tried to create meaning through melodramatizing her life, liable to feel she did not exist unless she was at the centre of a crisis which – despite protestations to the contrary – she liked the tabloids to chronicle because they made her feel significant. She tried to bolster her poor self-regard by associating with the famous.

Although Paula seemed to be aware of some of the ways in which her childhood affected her, she was powerless to avoid repeating the past. She saw parallels between the events that led to the end of her father's TV career and her own. In both cases they were exposed by the tabloids as having had affairs, and because they worked on family programmes were cast out. She glimpsed other similarities to her father, yet none of this knowledge helped her to avoid his mistakes.

This was partly because she persistently dressed her childhood in flippant humour. By converting it into a 'hilarious' tale she distanced herself from it. Patients in therapy sometimes recount childhood events without recalling the accompanying feelings. Paula's shallow glibness, evident in her autobiography, kept her at one remove from the emotions. As a result, despite all her awareness and vows not to do so, she did to her own daughters much of what had been done to her. 'I'd always thought that people should stay married forever no matter what it was like for them,' she wrote. Yet she was able to persuade herself it would be all right for her to leave her husband for another rock star, Michael Hutchence. She was probably blaming her external circumstances for an internal problem.

Sustaining any sense of coherence is a problem for all Disordered people, and a fundamental tactic for keeping reality at bay is called 'splitting'. The world is divided into good and bad, black and white, expressing a crude psychology in which there is only us and them, for and against. Knowledge of different aspects of our experience is kept in separate compartments to avoid incoherence and provide certainties. For example, a wife-beater may seem to his colleagues at work to be a mild, charming, generous fellow. Through splitting, the 'him' at work may be totally unaware of the 'him' that beats his wife.

Deep down in many Disordered people lies a paranoid conviction that others are hostile and malicious. Yet this can be reversed at a moment's notice by engaging in 'idealization', a conviction that others are benign and wonderful regardless of the evidence. Such processes are found in all three-year-olds, living as they do in a passionate world of extremes, but in adults they are perplexing.

'Denial' aids these mental operations, enabling the world to be cut up to suit inner reality. It is illustrated by the opening scene of the movie *Big Deal at Dodge City*. A woman is sitting on a bed doing a jigsaw when her husband comes in and notices that she has a pair of scissors. He asks what she is doing with them, and she replies that she is cutting the jigsaw pieces to make them fit. The Personality Disordered are constantly chopping up the noncompliant, complex and intolerably independent jigsaw of reality. It can be adapted to fit the needs of the moment but, like the woman using the scissors as a childish solution to her problem, in the end they are confronted with the incomplete puzzle of their lives.

Another way of getting rid of unwanted experience is 'projection' – attributing our own feelings to others. This is easily done by a person who has weak boundaries between Me and Not-Me. Instead of experiencing our own depression or anger, we simply attribute it to someone else and say, 'Why are you so unhappy?' or 'What are you so angry about?'

Equally, we can remove ourselves from the situation through 'dissociation'. At its most extreme, this is the strange sensation of being outside ourselves, looking in. It is often found in Wobblers, a pattern of insecurity very common in Personality Disordered people.

For such a fluid psyche or mind, the adoption of multiple per-sonalities is easy. If reality is unbearable, a fine way to evade it is to be someone else. We often feel unreal and 'unable to be myself', uncertain as we are of who that is. Paradoxically, we may feel most real when pretending to be someone else, and even if we don't gravitate towards acting as an actual profession we still make a good actor. We live our lives as if playing ourselves, rather than really feeling that is who we are. We recognize our passport photograph or face in a mirror as belonging to the person named on our birth certificate or passport, but for much of the time that person is someone we are pretending to be. This can make us a very adept mimic, able to switch identities with remarkable fluidity, just as small children can flit between one fantasy game and another.

Often witty and charming when we first meet someone, afterwards we leave an indefinable sense of something being not quite right. We are an impostor impersonating ourselves, expecting someone to tap us on the shoulder and expose us as the fraud we feel ourselves to be.

I have interviewed dozens of politicians, TV presenters and business entrepreneurs with these As If personalities. For example, the disc jockey Tony Blackburn told me in a television interview that 'I only feel myself when I am pretending to be Tony Blackburn, the disc jockey. I live for the daily two hours of broadcasting – you can forget the rest. I wish my whole life could be a live radio show.' A more recent example is the TV presenter Michael Barrymore, at the time of writing hoping to make a comeback after a scandal involving the death of a man at his home. Speaking of his difficulty in living a life without regular TV appearances and unadulterated by heavy drug abuse, he states that 'the only time I ever found true contentment or happiness is in front of an audience. The minute I came away from that environment and had to be me, that's what I couldn't live with.'

Jeffrey Archer, the novelist and former politician, whom I interviewed on TV in 1987, is another example. The interview took place shortly after he had won a libel case which, although exonerating him from consorting with a prostitute, had cost him his vice-chairmanship of the Conservative party. This was the second time that he had snatched disaster from the jaws of success, the first having been when he was declared bankrupt as a Member of Parliament and had to resign. I asked him why his career had been so full of boom or bust moments and, although he displayed little insight, one reply was revealing. He told me about a play in which all the characters had lost everything they cared about. 'The duchess faints,' he recounted, 'the prostitute swears, the soldier shoots himself and the one who's lost more than anyone else sits in the chair, looks straight at the audience and says, "Wonderful. Now I can start again." And I think there is a little bit of that in me. When I reach a certain point I need to be driven – I need a challenge.'

Perhaps he does need challenges, but this could also suggest that Archer has a need to be found out: his feeling of fraudulence becomes so overwhelming that he longs to be exposed. For living a pretend life is lonely and unsatisfying. There is no real (a word which Archer used again and again in my interview with him) contact between you and others because all you are offering is a false self, a carapace. You feel a desperate longing for something real to happen, of which being unmasked is an example. That may explain the huge pleasure Archer took in the wealthy man's 'Wonderful. Now I can start all over again.' It may have meant, 'At last I can give up all this pretence and be someone authentic, if only for a brief moment.'

At the same time, there is often a huge desire in the As If personality to play games and to sustain the fraud – to get away with it, as so many of Archer's characters do in his novels. This ploy, called 'imposture', can encompass revenge on people who live first-person lives and joy at having suckered them. Not far beneath Archer's exuberant, Roy of the Rovers exterior is an infantile rage. It briefly exploded during my interview with him when, upset by my repeated attempts to find out why he was so self-destructive, he embarked on a peculiar rant: 'I am by nature an enthusiast, I am by nature someone who wants to change things and do real things with my life, and when the petty, and the little belittlers go on decrying my feeble achievements – and many of them are feeble – I feel sorry for them because they're not getting up and achieving something themselves. And I won't become a cynic myself or become a belittler myself, running everyone else down to prove that I was really in the right. I was wrong [in his relationship with the prostitute Monica Coghlan], I made a mistake, I'm trying by hard work to get back and none of your psychological discussions on Channel 4 will change that – that is the way I am.' In a more recent television documentary, his rage was even more evident. When the interviewer asked him a difficult question he completely lost his temper.

Archer may pretend to be the Great Novelist or Politician or Playwright, rather than actually feeling himself to be that person. For the As If, proving to the world that their fantastical version of reality is true, even when they know it is not, may become crucial for their mental health. Alongside their desire to be exposed, the As If often also want to force the rest of us to accept their fraudulence as real, because otherwise they will be conscious of how fake they feel.

In bringing a libel court case against the tabloid newspaper which exposed his relationship with the prostitute Monica Coghlan, Archer was taking an extraordinary risk, since we now know that he was lying. What he may really have been doing was trying to prove to us that our world is a sham and that his is real. By all accounts, the air of unreality at the trial was striking. A 'fragrant' wife (as the judge famously described Mary Archer in his infatuated summing up) with a loving, dutiful husband were portrayed, yet most of the press knew perfectly well that Archer had had at least one mistress and that it was all too probable that he had used the services of a prostitute. That made it all the more satisfying to Archer to turn reality on its head by getting a verdict which supported his Walter Mitty existence.

Of course, all of us are playful and engage in As If pretence – it is a vital sign of emotional richness. Children love to make believe that they are someone else, and as adults we often feel at our most alive when telling stories and putting on funny voices. By identifying with fictional characters in films and novels we can experience our emotions more intensely than in reality. Once, when I split up with a girlfriend, I shed no tears until months later when watching the Humphrey Bogart and Ingrid Bergman characters part at the end of the film *Casablanca* – a piece of fiction was more real to me than what had actually happened in my life. Escape into fiction and a regaining of power over otherwise inescapable reality is a prime motivation of creative artists. In all the arts there is an element of controlling other humans by engaging their emotions and forcing them to feel those that are difficult or moving – most often, the ones that artists cannot cope with themselves.

In Archer's case, his novels and plays, with their cast of liars, conmen and charmers, are transparent attempts to live out in fiction what he would (or actually did) enjoy in real life; they are wish-fulfilments. He began scripting a play immediately after the Monica Coghlan story broke, and did the same as soon as his ambitions to become Mayor of London collapsed in 1999. In the play he even acted on stage the role of the accused, breaking down even further the barriers between his fantasy and his reality. Indeed, some of his words contain barely veiled accounts of his actual deceits, allowing him to sit back and enjoy the feeling that he has taken us in: not only did his lies remain uncovered in reality, but the reader has paid him money by buying the book that catalogues them.

One variant of the impostor type earns their living through criminal deception. This is the Machiavellian, 'false self' type – the psychopathic conman who assumes identities in order, for instance, to sell old ladies false insurance policies. Psychopaths are impulsive, greedy for sensation, prone to criminality and breaking norms of decent behaviour, and have a weak conscience. If we are psychopathic our emotions are not deeply felt. We do not experience guilt, remorse or empathy. Towards others we are promiscuous, manipulative, grandiose, egocentric, forceful and cold. In the popular imagination all psychopaths are drooling maniacs or cold killers, but in reality this is just the tiny minority who become serial killers or sexual criminals, some of whom are masters of disguise and take great pleasure in a game of evading detection. The great majority of the 2–4 per cent of Britons who are full psychopaths are not behind bars, and some of them occupy the most powerful positions in our society. For, although most Personality Disordered people have severe conflicts with authority figures at school and work, and personally unsatisfying professional and personal lives, the peculiar mix of traits involved means that a proportion are extremely successful.

The false self's tendency to define itself externally can make people-pleasing, especially towards authority figures such as teachers, examiners and employers, a central principle. That can make for excellent exam results and accelerated promotion. Such people may also go to fantastic lengths to try to make their fantasies come true, to confirm their narcissism or omnipotence in reality, using actual success to gain the adoration and power they were deprived of as infants. The performing arts – popular music, stage, screen and TV – are choc-a-bloc with narcissists desperate for the next fix of public approbation to over-ride their rock-bottom self-esteem. Their infantile sense of worthlessness is matched precisely by their surface need to be recognized as the opposite. Being acknowledged as Who They Are by a stranger in the street confers identity and temporarily compensates for the sense of invisibility to carers that they had in infancy.

The ones who are more omnipotent than narcissistic may become

214

power- and wealth-crazed politicians and businessmen, endlessly concerned to prove themselves, always restless and in need of further proof of their efficacy in succeeding over others or of a material wealth that disproves their inner poverty. No wonder we talk about promising, competitive individuals as 'hungry'; a great many literally hungered to be fed as babies. A particularly telling example is the Formula One racing team owner, Frank Williams.

Very unusually for a living famous man, an intimate and reliable portrait exists of him by the woman who shared his life, his ex-wife Virginia. In the preface, she claims that her purpose in writing the biography was not to make money or to settle scores. She wrote it 'as an exorcism' and in the hope that he would read it: publishing was the length to which she had to go in order to communicate with him. Whilst Williams has never told his side of the story, and hers inevitably cannot be wholly objective, it seems to have been done in a spirit of love rather than one of revenge.

The picture that emerges is of a man unfeeling to the point of sadism. Here was a husband who loved to drive her home late at night through country lanes with the headlights off. Virginia writes, 'I would shout at him to put the headlights back on. "I can go faster this way. I can see the lights of oncoming cars", he would reply, unmoved by my shrieks.' During their long courtship, she recalls, 'I knew I must behave as if I regarded our affair as lightly as he did, if I were to avoid him haring off in fright.' At first he refused to spend extended periods of time with her, but she gradually weaned him on to regular stays at her flat by doing his washing and cooking. However, she writes, 'he would seldom warn me that he was due to leave for a trip until the morning and would make endless excuses as to why he couldn't leave a phone number, so I was very far from feeling secure. I always hated him going away and often burst into tears, which he found hugely funny.' Typically, after a few nights 'suddenly, he would disappear for three weeks ... [he] liked the company of women but he also seemed terrified of the power they wielded'. He confirmed this himself in a rare confession of emotion after their marriage, saying to her that 'I've never been as scared of anything in my life as marriage.'

In accord with this, Williams found it as hard to give love as to receive it. Fifteen years passed before he gave Virginia a present.

When it happened, she 'was astounded. Frank had never given me anything in my life before. Present-giving was not a habit he had ever acquired.'

On the day of their wedding he rang the restaurant where they were supposed to meet first for lunch, to say he was detained by work and would see her at the registry office. As soon as the ceremony was over he said, 'Okay, I'm off back to the office. See you later.' When she had a son she asked what he thought. 'Oh, I've seen him. I feel sick,' was his only comment. Regarding the prospect of a daughter his observation was that 'I can't imagine anyone wanting a girl.'

Virginia learnt early on to expect sexual infidelity. 'I discovered something unwelcome – Frank was a terrible flirt', and before long she 'had no illusions that I was the only woman in his life'. When Williams's huge success came, 'Frank had become enormously attractive to women.' She believes that, for famous men, the lust of attractive women is 'a spoil of victory' to which they succumb and that 'alas, I knew that Frank was unlikely to be an exception'.

The psychoanalyst Anthony Storr gives a brilliant account of the mind of such men in his book *Human Aggression*. Storr writes that 'although they passionately long for love, they have so deep a mistrust that any really intimate relationship appears to them dangerous. Receiving love is humiliating or actually dangerous.' They may seek to insulate themselves through high achievement but, writes Storr, 'however successful he may be in the external world, love from another seems to threaten his independence and masculinity'. As a result, such men live in a perpetual dilemma. On the one hand, writes Storr, 'to deny the need for love is to enclose himself in a prison of isolation and sterile futility'; on the other, 'to accept love is to place himself in a position of dependence so humiliating that he feels himself to be despicably weak'. These observations are borne out in Williams's case. At their final parting Virginia asked him, 'Don't you care, Frank?' His reply was 'Ginny, emotion is weak.' She comments that 'he never discussed his emotions or feelings at all', so that as she left 'he just looked at me without expression'.

The explanation for Frank Williams's remote, cold personality is found in his earliest experiences. His parents separated shortly after his birth and he only saw his father a few times afterwards. He was never close to his mother, who was so out of touch with him, literally as well as emotionally, that, as young as three, he regularly wandered away from home, on one occasion being found by the police several miles away. Three was also the age at which his mother sent him to the first of a series of Catholic boarding schools. As if that were not sufficient maternal deprivation, many of his holidays were also spent away from her, farmed out to monks at a monastery. In effect he was like a child who had been taken into local authority care, and this is exactly the kind of childhood that creates the sort of person Anthony Storr describes. 'Human babies are, for a protracted period, peculiarly helpless and unable to fend for themselves,' he writes. If no one person meets their needs they end up desperately insecure and 'however competent or powerful a man may be in fact, he still feels at the mercy of anyone whom he allows to become emotionally important to him'.

Above all, Williams used cars to protect him from such dependence and to turn the lead of his pathology into the gold of success. Virginia records that they, much more than any woman (including her), were his true love. According to his mother, 'from his earliest days his heroes were drivers – people who could make wheeled machines do as they wanted'. He could not control the people in his chaotic childhood world, so he employed the controllable inanimate to replace the unpredictable intimate.

This account, of Frank Williams in particular and of Personality Disordered people in general, should not mislead us into believing that they are beings of a wholly different order from everyone else. All of us sometimes use the mental equipment described above and, when we are young adults or teenagers, more than half of us have Disorder either severely or mildly. Jokingly, but perhaps less so than we would care to admit, we may boast of our sexual or sporting prowess (grandiosity, narcissism). We may daydream about gaining total power at work or in bed (omnipotence). In intimate relationships we may confuse our own feelings with those of partners (projection) and initially blank out the truth when faced with intolerable realities (denial). All of us put on different faces to meet the faces that we encounter at work and play (false selves), and

217

we enjoy adopting personae when telling jokes or stories (imposture). Nobody could survive without keeping numerous facts about themselves secret from other parts of themselves (splitting). Truly, 'Humankind cannot bear very much reality', and normal, mentally healthy people use these mechanisms to bolster and embellish their illusions.

It is interesting that, as we get older, our tendency to be Disordered decreases. An American survey showed that one third of young people had severe signs of the illness and one fifth had it mildly. Yet between the ages of 25 and 59 the proportions dropped dramatically, and by the age of 60 only 11 per cent had the full illness (with a further 17 per cent having it mildly).

Whilst there are clear differences in the degree and amount of Disordered psychic defences employed by Personality Disordered people compared with those who are mentally healthy, we can see signs of infant deprivation and specific weaknesses of sense of self all around us. Watch people waiting for a long-delayed bus or for their supper when it is an hour late and you will witness hyper and depressive reactions, albeit concealed beneath an adult carapace. Whenever we are in a high state of expectation about something we long for – hearing an exam result, going on a date or day one of a new job – the strength of our sense of self and capacity to tolerate uncertainty, with its attendant risk of feeling deprived, are put to the test. Carefully observe the normal dances of dominance and submission to be found in most workplaces and many homes, including a casual cruelty and pleasure at others' suffering, a self-centredness verging on omnipotence and narcissism. The way we cope with adversities is hugely influenced by our earliest equivalent experiences. A baby waiting to be fed, or crying out to be held, or feeling too hot or too cold is in a state of extreme vulnerability, completely powerless and at the mercy of its carers. This was true of all of us once – and for all of us, too, there were times when the care we received was unempathic. The difference between Disordered people and the rest is only that the lack of empathy was greater for the Disordered and endured for longer, as illustrated by the scientific evidence directly linking early deprivation of empathic care with subsequent Personality Disorder.

The infantile causes of Personality Disorder

Children who received no empathic care at all during their infancies and who survived are rare. There are 39 well-documented cases of children who were either raised by wild animals or given no care from humans beyond the supply of food. Their language development is nonexistent, they have no social skills and their personality is totally Disordered. Where such extreme early deprivation is followed by nurturant care there is some improvement in speech, intelligence and social skills as a result. However, the care does not reverse the damage to the self. Invariably, such cases are severely Personality Disordered with a weak sense of self.

Systematic study of the impact of infant deprivation began after it was first noticed that orphans in impersonal institutions suffered physical and mental harm. Even in sanitary and hygienic orphanages, the mortality rates were far higher than among children raised by their parents. Institutionalized infants suffered from 'hospitalism', comprising lethargy, failure to develop the most basic social and mental skills, and extreme susceptibility to infections and disease. No infants thrived under these circumstances and all were damaged permanently, suggesting that if there is such a thing as genetic resilience it cannot withstand the absence of care. However, infants in orphanages where relationships with individual carers were encouraged did far better. It was not the sensory deprivation of the cribs with no toys or stimulation that damaged the neglected babies, nor was it simply the absence of their mother. What was critical was having someone who knew the child intimately and understood his or her specific needs, capacities, idiosyncrasies and vulnerabilities. Babies who had this experience could survive a lack of sensory stimulation or could cope without their mother. So long as someone was empathic, the infant could always develop to some extent.

That these early experiences are linked to later Personality Disorder is strongly supported by the fact that at least half of men and one third of women raised in children's homes suffer from it. Most revealing of all, the younger the child when poor parenting was experienced, the greater the likelihood of subsequent Personality Disorder. In one study, children were examined when in care and then interviewed years later in adult life. Many bad things had happened on the way to adulthood, from poor education via lack of supervision to being bullied, but one environmental experience above all others predicted whether they would be Personality Disordered: disrupted parenting before the age of two.

What is true of institutionalized children is also true of those raised by their own unempathic parents. John Ogawa and his American colleagues followed 168 children from birth to the age of nineteen, measuring the quality of the care they received throughout childhood. The sample, all of whose parents were at high risk of being depriving or abusive, had been selected to test whether early maltreatment correlated with problems in later life. In particular, the impact of early parental care on dissociation was measured. A distinctive sign of Personality Disorder, dissociation entails a variety of amnesiac mental tactics for evading painful realities, like not noticing or disallowing or forgetting or refusing to acknowledge. Dissociated people also escape from the present by becoming distantly absorbed in a single aspect of the inner or outer world, for instance gazing hypnotically at a pattern in the wallpaper or disappearing into a fantasy in the company of others. There may in addition be depersonalization, in which events are experienced as if by a third party disconnected from one's own body or feelings. At its most extreme, the dissociated person may develop different sub-personae into which they escape – in some cases this is the precursor of schizophrenia.

For the first time, Ogawa demonstrated that the degree to which care was neglectful or abusive or disharmonious before the age of two predicted whether a person would be suffering from dissociation seventeen years later. Genetic factors appeared to play no part: the baby's temperament after birth or at three months did not predict how he or she turned out. An accompanying report from the same study also proved that having a Wobbler pattern of attachment at two years old predicted later dissociation, and that it was the quality of early care which predicted both Wobbling and later dissociation. Despite the fact that an enormous number of influences intervened between the experience in infancy and the measurement of the nineteen-year-old personality, its substantial impact was proven. Those in the sample who were only maltreated after infancy, whether between the ages of two and four or in middle or later childhood, were significantly less likely to suffer dissociation at nineteen than those who suffered maltreatment in infancy: the earlier the maltreatment, the greater the likelihood of a symptom of Personality Disorder.

Ogawa was also able to explain why some of his sample were more likely than others to develop dissociation when faced with traumatic environments after infancy. A strong sense of self was crucial. If a child had developed a strong one early on, through empathic care, he was far less likely to become dissociated when faced by traumas in later life. Children who first suffered maltreatment after the age of five were much less likely to be dissociated at nineteen if they had an early strong self rather than a weak one. Being uncertain of who they are, the weak-selfed require less severe and less frequent trauma to make them doubt their reality, and, because their psychic boundaries are fragile, it is easier for them to make a dissociative escape from intolerable realities. The stronger-selfed can manage to remain themselves. This is suggested by another study which showed that twenty-two-year-old adults had a better quality of relationship with partners at that age – they had more insights about themselves and were more respectful of the other's autonomy – if their mothers had been sensitive in caring for them during the first year of their lives. Whether sensitive or insensitive, the impact of care is laid down as electro-chemical brain patterns, evident in either a strong self or Disorders of it.

Virtuous and vicious circles of early mothering, genes and Personality Disorder

Ogawa also checked to see if genes could be responsible for dissociation: inherited, temperamental traits were measured soon after birth and again at three months of age. These traits were found not to predict dissociation at age nineteen at all, suggesting that genes played little or no part in causing it. Studies of identical twins also suggest this is true of many Personality Disorders – no genetic influence was found in one study of borderline personality, for example. By contrast, strong evidence (from over twenty studies) exists to show that, compared with adults suffering other kinds of Personality Disorder or other mental illnesses, Borderlines are more likely to have suffered childhood maltreatment – in particular, loss or separation from parents, being unwanted, and abuse and neglect from both parents. In one large study, 84 per cent of Borderlines had been subjected to some form of maltreatment from both parents, compared with 61 per cent of patients with other Personality Disorders. None the less, at least in theory, the cause of Personality Disorder could lie in the child's inherited temperament. An infant could be born with difficult traits that would cause her parents to become unempathic. If she was born with a great reluctance to make eye contact, or was constantly irritable, or forever crying and never sleeping, a parent might give up the struggle to make contact and to understand life from that baby's point of view. It is true that a baby has no physical power over her parent and cannot deliberately use her mind to influence the parent. But because most parents are highly motivated to do their best to keep the baby happy, a difficult baby is particularly distressing because the parents feel they are failing her. On top of this, it is also perfectly possible that babies with genes that make them difficult are more likely to have the sort of parents who themselves have a genetic tendency to be unempathic, a double whammy. Not only is the child born disinclined to connect with her parents, the parents are also that way inclined themselves: poor relating could be heritable from both sides.

These ideas have been investigated by measuring the difficultness of infants soon after birth and relating this to the way the mother subsequently behaves and the infant turns out. As noted in Chapter 4, the main finding of such studies has been that it is usually the degree of maternal empathy, not the difficultness of the infant, that determines what happens.

In writing these words, I am acutely aware of how upsetting, enraging or depressing they could be to the reader who is a mother and feels she has failed her infant. Yet I bow to no one in my appreciation of how difficult it is to care for small babies, and stirring up guilt or anxiety is the very last thing I hope to do. In fact, one of my objects is to make more people aware that most women are unlikely to experience a greater threat to their mental health than the first few months of the life of their firstborn. Between 10 and 15 per cent develop a full-blown major depression soon after the birth and about 25 per cent have done so by the end of the first year. For the remaining women who do not become actually mentally ill, caring for the infant in the earliest months normally produces dysphoria, a state of low grade, depressed mood allied to total exhaustion. In a recent survey of 1,000 British mothers, over half said their exhaustion left them in 'a state of despair' and highly irritated by their babies, four hours' sleep a night being the average for those with small ones. Four-fifths of mothers of under-twos said the infant had placed their relationship with partners or husbands under 'immense strain', with rows commonplace; and two-thirds of the women had been 'completely put off sex'. Whilst at times they doubtless also feel an unprecedented sense of fulfilment and achievement, until the infant has settled into stable patterns of sleeping and eating on a daily (and especially nightly) basis they will be liable to terrifying feelings of loss of control, periodically violent impulses towards their baby and chronic desperation.

Most people imagine that postnatal depression is a largely or purely biological misfortune, the flipping of a hormonal switch, but there is no evidence whatever that it has a primarily physical cause. Whilst all women's hormonal levels alter during pregnancy and for a few weeks after the birth, no difference has ever been demonstrated between the hormones of mothers who become depressed and those of mothers who do not. By contrast, it is possible to predict with a fair degree of accuracy which mothers are most likely to become postnatally depressed by asking them about their psychological histories: it is the kind of childhood they had and their current circumstances, not hormones or genes, which predict postnatal depression. In most cases, it is the combination of these with the sheer hell of trying to meet the needs of the infant which tips them into depression. The fundamental problem is the total dependence of the baby, twenty-four hours a day, resulting in an equally total loss of autonomy in the mother. The great majority of mothers do not have someone else there at all times to help them out when the grinding relentlessness of meeting the infant's needs becomes too much. Whereas in pre-industrial revolution societies people lived in extended families and there would always be sisters, mothers and other relatives on hand, all too often when the exhausted modern mother of a tiny infant is at the end of her tether there is no one available to share the burden.

Having money certainly reduces the risk of depression because it buys mechanical support (cars, washing machines) and the human equivalent (maternity nurses, nannies). In the definitive study of the subject, whereas nearly one third of the mothers of an under-six-year-old from poor homes suffered from full-scale depression, only 4 per cent of middle-class women did. But it is not only a matter of money. Women who breastfeed are at greater risk, since they get less sleep and producing the milk is tiring; likewise, women who lack strong intimate relationships, putting divorcees and single mothers in particular danger. So are those whose own mother died when they were young; and, if she is still alive, the risk is greater for those who have a bad current relationship with her or who did so when small (and are insecurely attached adults as a result). The sheer number of children is also a factor. Having three or more under fourteen, especially if one of them is under six, increases the risk. Even if the mother has survived the firstborn, trying to cope with a second or third newborn offspring is nightmarish when there is little support. The increased risk of having more than one child to care for is neatly illustrated by the simple fact that significantly more mothers who have twins, rather than singletons, are depressed.

All these factors, mostly predictable from before birth and many of them preventable if our society were better organized, affect whether the mother becomes depressed. But what about the baby – surely his difficultness is important too? Any mother who has had more than one will agree that they are all different, and this could provoke despair. In particular, during the first few weeks and up to three months about one fifth of babies cry more than three hours a day for more than three days a week; it is known as colic. In such cases, crying increases from about two weeks after the birth, reaching a peak in the second month. The infant has prolonged bouts of intense crying that are resistant to soothing, including feeding, and is fretful and fussy in between. Typically, these bouts occur in the early evening, precisely the time when the mother is at her most exhausted. The infant often clenches his fists, flexing his legs over his stomach and arching his back; he has a flushed face and makes grimaces; his stomach may be hard and distended, and he often has wind and vomits. Perplexingly, the crying bouts seem to have no obvious trigger and may end suddenly and spontaneously.

If ever there would seem to be an illness with a physical cause it is colic and yet, despite decades of research, no causal factor has been identified as present in colicky babies and absent in those who don't have colic. On the contrary, the fact that it seems to be rare or even unknown in some non-Western cultures suggests it is not a medical condition, like a bacterial infection. Nor has the problem been linked to the kind of feeding method used, breast or bottle. The most respected authorities estimate that a physical cause exists in only about 5–10 per cent of the one fifth of babies who get colic. What is more, it is becoming increasingly clear that most, if not all newborns develop some symptoms of colic, and that the only distinguishing feature of babies labelled as having the illness is the severity and frequency of the bouts. Full colic is the extreme end of a universal spectrum. For the newborn it is probably a normal part of adjusting to using their stomach to digest food – having spent nine months using the placenta for nutrition – and to other peculiarities of life beyond the womb, like the need for stomach bacteria. Whilst there may be substantial differences in the degree of sensitivity newborns display to these new conditions, the evidence suggests that it is how the mother reacts to the baby's inevitable periodic distress that is critical.

The first few weeks are a very delicate period in the mother's life. She is emotionally fragile, vulnerable, yet the need to fit into the infant's patterns feels like permanent jet-lag, with her sleep patterns going haywire. Worst of all, she is having to expect the unexpected as regards the baby's patterns. Until the age of three months, and sometimes beyond, just when the mother thinks she has established a routine of sorts, the baby is liable to break it – no wonder the helplessness and despair of depression are provoked. None the less, if she has adequate support and is able to tolerate the chaos, a virtuous circle can develop, in which the infant senses her overall calmness and has its needs met with a good-enough rapidity and perception, so that by six or eight weeks he has settled into reasonable gaps of two to four hours between feeds and sleeps for much of the night, requiring one or two feeds.

But if the mother is unsupported and at risk of postnatal desperation, a vicious circle can develop – about as horrible an experience as it is possible for a human to have, because the mother's mental pain is exacerbated by the sense that she is failing her infant. In this situation, when the baby displays colicky signs the mother is either unable to remain calm or simply ignores it, leaving the baby to cry, perhaps encouraged by a childcare manual suggesting she introduce strict sleeping and feeding regimes from birth. (In fact, the imposition of such patterns should not begin before two months and should be done only gradually.) Picking the baby up during the first two months has been proven to reduce crying – babies whose mothers carried them for three hours a day or more cry less at three and six weeks than those who do not. Close contact with mothers also promotes secure attachments, proven experimentally. Starting at birth, one group of mothers, randomly selected, were asked to use a sling regularly to carry their infant around with them, doing so at least 2–3 times a day, whereas another were not. At 13 months, twice as many (80 per cent) of the sling-carried babies were securely attached compared with the sling-less (40 per cent). What is more, babies whose mothers feed on demand and impose no rules before two months cry less at that age than mothers who do impose rules (but by four months the gradual introduction of regulated sleeping and feeding do help). However, once the infant has developed full-blown colic, often absolutely nothing works and the mother has to just ride out the storm. Carrying will not necessarily make any difference to the crying but it probably does give some solace.

The more that the mother can hand the 'screaming brat' – as he will seem at such moments – to someone else and grab a much needed nap, the better for all concerned. But if none of this is possible, the infant becomes increasingly distressed, the mother becomes more depressed, making her less empathic, leading to still further infant distress and so on. By two or three weeks into the baby's life a hellish scenario is developing in which the mother is getting very little sleep and most of the baby's waking moments seem to be filled with grimaces and screams. The vicious cycle is established and unless it is broken, it predicts problems later:

three-year-olds who had infant colic are more liable to suffer from hyperactivity, temper tantrums and sleep disorders.

Given all this, it is little wonder that there is a direct connection between the degree to which a baby has colic and the mother's mental health. Fully 50 per cent of mothers with extremely colicky babies are liable to have a mental illness. This falls to 25 per cent of mothers of babies with moderate colic and to just 3 per cent of mothers whose infant has minimal signs of colic. But which comes first – the colic or the mother's mental illness?

Whereas no physical pathogens (germs) have been shown to cause colic, the mother's history has been proven to play a part. In one study 1200 mothers were interviewed during pregnancy and again when their babies were three months old, at which point one third of them were diagnosed as severely or moderately colicky. Having a good relationship with a partner before the birth protected against colic. Also, if during the pregnancy the mother experienced a lot of stress, felt socially isolated and anticipated needing a lot of help after the baby was born, her child was more likely to have colic. For example, 25 per cent of the women who had no symptoms of stress during pregnancy subsequently had a colicky baby, whereas 70 per cent of babies did so if their mother exhibited five or more symptoms while pregnant. Since mother and baby are connected via the placenta, it is highly probable that the physical effects of this stress on the mother, such as raised levels of the stress hormone cortisol, are directly passed into the baby's bloodstream, perhaps creating greater vulnerability to colic.

Other studies show that the mother's relationship with her own mother and the kind of childhood she experienced affect whether the circle of mother–infant relationship is virtuous or vicious, the latter inducing colic. When asked during pregnancy or a few days after the birth, mothers who recall distressing childhood memories of their own, or expect a lack of support or excessive interference from their mothers, are more likely to have colicky infants. There is a real likelihood that some colic is transferred between generations not by genes but by problematic relationships between grandmother and daughter. At its simplest, this may consist of the grandmother teaching her daughter to care for the newborn in ways that exacerbate the colic, such as discouraging her from picking the infant up or from feeding on demand during the early weeks. Other grandmothers may be unable to provide support, reminded of their own grim experience of a colicky baby, leaving their daughters vulnerable to exhaustion and isolation. Alternatively, if the grandmother is still 'enmeshed' with her daughter, unable to allow her to function separately as an adult and still treating her as a child, she may interfere so much that her daughter is made anxious or becomes furious, preventing her from riding out the signs of colic that all babies show. All of this may be made worse by the fact that both mother and grandmother suffered unempathic care themselves, leaving them ill-equipped to provide it, reproducing the vicious cycle.

Of course, colic is not the only problem the baby can present to the mother. Immediately after birth, most babies are not difficult eaters or sleepers, not particularly moody, floppy, passive and so forth but they do vary in these respects. The reason has little or nothing to do with genes, as a fascinating Italian study of twins in the womb showed: variations in the prenatal experience are very considerable. Combined with the difficultness of the actual birth, this explains why newborns' temperaments differ, insofar as they do. But the crucial finding is that, largely regardless of what a baby is like at birth, what determines what they are like at six months or twelve months is how their mother reacts to them. Nearly always, a difficult baby will be made less so if the mother is empathic and the converse also applies: an easy, relaxed baby will be made difficult by unempathic care. When mothers who have a 'difficult' infant are given help in learning how to care for them, they soon cease to regard their baby as a problem because its feeding or sleeping soon become better regulated.

Another way of examining the role of maternal care is to see how infants of prenatally disturbed mothers fare. Numerous studies of infants born to parents who are liable to be unempathic because they are depressed, abusive, alcohol abusing or cocaine using show that they are far less likely to thrive. It is estimated that at any one time there are no fewer than two hundred thousand children with heroin-addicted parents. Easy babies of such parents are far more likely to be made difficult, and subsequently to have Disordered patterns of relating, than babies of normal parents. In particular, it is now clear that children of severely maltreating parents are more likely to be Wobblers, and that Wobblers are more liable to become Disordered in adult life.

In the case of infants of depressed mothers, the unempathic care has been shown actually to change the infant's brainwave patterns from healthy to abnormal. Long-term damage to the right brain occurs regardless of the infant's initial temperament. When the brainwaves of infants of depressed mothers were measured shortly after birth and again at later ages, the patterns had changed in the offspring whose depressed mothers had been unempathic. This occurred regardless of the pattern with which the infant was born. Interestingly, even if the mother recovered from her depression in the child's second or third year, the evidence of the damage to the child's brain was still there. Pathological patterns of brain electro-chemistry are the physical and enduring effect of unempathic care.

Above all, if there are any difficult newborns who drive their mothers to mental illness, it only happens because the mother was already vulnerable. Thus, it has been shown that if a newborn is especially floppy, mothers at high risk of depression are more likely to develop the condition. They are particularly at risk of being pushed from gloom to despair by a difficult baby if they lack the social support provided by a partner or friends. In this very limited sense, infants' neonatal temperaments can cause parents to become unempathic by triggering depression.

Of course, there may be such a thing as the baby from hell. Perhaps one in two thousand are born with only a minimal capacity to relate to their parents or anyone else. As the years pass it becomes apparent that, whoever had been their parents, these traits would have developed. No amount of empathic care would have changed them fundamentally and, given enough bad luck, almost any mentally healthy parent could be severely emotionally disturbed by such a child. Yet even in these cases, such as the illness autism, how parents respond makes a significant difference. Autistic children whose parents are helped to care for them in ways that are specifically attuned to dealing with their emotional deficits fare far better if the training occurs while the child is still only two or three. Usually the child will finish up being cared for by the state, but where the parents have exceptional commitment, resources and professional help it is possible for an autistic child to lead a relatively normal life.

Overall, there is no doubt that newborn babies pose a serious threat to the mental health of many mothers. As a society we do far too little to counter that threat. None the less, it is worth recalling that the average baby's emotional state is almost completely dictated by the minute-to-minute, hour-to-hour, month-to-month level of empathy of its carers, whereas the average carer's capacity for empathy is not controlled by an infant's character. Two simple facts highlight this power ratio. First, three-quarters of babies have been hit by their parent before the end of the first year. Infants cannot hit parents. Secondly, the infant age group (under two) are four times more likely to be murdered than any other. Infants cannot murder mothers.

Later childhood scripting of Personality Disorder

It is unempathic care, then, not genes, which creates a weak self biologically anchored in brain electro-chemistry, but subsequent care can alter the brain's patterns. It can have an effect on whether the potential to become Personality Disordered is fulfilled and what particular form it takes – the degree to which it is narcissistic, addictive, psychopathic and so forth. So, to be precise, deprivation in infancy does not cause Personality Disorder; rather, it creates the vulnerability. The type of later childcare and family script that are experienced affect the extent to which the potential remains dormant or finds expression, and so does the kind of society in which the child grows up.

Of these various experiences after infancy, neglect and physical and sexual abuse in childhood are the greatest predictors of Personality Disorder. When asked in adulthood, about half of Disordered people say they suffered abuse – far more than non-Disordered people and more than people with other kinds of mental illness, like depression. People who suffered childhood maltreatment are four times more likely to have a Personality Disorder than people who did not. Indeed, it is possible that a few Disordered adults emerged from infancy with strong selves, only to have them weakened by atrocious subsequent childhood experiences.

Symptoms of Disorder, like dissociation, are ways of coping with intolerable circumstances. If a child is being beaten or sexually interfered with, dissociation is a good way to distance himself from the experience; so are many of the other defences commonly found in the Disordered. Denial ('This is not happening'), omnipotence (saying to oneself 'I can do anything I choose' when feeling power-less) and narcissism ('I am perfect' when feeling worthless) will help the child evade the feelings of physical pain, fury, self-hatred, sad-ness and humiliation that abuse activates. Sub-personalities, which may develop into full-blown schizophrenia, are the ultimate escape mechanism ('I am someone else, so this abuse is not happening to me'). Interestingly, among survivors of sexual abuse the age at which it was committed affects the number of sub-personalities that subsequently develop and the extent to which the self is fragmented. Children abused before the age of six have much larger numbers of sub-personalities and more chaotic selves than those abused at later ages. Similarly, the earlier and more severe the abuse suffered, the more likely a boy is to become a violent man. All of this strongly supports the idea that, since our ego is weaker when young and less able to cope with adversity, the impact of early experience is greater.

The life of the film director Woody Allen well illustrates the role of later childhood abuse in fulfilling a potential to become Personality Disordered that was originated in infancy. It is not only the characters that he writes for himself in his films that are Disordered; he exhibits many of the symptoms in real life.

Ordinary contact with ordinary people is painful and disturbing to Allen. Rather than exchange glances or a word in a lift with a stranger, he will turn to the wall or bury himself in a newspaper. He shrinks if he is physically touched by acquaintances. He has known for decades the members of the jazz band with whom he plays every week, yet he eschews all but the most functional social exchanges with them. The performance of comic shows is his ideal form of companionship. 'I go up on stage, I tell my jokes, they laugh, I go home,' he relates, for he regards others with mistrust and suspicion.

From as young as he can remember, through to his early forties, he was unable to go to sleep without a night light on and until he had searched through the whole of his home to check that 'there were not any enemy there out to get me'. He is scared by any form of transport that he cannot directly control, be it an aeroplane, lift or train. He has a terror of being kidnapped and of being poisoned, unable to eat food unless it has been carefully vetted by an elaborate ritual. He will not use a bath – only a shower – for fear of dirty water. When he was married to the actress Mia Farrow and stayed for the first time at her country residence, he returned to New York because there was no shower. A special shower room had to be constructed exclusively for his use, and he always took his own food with him for the visits.

For Allen, relationships with women become painful when the sexual element grows into dependence and intimacy. Women who inspire these feelings in him are soon liable to be treated with malice and condemned as mad, bad or stupid. Stable, enduring love has been elusive to him, both his early marriages foundering within weeks. His comment about the actress Diane Keaton, whom he dubbed 'the coatcheck girl', illustrates his need to feel superior towards his wives: 'When I first met her, her mind was completely blank.' Of his current marriage to Mia Farrow's adopted daughter Soon-Yi, he said, 'The inequality of my relationship is a wonderful thing. The fact that I am with a much less accomplished woman works very well.' Thus far, this relationship seems to have been the most stable.

His relationship with Farrow was the first in which he felt truly loved. Initially he was able to keep her at arm's length despite the unprecedented sense of being nurtured, saying that 'a relationship is always better if you do not actually have to live with the other person'. They continued to live apart, but gradually he was drawn into domestic commitment. Farrow had his first child but he felt far more involved with Dylan, a girl whom they had adopted together. Until then he had been voluble in his dislike of children but, knowing he was most beguiled by blondes, Farrow went to some lengths to find a fair-haired infant to adopt. For the first time in his life Allen showed real passion for a child, a passion which famously became destructive and distorted.

He lavished such massive attention and love on Dylan that the other children felt excluded. Sometimes he would chew her food for her so that it would be easier to eat. He encouraged her to suck his thumb and was witnessed massaging sun cream into the crack of her buttocks. When the authorities were brought in to investigate, seven-year-old Dylan claimed that Allen had touched her genitals on two occasions. The detective responsible for the criminal case

232

declared unequivocally that there was sufficient evidence to pros-ecute, but would not do so for fear of the damage such a case would do to Dylan. The child psychologist whom Allen himself appointed to assess the relationship labelled it 'inappropriate'. The conclusion of the Supreme Court examination of his plea for custody of the child was that Allen was a 'self-absorbed, untrustworthy and insensitive' father.

It is likely that none of this would have become a matter of public record had Allen not engaged in another questionable relationship – that with Soon-Yi, whom Farrow had adopted whilst married to André Previn. Then a needy nineteen-year-old, it seems that Soon-Yi deliberately sought to seduce him. Allen is genuinely unable to comprehend why succumbing to her advances should be regarded as inappropriate. Whilst not the girl's father, his role in the household made him an authority figure; yet Allen consistently represents it to this day as a normal sexual liaison. Equally strangely, he was unable to discern the hurt it would cause Farrow, his partner of nearly ten years, for him to have an affair with her daughter.

It might seem surprising that a man who had had thirty years of psychoanalysis could act with such minimal understanding of the impact of his actions. But Allen is highly skilled at keeping others at bay, limiting both his knowledge of their feelings and theirs of his. A friend described him as 'one of the most insensitive people I know' and of his several psychoanalyses he said there had 'not been one emotionally charged moment'. When one analyst persistently pressed him to reveal more and to cease withholding his feelings, Allen abruptly terminated the treatment.

Since the age of five Allen had immersed himself in feature film fiction, enjoying a secondhand life through the characters on the screen of his local cinema. Alternative personae have been the principal way of living his real life as well. Working seven days a week and making a film every year for the last twenty-five, he has devoted most of his existence to the creation and writing of the parts that he plays in his films and the narratives of these characters' lives. For that small part of his life when he is not engaged in these fictions he keeps a distance from others through living his life As If he is himself, acting parts written by his immediate social environment. In 1996 he permitted a documentary film crew to follow him for sixteen hours a day for six weeks, complete with a radio microphone attached to him to record every utterance. After seeing the film a close friend was amazed that he had managed to act the part of a reasonable, decent, warm man for all this time without a single slip-up. He was acting a self that he wanted to portray, presenting it as his real self.

What it is like to change personae constantly is described in his film *Zelig*, summarized by his biographer as follows:

> Set during the 20s and early 30s, the story purports to be a real documentary about Leonard Zelig, a man who so badly wants to be liked that he tries to fit in everywhere. Insecure and anxious, he can't help assuming the personality, even the appearance, of people he meets. For example, contact with a black musician makes him change into a black musician; conversation with a psychiatrist transforms him into a learned doctor, and so forth. As a consequence of this miraculous talent, Zelig the human chameleon quickly becomes an international celebrity feted with ticker-tape parades and merchandised with board games and dolls, songs and dance crazes. At the same time, fame takes its toll and 'the price he paid was being an unhappy, empty human being', Woody explained.

Allen's biographer believes this to be a straighforwardly autobio-graphical film: 'Like Leonard Zelig, Woody Allen was a famous figure who, forced to battle the beast of celebrity, continued to feel like a poor lost sheep once known as Allan Konigsberg.'

If Allen does have an imposturous personality, there have none the less been some constants. From as young as he can remember he has been 'pessimistically depressive'. His first visit to a therapist at the age of twenty-three was triggered by 'a continual awareness of seemingly unmotivated depression'. Aged fifty-one, he was to say that 'not a day goes past when I do not seriously consider the possibility of suicide'. However much he achieves, he feels such an intense sense of failure that 'everything dissatisfies me'. When he discovered the technical term for an inability to experience pleasure, anhedonia, he immediately wanted this to be the title of his next film. His lack of pleasure and lurches between workaholism and depression are strongly reminiscent of the hyper and depressive coping strategies of the infant who suffers unempathic care.

Typical of the successful Personality Disordered, as part of their defence against feeling impotent and worthless they charm, bully or manipulate the world into accepting their fantasies of their own power and status. That is not to say that the high-achieving Disordered person does not earn his glory. Allen has been making films full-time, without weekend breaks, for a quarter of a century. But he is the first to say that his workaholism is a defence, saying that he works hard 'so I don't get depressed'. Earlier in his life he was motivated more by a simple desire for fame. A childhood friend says that 'he had this powerful need to be recognized' and Allen recalls having wanted to be famous at the age of thirteen. At sixteen, he was thrilled to see his name in a national newspaper for the first time when a one-line joke was published by a columnist. It was only in the last two decades of the twentieth century that he came to loathe the attention that fame brings.

The childhood experiences that caused this complex and Disordered personality illustrate the importance of events after, as well as during, infancy in fulfilling the potential for Disorder created during it. Allen was born and raised in lower middle-class New York. His father Marty had a peripatetic career until he finally settled into a job at a nightclub. He was a benign, laid back character, in marked contrast to his wife, Nettie, who was liable to fly into a temper at the slightest provocation. As a result, there was a state of almost perpetual warfare between husband and wife. For most of their son's childhood they lived in very cramped accommodation, and Allen recalled their daily disharmony being 'there all the time as soon as I could understand anything'. Although he had a sister eight years his junior Allen was the particular focus of his mother, who was probably depressive as well as aggressive. According to Jack Freed, a boyhood friend, 'His mother had a hot temper and was always taking a whack at him. Whenever he got her goat, she'd start howling and yelling before taking a good swipe at him. If my mother hit me that hard, I'd have run away crying, but he never cried. He had an amazing ability to restrain his emotions. His mother couldn't control herself at all.' Another boyhood friend described her as 'extremely naggy. Sometimes Woody couldn't take it and would say very cruel things.' Years later, in 1986, Allen was to confront his mother about her use of corporal punishment. After a great deal of testy refusal to accept that her care had been in any way inadequate she eventually acknowledged that 'I wasn't that good to you because I was very strict, which I regret. I was much sweeter to Letty [his younger sister] than I was to you.' So in his family script, Allen was the recipient of his mother's negativity.

Possibly the only way he might have pleased her would have been through conventional achievement. She was incensed by his refusal to take school seriously and impressed upon him the importance of work and success. 'Don't waste time!' she would yell. She sent him to a Hebrew school and from a young age he was indeed immensely industrious – but only in activities he had chosen himself, such as learning magic, playing the clarinet or selling jokes to newspapers.

According to his mother, he had a sweet, happy-go-lucky temperament until about the age of five, only to become sour and antisocial overnight. His sadness made no sense to her. He became obsessed with death, and as young as six years old tried to imagine in minute detail what it would be like to die, which left him desperately depressed.

From this brief account we could make some plausible guesses at linking Allen's childhood experiences with his adult personality, taking into account the evidence of the role of family scripts and parental aggression in causing childhood depression. But the most important factor may have been his very earliest care. Precisely what it was like during his first six months is not known, but there are some strong clues. His biographer describes his infancy thus:

> Shortly after the baby's first birthday, Nettie found work as a book-keeper for a Manhattan florist and began travelling back and forth to the city every day. Her son was tended by a succession of caretakers, mostly ill-educated young women who desperately needed the money and were not terribly interested in the finer points of early childhood development. As Woody later recollected, they invited friends to the house and sat around gossiping all day while he played by himself . . . In the evening when his mother returned from work, she had little time for reciting bedtime stories. When he got on her nerves, which was frequently, she wound up spanking

him. As a result, he grew up believing that from the cradle he had been unwanted. Nothing would ever convince him otherwise.

Whilst we cannot be certain from this account how good a relationship he had with his mother during the first six months of his life, the likelihood is strong that it was poor. Studies of mothers who share Nettie's personality – aggressive and abusive to their offspring, as she is known to have been since Allen was one year old – show that they are rarely able to be empathic to their offspring before that age.

Given that a strong potential to develop Personality Disorder was created by this type of infancy, Allen's subsequent abuse and neglect ensured that the potential was fulfilled. Such a rancorous home may explain his thin-skinned dislike of contact with humanity and his enduring expectation that other people will be hostile. Repeated physical abuse at his mother's hands would have reinforced this feeling, as well as making him an angry person with the potential for malice and destructiveness that he was later to show to his lovers. Such unhappiness may have driven him to escape into private fantasies and to live much of his life through fictions, not just as a fanatical movie-goer but in his second-by-second experience of himself, dissociating himself from an intolerable present and encouraging As If, imposturous living. His mother's aggressive–depressive negativity towards him may account for his depressive tendencies. No praise, no unconditional love, no simple warmth and affection are a recipe for low self-esteem. Feeling so unimpressive and worthless could also fuel a desire to be recognized through fame. Her sense that nothing he did was ever good enough could explain his joylessness, his inability to experience pleasure and his impregnable sense that he was a failure, however many Oscars he was awarded. That the only possible means of gaining her approval was achievement could help to explain his drive to succeed and to work hard.

Society's scripting of Personality Disorder

Woody Allen's life shows the importance of experience after early infancy in fulfilling a potential to become Disordered. It may also illustrate the role of society in fostering or suppressing it. The number of people suffering from Personality Disorder varies from nation to nation and between historical periods. Someone who had experienced Allen's childcare in France would be less likely to have been so Disordered. New York and Los Angeles probably contain the highest concentration of Disordered people anywhere in the world, and North America is the most Disordered continent. Riddled with crime, it has fully four times more convicted psychopaths per head of population, and they are far more likely to exhibit callousness and lack of empathy and to be glib, and prone to superficial charm and to grandiosity. This is likely to be true of the normal population. Individualist American culture, largely through the effect of television, strongly encourages grandiose and selfish behaviour; cultures in developed nations which prize community and collectivism, however, such as Sweden (which has half as much Disorder as the USA) and Japan, with their lower crime rates, have the lowest proportion of psychopathic people in the general population in the developed world. In the even more collectivist nations of India and China, surveys show that only one in one thousand suffers psychopathy, compared with twenty to forty per thousand in much of the developed world.

Such large variations between nations strongly suggest that genes play little role in causing the illness. Japan is a particularly telling example of the way that infantile experience and social order together, not genes, determine its prevalence. Japan is one of the few developed societies in which empathic infantile care is regarded as indispensable. Until very recently, almost all Japanese infants were raised exclusively by their biological mother, with tremendous attention paid to meeting their every need. If there was ever a society organized to provide the kind of empathic care which should create a strong sense of self, this is it. Given that Japanese children have so much better-quality early experiences than the average American infant, half of whom are given over to often inadequate substitute care before the age of one, they should also have much lower rates of Personality Disorder. And so it seems: certainly, even today, Japan has very low rates of crime and drug abuse amongst its young people compared with America. However, Japanese mothering cannot take all the credit. It is likely that the social system, too, reduces the probability of those relatively few Japanese who suffer infantile deprivation or abuse, and who therefore have the potential to be Disordered, actually expressing it. Japan provides a sense of order and connectedness that is largely absent in American society. As this order has gradually diminished over the last quarter of the twentieth century, with the colonization of Japanese culture and *mores* by America, so crime and drug abuse among the young have increased. However, their very solid early infancies may mean that they never develop American rates of Disorder, especially psychopathy.

The most fundamental of the numerous destabilizing trends that scar America is the shift from a collectivist to an individualist society. In the collectivist system, identity is ascribed on the basis of family ties preordained by birth. People put the group's interests before their own, and the erasure of hedonistic and selfish desires is valued. The cardinal virtue is sensitivity to the potential impact of one's individual actions on others. The collectivist teenager is discouraged from cutting loose from the family, which is often large. Obedience, reliability and approved conventional behaviour are fostered, and rebellion is quickly stifled.

By contrast, in the individualist cultures to be found to a greater or lesser degree today in most developed nations, but epitomized by America, identity is achieved through education and occupation, in open competition in a supposedly meritocratic system regardless of gender, class and race. In such societies, the self is defined by reference to inner feelings and thoughts rather than to external and preordained social roles. The goal of the individualist is to express himself, whether through hedonism, achievement or consumerism. In order to realize the achieved self, he must break away from ascribed family roles. The adolescent must seek out new ties and become part of many networks – school, university, profession – to which he has divided loyalties. Families are small, and parents see their job as being to foster self-reliance, independence and creativity.

For those who had empathic, unabusive childhoods, and therefore have a strong self, the individualist society presents an opportunity

239

for self-expression never previously witnessed throughout history. But for the large numbers of children with a weak sense of self, the cornucopia of choice and constant demands for self-definition are severely destabilizing. As they grow up they are required to face the tasks of life without the aid of accepted, durable traditions. The strain of making choices from discordant standards and goals besets them at every turn. These competing beliefs and demands, inflated by aspirations stimulated by consumerism, reduce what hope there may have been for a weak-selfed person to develop internal stability. In a society where families are increasingly falling apart, schools are constantly being changed as parents move home, and grandparents, teachers and church leaders no longer provide an alternative, children have no stable models. Where once there were powerful social institutions and conventions that could vitiate the effects of Disordering childcare, today there are many fewer. This has been proven in a study which measured increases in anxiety levels among American young adults and children between 1952 and 1993. Anxiety levels rose steadily during that time, to such an extent that the average level of a normal child today is the same as that which was found in children who had been sent for psychiatric treatment in the 1950s. The causes of the rise were shown to be the increase in social disconnectedness, arising from such trends as high divorce rates and more people living alone, and the rise in feelings of being under threat, such as from violent crime. Just as depression has increased up to tenfold since 1950, so it is with anxiety, and the fabled American way of life seems to be playing a big part. What is true of these mental illnesses is almost certainly true of Personality Disorders, such as addictions, as well.

Unfortunately, the disturbed children most in need of structure are precisely the ones most likely to abuse mind-blurring drugs and alcohol. For those with a strong self these explorations are not dangerous, but for a weak-selfed person desperately in need of reliable and exciting sources of satisfaction stimulants seem a godsend. Those who least should be taking drugs which reduce realism and encourage childish fantasizing are those most attracted to it. One study, which followed 101 Americans from the ages of three to eighteen, illustrates this well. Those teenagers who had used drugs, but only moderately, were actually the healthiest at eighteen.

Both frequent users and those who had never taken them at all had significantly worse relationships with their peers and higher levels of emotional distress such as depression and anxiety. Right from the beginning of the study, these two groups had both been more disturbed and suffered bad relationships. The way their parents had related to them caused their pattern of drug use, whether they abused them or abstained. Compared with those of occasional users, abusers' parents were observed to be cold, critical, pressuring and unresponsive to their child's needs. In the case of the abstainers, their fathers had been autocratic, authoritarian and domineering. The researchers actually concluded that moderate marijuana use was an indicator of mental health.

Nowhere are the risks for the weak-selfed more startlingly evident than in America. Because of its enormous gulf between rich and poor, one quarter of the population live in almost Third World conditions, with minimal health care – the more unequal the state, the greater the health deficit of the citizen. Slightly less than 2 per cent of the 260 million population are imprisoned by the state at any one time, mostly in appalling conditions (as it happens, about the same percentage as were incarcerated in the old USSR; where once that totalitarian regime was the focus of attention from the human rights movement, now it is the US prison system that is being targeted by Amnesty International). Nearly half of American infants under the age of one are placed in some kind of daycare, much of it of a kind liable to induce insecure attachment. These are conditions for creating widespread Disorder (as well as the Wobbler pattern of attachment) and, not surprisingly, the American version of individualism is very different from that found in Europe. The selfishness, greed and shallowness that were healthy adaptations to the tough conditions of much of the world for much of history are also appropriate in modern America, and are valued there. By contrast, European individualism is based far less on historical precedents, and is encouraged within a civic, collectivist context.

If a comprehensive study were carried out following several thousand people from birth to death (as opposed to the 168 in Ogawa's study, the best that currently exists), I believe it would demonstrate the crucial role of unempathic care in infancy in causing Personality Disorder. Very few of the most severely deprived infants would not develop it. Virtually none of the infants who had very empathic care and strong selves would become Disordered, even in the face of severe subsequent maltreatment. However, if the self was only strongish, the more unempathic the care the more symptoms would arise, and the more that subsequent childhood abuse would trigger them.

Studies would show little or no significant differences between the genes of the Disordered and the non-Disordered. Nor would genetic differences between races or nations be found to influence rates. Above all, in comparisons between developed nations the degree to which a social group has an American style of advanced capitalism would predict the extent to which those who had been made vulnerable to Disorder by their childhoods would fulfil that potential.

The practical implication of such findings would be to focus on creating a more child-friendly society in which mothers and their infants were given every possible assistance and encouragement, especially during the first six months. Unfortunately, a major impediment to such a change may exist: on the basis of meeting many of them, I have come to the conclusion that a disproportionate number of leaders and chief communicators in our society suffer from Personality Disorders, or symptoms of them; they are often unable to acknowledge the dependence of children on parents or increasingly, of citizens on states.

Many of the traits that accompany Disorder are also an advantage in reaching positions of power. Being a chameleon, with the self-monitoring, game playing distance which often accompanies dissociation, has been shown to enhance career success in organizations. If concealed well enough, an omnipotent drive to control others can motivate the industriousness that is so vital to success. Few people are prepared to work in the evenings and weekends, but people such as Woody Allen or Frank Williams who find intimacy painful actively prefer work to domestic life. Ruthlessness is easier if you lack empathy for the emotions of others, as borderline people often do, and being ruthless is usually necessary if you are to reach the very top. These traits are mostly developed in early infancy as ways of coping with unmet emotional needs. The Disordered are the last people to want to acknowledge that they have such needs or that others have them, because it would make them feel lonely, angry and depressed.

There is no obvious solution to this problem. To run a large business or government department requires extremely hard work, and it may even be that the Disordered are the best equipped to make what to others would be a sacrifice of their personal lives. Perhaps the best hope is to persuade leaders that creating a child-focused society is actually the most cost-effective way to manage our affairs. Certainly, there is good evidence that spending money to help parents do their job better saves a great deal in the longer term. One American scheme proved that for every dollar spent in improving the quality of care and education of a group of deprived children in early life, six dollars were saved, in terms of reduced expenditure on crime, mental health services and unemployment benefits, when they grew up.

The link between our infantile care and subsequent sense of self is ghostly: it is there, but we cannot see it. Yet it has played a vital role in the chain of connections that explain us. By the end of childhood, not only do we have a more or less strong sense of self but the security of our relationship patterns varies too, as do the weakness or strength of our conscience and role in our family drama. In theory there are dozens of permutations, but in practice some combinations are more common than others because parents, whether responsive and empathic, or the opposite, tend to stay that way. People with weak selves and Personality Disorders as a result of unempathic care in early infancy are also liable to be insecure Wobblers, and both are liable to have a weak conscience because their parents consistently maltreated them throughout childhood. Conversely, the strong-selfed tend to be secure and to have a benign conscience, the most mentally healthy that it gets, because their parents were always responsive and authoritative.

Of course, there are plenty of exceptions to these rules. Some of us had empathic infancies followed by insecurity-inducing toddler years followed by authoritative subsequent care. Parents can change. They can get depressed, go back to work, fall out with each other – any number of shifts can mean that the care we receive differs from one stage to another. Futhermore, as parents we all vary in the degree to which we find different stages of childhood appealing. Some cannot get enough of tiny infants, endlessly intrigued by the strange psychic place that they inhabit, and find the fascination with facts and the rumbustiousness of middle childhood less congenial. Others are left cold by infancy and adore toddlers, with their vivid fantasy lives. Still others do not really connect with children until adolescence. Many parents are so wrapped up in the adult world that they cannot easily plug into childhood reality, finding it repetitive and unstimulating. Their response is to accelerate the pace at which childishness is replaced by adult capacities – a concern with education, education, education that can diminish creativity.

Having reached the end of this chapter, after completing the audit below (if you have decided to audit chapter by chapter) you will be ready to apply the evidence of this book to yourself: the subject of Chapter 6.

Auditing your sense of self

If you do have a weak or weakish sense of self and symptoms of Personality Disorder, one of the symptoms may be a lack of self-knowledge which will make it hard to realize that you have these problems. One way around this may be to think of someone who you suspect does have them.

Picturing this friend or acquaintance or colleague, ask yourself if they are especially prone to any of the psychological defences described in the section on Personality Disorder (p. 193).

- *Omnipotence*: Are they a 'control freak', or someone who likes to feel they are powerful and dominant, or someone who hates being bettered in arguments, or given to fantastical ideas of what they can achieve?
- *Narcissism*: Do they find it hard to talk about anyone other than themselves, only really relaxed when the subject is them or a matter about which they feel expert? Do they tend to exhibit an inflated sense of their own importance?
- *Denial*: How do they cope with unpleasant truths or bad news? Do they tend to forget they told you they were going to do

something for you, or not to recall awkward facts or acts from their past?

- *Projection*: Are they liable to attribute feelings to you which you do not have – do they pressurize you to feel things you do not feel? Do you tend to leave their company in the grip of uncomfortable emotions or a sense that you have been emotionally ripped off?
- *Dissociation*: Do they sometimes feel very 'out of it', as if they are not really there, perhaps gazing off into space, not having noticed that several minutes have passed, as if in a dazed, shocked state? Do they seem to find it difficult to concentrate on what you are saying to them? Do they have sub-personalities – are they very different people in different contexts?
- *Imposture*: Are they skilled at imitating others' voices or gestures? Do they enjoy acting, perhaps seeing life as a game?
- *False self or Machiavellianism*: Are they devious, so that you have witnessed them put on an act deliberately to deceive someone? Do they prefer to skate on the surface of life, artificial and superficial in their intimacies?

You can see that the person you are picturing has some of these characteristics, but do you not as well?

Another way of trying to penetrate your possible reluctance to see these things in yourself is to consider how 'normal' you are. Psychiatrists have four main criteria for defining Disorder. Try these out on yourself:

- Do my moods often seem at variance with what is generally considered normal. For example, do I feel things more keenly than most? Do I have spectacular, unpredictable rages?
- Are my thoughts often of an unusual kind? For example, do my colleagues think that I am not just a lateral thinker but that my ideas often seem downright weird, likewise my friends?
- Are my relationships unusual? Is my love life distinctly odd by most people's standards? Do I dip in and out of friendships exceptionally often?
- Am I impulsive, perhaps prone to sudden bingeing on food or shopping, perhaps liable to lash out for no reason that

245

even I can fathom, perhaps given to astonishing U-turns in
my decisions?
- Have one or all of these tendencies been present for as long as
I can remember?

A way of checking whether any of these apply to you would be
to ask the most normal person you know to go through this list. But
be aware that you may tend to seek out the company of abnormal
people, so don't simply ask your best friend.

DID YOU HAVE EMPATHIC CARE AS AN INFANT?

Even more than the causes of your pattern of attachment, identifying
what created your sense of self depends on other people's accounts
of your early infancy. How empathic your care was at that time
may be hard for your mother to think about accurately, especially
if she provided the sort that results in Personality Disorder – for
instance, if she was depressive or drank a lot. The best approach
will probably be to ask what sort of baby you were – a good sleeper?
a good eater? prone to crying? liable to illnesses? If your mother was
herself Disordered, she may find it easier to recall what you were
like than how she related to you, but in doing so she may convey
how much she tuned into your unique needs. If she attributes a lot
of negative qualities to you in your infant days, that may be a clue
that she was not in a good frame of mind herself. Of course, it could
be that you were a Baby from Hell, but this is very rare – it is more
likely that she is projecting on to the infantile you the way she was
feeling at the time.

Another important clue to what went on in your first six months
can be gleaned from refering to the questions in Method 1 of the
audit at the end of Chapter 4 (p. 181). For example, if your mother
suffered a serious problem shortly before or after your birth, like
the death of one of her parents or a divorce, that could have severely
diminished her capacity for empathy.

Easier to recall is whether you were abused or severely neglected
in later life, although even this can be difficult to admit to if you
feel protective of your parents. If you do recall being belittled,
ignored, attacked or humiliated, this is a strong clue that you

were also maltreated in early infancy. The sort of parent who does this to older children is also more likely to provide unempathic infantile care.

LINKING YOUR SENSE OF SELF TO YOUR EARLIEST EXPERIENCES: MORE PSYCHO-ARCHAEOLOGY

Direct, conscious memory of early experience is inaccessible, for the simple reason that it predates our acquisition of language. Once we can label events with words it is far easier to recall them, but from the time before that we have only a flow of sensations which we are unable to order or control with the thought that language makes possible. For this reason, the search for evidence about your earliest experiences in your present-day life must focus on the nonverbal. Two very basic areas are your attitude to food and to time:

- *Food*: Would you call yourself a hungry, greedy person, prone to overeating, insatiable? If so, this could be because you were not fed when you needed milk as an infant – your mother was not attuned to what you wanted. Equally, having been cared for like this could have made you someone who denies hunger pains altogether, so that you don't eat enough or are a very erratic eater.

  Another clue is your attitude to being fed by others. Perhaps you are exceptionally uneasy about being kept waiting for food, or always have to be the cook, or are extremely particular about what you will eat, having very strong dislikes of whole categories of food – no meat, say, or no fish or dairy products. Control over food is at the core of anorexia, but to a greater or lesser degree many of us have a strong desire to determine what or when we eat. This dates back to patterns of breast or bottle feeding in our first few months.

- *Time*: Extreme punctuality or its opposite, or very strong feelings about these in others – passionate dislike of their pattern, be it punctilious or not – may tell you that you were kept waiting a lot in infancy by unempathic care. The infant's sense of time exists as a sensation of hunger or non-hunger, feeling cold or having that feeling removed by nurture, feeling lonely or the end of that feeling, and so on – it is very basic. But we all continue to experience these things as adults. How you react to being kept waiting, to minor illness or aches and pains, to living alone or not, are all heavily influenced by your earliest experiences.

You may detect signs that you suffered unempathic care if you are intolerant of being kept waiting, either by others or by your body. Extreme impatience at the smallest frustration, like a cold that does not clear up immediately or a delayed train, are the most obvious signs. However, some deprived people deal with it by becoming extraordinarily immune to frustrating events. Through denial or dissociation they ignore their bodies' signs that they are in discomfort. Yet these same people may amaze you occasionally by infantile bursts of rage at apparently insignificant events.

The strongest single sign that you had unempathic early care is if normal dependence on others, or of them on you, is your worst nightmare. To have to rely on someone else for love or food or sharing of recreation is hell for a person who had unempathic care, because it re-creates that very early experience of being kept waiting.

This completes the emotional audit which this book offers. It is now time to put it to use.

Chapter 6

BE YOUR OWN SCRIPTWRITER

They fuck you up, your mum and dad.
They may not mean to, but they do . . .

If you have completed the emotional audit at the end of each chapter, you should now be able to define anew your different attributes and their causes: your role in the family drama; your punitive, weak or benign conscience; your Avoidant, Clinger, Wobbler or Secure pattern of attachment; and your more or less weak or strong sense of self.

These characteristics have become laid down as patterns of electro-chemistry in your brain. In general, the earlier they were established the harder they are to change, should you so wish. But they have not been set in concrete: there is much that you can do to colonize your past and, in conquering this territory, change your present.

At the end of this chapter you will find an exercise that puts your emotional audit to work on your behalf: writing a short story with yourself as the central character. But before you attempt this there are a number of other ideas that need considering.

Change occurs through the particular form of self-knowledge known as insight. There are many ways to gain it, but self-expression is the thread that runs through all of them. In the first place, you probably already talk about yourself sometimes to colleagues, friends, family or lovers, wondering what makes you tick. You may also ask them what they think about you and every so often they will offer their thoughts unprompted – whether you like what they have to say or not. Such talking may lead to insight, but often it does not. Indeed, paradoxically, being honest about yourself in talking with intimates can actually be a way of avoiding the truth. Putting some thoughts into words places them outside you, so that they can no longer apply to yourself.

Another option, increasingly common, is to talk to someone who is professionally trained to listen, such as a counsellor or a therapist. Although not specifically trained to carry out the emotional audit in this book and to apply it, they can be incredibly helpful. But here be dragons – it can be difficult to match the problem with the therapist.

Another possibility is to express yourself through artistic activity, for instance writing or painting. Again, this can result in vivid self-realization but it can also be a way of not knowing. The literary exercise at the end of this chapter offers a way of using writing to achieve insight. For, whatever the means, it is insight that you are seeking.

Insight

While making a television programme about Mark Chapman, the man who shot John Lennon, I met a psychologist in New York who believed that it was possible, through hypnotism, to delete past bad experiences from the mind and replace them with good ones. In the case of a sexually abused patient whom he had hypnotized, for example, he claimed to have returned her to the moment just before her first memory of being abused. She described her father coming into her bedroom and sitting on the bed; then, instead of talking about the abuse that had actually happened, the psychologist told her that her father had simply read her a story and kissed her goodnight. When his patient came out of the trance she recalled her father as having been kind to her, not abusive; and so, the psychologist claimed, she was liberated from the effects of maltreatment.

I know of no firm evidence that such a cure is actually possible, but even if it were I would doubt its desirability. The key to changing the impact of our past on our present is not suppression of reality but insight. Being able to picture ourselves, to analyze what we are like, to see ourselves as others do, to evaluate our motives – these are the capacities that separate us from other species. Most valuable of all is the realization that what is happening in the here-and-now is an expression of the there-and-then. When such moments occur they can change who we are.

Simply understanding why we do something, or remembering incidents from our past, may be helpful but in themselves they are no more than an intellectual apprehension. For instance, I have some plausible theories as to what has made me smoke while under pressure to finish this book. Both my parents were smokers, I was not breastfed and I clearly have elements of what Freud called orality, probably dating back to my earliest infancy. But merely knowing this is not sufficient, as I realize all too well, reaching, as I am about to do, for another pull on a cigarette before writing the next sentence. I lack true insight into how I am dealing with the past by this act in the present. Although I am a psychologist with a more than passing interest in such matters, I have no more than a cerebral understanding of the possible causes.

For insight to happen, there must be both emotion and re-experience. When the brother of my patient Mrs B came to stay with her, she caught herself talking like her mother to him. He had left behind the sandwiches that she had made for his lunch, so she called him on the telephone and offered to bring them to the office 'because he ought to eat'. The moment of insight came when she realized that this was quite unnecessary and that her brother had been actively discouraging her from doing so: it was her mother who would have taken him his lunch. What made this an insight, rather than mere understanding, was the aliveness of the connection between past and present at the precise point when she was about to insist on taking the sandwiches to her brother. There was a surge of comprehension of the actual reality of her mother's experience of such matters, and of its existence in Mrs B. After that moment she was better able to see her mother's 'little voice' in her, and her own wishes, as separate entities.

Another example is a female patient of mine who was torn between two men. Having decided that she was ready to have a baby, she was unsure which to marry and to father her child. She had lived with one of them for many years and, although they were still very close, they rarely had sex. With the other she had rich and varied sex but was less emotionally intimate. Up to this point in her life, aged thirty-three, she had never seen the need to make sacrifices in her sex life. While she lived with her steady partner, if she found herself attracted to others she simply had a covert affair. Ambitious and highly motivated by her career in criminal law, she rarely went home before 10 p.m. and spent much of the weekend reading up her cases. Although she greatly enjoyed her work, she was dimly aware that it was at a cost. For most of her waking hours she had to adopt a responsible attitude, constantly monitoring her thoughts and feelings, concealing her true ones and exercising tremendous self-control.

There were three key moments of insight that enabled her to make her decision. The first was when she realized that the perceived choice between the competing virtues of the two men was wholly spurious: the real issue was her own psychology, not theirs. By endlessly agonizing over their differing merits she was distracting herself from looking inwards. The second revelation was that affairs were the one area in her rather solemn existence where she could let go and give her lively, playful self full rein: she was working too hard and it was distorting her love life. But the final, and most vital, insight linked the other two to her childhood. Affairs were an insurance against settling down and having children, something for which she now felt a strong desire but which terrified her because she had witnessed what happened to her parents: they had divorced after a loveless and empty life together. Unconsciously convinced that if she married and had children her life would become equally dreary and it would also end in divorce, she had let affairs put off that evil day.

As so often in therapy, none of these three insights was a Damascene moment that only required to be experienced once. She had to rediscover them several times before they really sank in. But finally she was able to see that she wanted to spend the rest of her life with the man she had always been with, and that this would not mean having to suffer the living death that was her parents' union.

Insight is not a luxury, not self-indulgent navel-gazing; it is indispensable. For Jeffrey Archer, its lack has proved disastrous. On hearing the news that the prostitute Monica Coghlan had exposed their relationship in a tabloid newspaper, which had also printed pictures of an attempt to buy her off, he told me that he felt depressed but immediately got down to work: 'You just carry on – that is what you do. You just carry on.' Pressed to explain why he kept slipping up in his career, he could only say, 'I was a bloody fool.' But why? 'Because I was naïve.' Why, Why? 'Because I am. Most enthusiasts are.' Pressed still further, he blamed his 'naïvety' for not realizing that there are 'evil people out there and you have to face that fact'. With so little insight into why he kept sabotaging his career, I asked, was there not every likelihood of it happening again? Given his criminal conviction for perjury in 2001, it is ironic that he replied, 'Well, we'll meet in twenty years' time and I hope you will be proved wrong.'

Oddly enough this interview was not the first time I had met Archer. That was in 1965, while I was at school in Kent. He was then a trainee physical education teacher at another school nearby and was brought over to mine to increase our fitness. He was endearingly enthusiastic, almost childishly so – in fact, a wonderful PE teacher. Perhaps his life would have been happier if he had stuck to that. Certainly, had he undergone the emotional audit in this book and put it to use he could have been spared a great deal of misery, for the similarities between his past and his present could not be clearer.

Archer's tendency to lie about his past has been documented by his biographer, Michael Crick, and I have firsthand experience of his wish to excise aspects of it that no longer fit with the way he would like to be seen. Chatting with him before the interview in 1987, I warmly recalled how helpful he had been as a PE teacher. But he simply looked blank at me, making no response whatever, as if I had not spoken. Archer made no acknowledgement of the fact that he had once been a teacher. But his tampering with the truth goes beyond repression. Whilst his ever-loyal wife, Mary, prefers to characterize him as a fantasist, he is also a conman and liar. From early in his career he misrepresented his curriculum vitae, such as falsely claiming that he had obtained a degree from the University of California. There have been several government investigations into his financial affairs, from his dealings in Anglia Television shares, when his wife was a member of its board, to the destination of donations to charities for which he was responsible.

His conviction for perjury finally proved that he had lied about the Monica Coghlan affair.

How similar all this was to the life of his father, William. He claimed to have been a captain in the Royal Engineers during the First World War but Michael Crick proves that he spent it 'not in the trenches, but in the courts. He was exposed as a compulsive liar, an impostor and a conman, pursued by police in three different countries'. As intransigent an adulterer as his son was to be, William made his living by low-grade frauds. While working as a mortgage broker he was charged with using false pretences to obtain money, and he received convictions for fraud in both the USA and Canada. As his son was to do, William committed perjury on at least one occasion, and he was equally dishonest about his curriculum vitae. He presented himself as having been educated at Eton and Oxford and, having discovered that another William Archer had received a war medal, Jeffrey's father claimed to be the dead man and through this impersonation stole considerable esteem.

The idea that Jeffrey inherited an antisocial gene from his father has been touted in the press as an explanation for these similarities, but it is far more likely that the care he received as a child is responsible. I believe that, with insight into the way his past had affected his present, he could have avoided repeating his father's confidence tricks and imposturous life. How could this insight have been achieved?

Self-revelation to intimates

Considerable insight can be gained from our relationships with others. At its crudest, this may consist of someone telling you an uncomfortable truth. Perhaps a frustrated friend can no longer restrain herself from blurting out that your constant whining about your lover's tendency to let you down is a bore, and that you must face the fact that he only gets away with it because you are scared to leave him. Maybe you have always been a bully, and a usually compliant friend suddenly points this out. Your child may be extremely reluctant to go to school, and one day your best friend is unable to restrain herself from pointing out that this is because you don't really want the child to leave you at home alone. But far from resulting in insight, this kind of home truth is often the cause of a major row and a fracture in your relationship with the truth-teller. Rarely can we stand it if someone destroys our illusions – we can turn very nasty as a way of avoiding the depression or fear from which those illusions insulate us.

It is far less dangerous when we open up to others, perhaps after a bad day or when feeling unaccountably down. In such moods we can be briefly accessible to the truth about ourselves. Realizing that our unhappy state could be connected with our behaviour or beliefs, we are more prepared to question them. A friend who provides a sympathetic ear and does not try to impose solutions can help, simply by being there. Insight from these exchanges tends to happen because it has been bubbling just beneath the surface for some time. So long as we do not become depressive – irrationally blaming ourselves for what is not our fault – we can suddenly perceive a pattern of self-destruction in our misfortune or unhappiness and grasp that alternatives are possible, that volition is an option. At such moments, there is a potent realization that it does not have to be that way and that certain actions will bring desired change. We make a mental note to be different next time, perhaps shed a few tears, have a few drinks, and move on.

People vary considerably in the extent to which they are introverted or extroverted, and in how suspicious of others or trusting. Whilst most do confide in intimates about their darkest fears from time to time, quite a few do not, either because they have no faith that another can understand or care about them, or because they are fearful that the confidante will use the confidences against them. Depressed people are very liable to be isolated by their negativity, rendered withdrawn and tongue-tied by their sense of powerlessness. Studies of depressed mothers with low incomes, for example, show them to be far less likely to have any close confidantes than similarly disadvantaged women who are not depressed.

However, confiding per se, does not necessarily result in insight. Those depressed who do talk about themselves, are liable to do so in such negative and obsessive terms that they rapidly alienate their audience. They may reveal far too much about themselves, in too personal detail, inappropriate for the social context or degree of closeness to the listener. They tell new acquaintances all about their dreadful marriage or impossible children, very offputting. The conversation is so littered with despondency that the listener leaves their company feeling dreadful.

As a result, they have trouble making new friends, only serving to confirm their feeling that they are stupid, ugly or otherwise socially unacceptable. Their self-revelation is not a search for truth but an attempt to offload their depressed feeling.

The Personality Disordered, too, are liable to confide in self-defeating ways. Narcissists are only happy talking about themselves, regarding others as merely an admiring audience, a mirror in which to preen. For much of the time they allow little or no communication from others to them. They are like a two-way radio jammed on transmit and unable to receive, so no knowledge gained from outside their self-preoccupied universe is possible.

Another impermeable group comprises the 10–20 per cent of people known as Repressors, who edit out the negatives in their past, present or future lives. Their relentless glass-half-full optimism seals them off from learning anything unpalatable about themselves.

Whilst these various groups may confide ineffectually, there are also generational differences in openness. Detailed studies of differences between the generations show that younger ones increasingly use psychological terms, rather than their social role, to describe themselves. Young people are much more prone to self-analysis and self-revelation than they used to be. Whilst there is still some way to go before British youth are as prone to this as their American counterparts, therapy culture is taking hold – for instance in the innumerable Reality Television programmes such as Channel 4's *Big Brother*, in which young people expose themselves to public scrutiny and analyze themselves and each other for weeks on end. That we are more open can only be for the good, insofar as it marks the end of the buttoned-up self-repression of older generations, dating back to Victorian times. The trouble is that exhibitionistic self-revelation is in danger of being confused with honesty, obsessive self-preoccupation mistaken for insight, passionately earnest sincerity conflated with the more honest truth-seeking of authenticity (a particularly American problem that Lionel Trilling acutely described in his book *Sincerity and Authenticity*).

Of course, most of us sometimes use honesty about ourselves as camouflage. Whether intelligent or stupid, humans tend to be tremendously cunning in deceiving themselves and others. From time to time we try to keep control over an uncomfortable idea, and reduce our vulnerability to sudden ambush by voicing it. We may do it semi-jokingly, as in 'I know I'm a terrible bully but ...' when about to be tyrannical or 'That's just typical of me – always putting my foot in it' when we have just done so. Alternatively, feeling upset and angry, we may embark on a stream of self-consciousness, offering a rapid-fire self-analysis that is forgotten as soon as spoken (users of cocaine are especially prone to do this). Where it most closely counterfeits insight is in the penance, the ponderous, agonized confession to the close friend that has no effect whatever on subsequent behaviour (sympathetically satirized in *Bridget Jones' Diary*). This strategy is particularly common on Reality Television, in which a tearful or repentant contestant is ostensibly speaking to a new-found friend on the show, but has actually got the viewer's affections in mind. The aim is to seem honest and self-aware, but the intention is ingratiation.

Self-revelation as a way of not knowing

A recent symptom of the misuse of openness has been the spate of newspaper columns and autobiographies which cross the boundaries between private and public in unprecedented fashion. Some of these confessionals may be motivated by the desire to break damaging taboos against public discussion of unpleasant experiences, such as the impact of cancer or of Alzheimer's disease. But in other cases it can be difficult to judge where exhibitionism and the pursuit of wealth and status end and legitimate public education begins. In other cases still it is harder to see what the purpose was, other than a paradoxical desire to avoid the truth by seeming to tell it.

Some years ago, the journalist Kathryn Flett used her *Observer* Sunday newspaper column to chronicle the collapse of her marriage. Subsequently, she wrote a book which took us on the agonizing journey that has been her adult life. Whilst she might have created a work of fiction based on herself, as so many novelists do, it may have been crucial for her to make us feel what it was like to be her in reality: a novel's pretence that it had happened to someone else would not have satisfied that need. This way we must meet her, aged thirty, single and filled with insecurities and the longing for love. We must embark on her courtship and marriage to Eric. We must share the experience of his subsequent rejection of her, and her nervous breakdown when her next boyfriend does the same.

Employing one of the most basic techniques in psychoanalytic therapy may help to explain Flett's unconscious goal in writing her book as fact rather than fiction. The analyst constantly examines how the patient is making him feel, and quite often this is a guide to the patient's unwanted feelings. For instance, the analyst may find himself feeling unaccountably impatient or angry towards the patient. Further investigation may show that this is what the *patient* is really feeling. Instead of having these feelings directly, the patient has used various tactics, such as offering only tantalizingly incomplete details of events or covert insults, to provoke them in the therapist. By being aware of how the patient is making him feel (known as counter-transference), the analyst can discover what is in the patient.

If the reader of Flett's memoir is the therapist and the book is the patient, then its literary techniques and its content are the manipulations by which the patient transfers unwanted emotion into the reader (therapist): how Flett makes *us* feel may tell us what *she* was feeling when she wrote the words. For example, at several places in the book it is hard for the reader not to feel that Flett is a pretty monstrous human being. That may also be how she felt about herself at the time. This perception of herself may be one she prefers to keep at arm's length by writing a book about it and inducing the feeling in others. She has lost herself in the activity of writing in order to avoid the truth about herself, even though the project appears, on the surface, to be a quest for honesty.

Although I have never met Kathryn Flett, I am assured by people who know her well that she is not as selfish as the person she portrays and there is no doubt that she is a highly intelligent woman. She is aware that she was insufferably bossy, manipulative and controlling in her marriage. Yet, instead of believing this behaviour to have been the reason that Eric left her, she develops the theory that he did so because he was a commitment-phobe, that deficiencies in his, rather than her, psychology destroyed their marriage. She offers an impressive analysis of how her unhappy childhood (divorced parents, wife-beating father, depressed mother) has affected her, but, because it seems to be experienced cerebrally and not emotionally, it does not change her behaviour.

There is a strong sense that the self that is struggling to overcome its pathologies suffers from them. As R.D. Laing put it, 'We are the veils that veil us from ourselves': Flett can only view through the distorted lens of herself. This emerges in the ways she justifies her behaviour to herself, such as her unilateral use of others' intimate lives for her journalistic ends.

The other members of a therapy group that she attends are furious when she writes about them in her newspaper column, and she observes that 'apparently this piece had been some kind of betrayal, although I failed to see how'. Only lack of insight could explain how such a manifestly intelligent woman could fail 'to see how' this was a betrayal. Given that her husband features so heavily and in such embarrassing detail, some people might suppose that her motive in writing such a personally revealing book is revenge; but rationalizing a horrible experience seems more likely. Despite her perceptiveness she dismisses Eric's concerns about her writing about him: 'Eric tried to explain, quite candidly, about how being written about every week in a national newspaper had made him feel. But he didn't tell it all (well, he wouldn't, would he). And anyway, listening to him, full of the same old predictably unemotional jargon . . . I suspected that he hadn't really dealt with any of it. He probably never would.' This sums up Flett's predicament. She is able to acknowledge that her writing about Eric against his will could be a problem for him. But she quickly rationalizes it by dismissing him as incapable of insight, an incapacity which is actually one of *her* greatest problems but is now attributed to him.

Flett's memoir is a fine example of the increasingly common use of newspaper columns by the Personality Disordered to dispose of unwanted emotions and to seek control of them by making them public. Julie Burchill performed this trick weekly in a *Guardian* column, and in her autobiography, tried to throw us and herself off the scent of her Disordered state by diagnosing herself as a psychopath – perhaps thereby enabling her to continue to behave like one. Another example is the TV presenter Anne Robinson's heart-stoppingly revelatory autobiography, with no personal detail, however damning and intimate, left uncovered. If she has revealed herself then no one can catch her out through this knowledge – no one can shock or suprise her with it.

Blaming parents

One important question is this: what, precisely, is it that we are supposed to be having insights about? Realizing something sensational about one's golf swing or tennis backhand may be pleasing, but it will not alter a persistent tendency to fall for the wrong man or to feel depressed. As readers will have gathered by now, it is my contention that the most important insights concern how we were cared for by our parents in early life; but it is vital that this should not entail blaming them.

Anne Robinson writes, 'Alas, a mother's best is rarely enough.' Of her own mother's failings, she leaves no doubt: 'Her best intentions were constantly thwarted by unintentional abuse. She loathed bullies while being a disgraceful bully herself.' Robinson is emphatic about the consequences of this early care. She married a man who was like her mother, and she followed her mother into alcoholism. Robinson concludes that 'She [her mother] was part bully, part magic. I have felt comfortable in the company of bullies, monsters and madness ever since.'

But apart from the money and publicity that Robinson gained from these revelations, where did they get her? Does it actually help to blame our parents? Of course it does not – blame gets us nowhere. But insight can get us everywhere.

Most parents do their best for their children within their limits. Philip Larkin's poem may begin, 'They fuck you up, your mum and dad'; 'But', the second verse points out, 'they were fucked up in their turn.' Given that our parents largely could not help what they did, blame is inappropriate. The second reason is that, if we are blaming, we may have understanding but we lack insight. Blame means we are still feeling rage, unable to exorcize it through self-awareness. What our parents did to us is not digested; we are simply continuing to experience them as, even now, maltreating us. They live on inside us.

But if blame does not work, that should not prevent careful appraisal of the way they cared for us. Although most of us did not have an alcoholic workaholic for a mother and become one ourselves, like Anne Robinson, there is no one who did not suffer to some degree at the hands of their parents. If someone tells us they had a perfect childhood, we can be certain they are in denial. Tales of a childhood idyll are self-deception. Studies of people making this claim show that they cannot recall specific memories from early life. They pay a heavy price for blanking out all the bad memories, tending to be very insecure in relationships and more prone to all manner of mental illness.

The difficulty about facing the truth of how our parents really cared for us is that we feel very protective of them: we love, as well as hate, them. Furthermore, if we denigrate them we often feel we are denigrating ourselves, because they are a part of us. Instead, we say, 'What's the point? What's done is done.' This response is widespread despite the modern shift towards a culture of complaint in which we are increasingly given to seeing ourselves as victims. Yes, there is a small minority of people who use parents as an excuse for their own failings, but most feel safer denying the bad and emphasizing the good when thinking about their parents.

This needs to change. Truly, those who forget the past are condemned to repeat it. Gaining a realistic view of how our early care is being relived in our minute-to-minute, here-and-now experience is invaluable. Therapy can be the best way to achieve this.

Therapy and insight

Since much of what motivates us is unconscious, it stands to reason that other people will be better equipped to provide the insight that we lack. Friends or lovers can be enormously helpful, but their view of us is inevitably prejudiced and distorted to some extent by our relationship with them. A therapist should be able to provide a more objective view, as well as being expert in the stimulation of insight. At its best, therapy enables us to talk to someone else about our past and present life in a way that changes who we are. There are various types of therapy, but unfortunately no single approach currently offers the audit presented in this book. I briefly outline what the main ones can offer.

The psychoanalytic tradition, started by Freud but now far advanced from his original theories, covers a huge variety of different approaches, quite apart from the hugely varying aptitudes of individual therapists. Whilst psychoanalytic therapists do have the most to offer for many kinds of problem, there are some frustrating shortcomings. One is that they are liable not to give much away, sitting silently as we grapple with our inner life and refusing to offer guidance on how to sort out specific dilemmas. Even more problematic is that most of them tend to focus too much on what we made of what was done to us, rather than on what actually happened. I can illustrate this with an anecdote from my time working in a therapeutic community in the 1980s.

A five-year-old girl was sent to me for an intelligence test. She acted in a highly sexual manner, such as asking if I wanted her to touch my penis and rubbing against my leg in a suggestive manner. I reported all this to my very old-fashioned Freudian psychoanalyst supervisor, expressing alarm that the girl should have known the words for adult sexual acts and should have been so sexualized at all, let alone with a strange adult. Surely something abusive must be going on – how else could she be aware of such things? The psychoanalyst rebuked me. Had I not read Freud? What did I expect from a five-year-old girl at the height of her Oedipus complex in the presence of a man? Her mind would be full of such fantasies without any prompting from adults. The analyst still believed in Freud's idea that we are all born with a genetic tendency to desire our opposite-sexed parent when we are between three and six years old.

This kind of thinking is now much less widespread than in the 1980s because of the large body of evidence proving that sexual activity between parents and children is caused by the parent, not the child. But unfortunately, many analysts still have a strong conviction that what is 'in' the child is responsible to a large degree for what happens. It dates back to a critical change in Freud's thinking at the end of the nineteenth century. On 21 April 1896 he presented his

262

Seduction Theory, which postulated that sexual abuse by parents was the major cause of neurosis, to a horrified Viennese psychiatric community. But, not surprisingly, having correctly located the cause of sexual abuse in the adult, he was greeted with an icy reception. Soon after his lecture, he reported that 'I am as isolated as you could wish me to be: the word has been given out to abandon me, and a void is forming around me.' The reason for his subsequent retraction of the theory, finally made in 1905, was not professional ostracism but a genuinely held new conviction. On further analysis he had come to the conclusion that the seducer was not the parent but the child, relocating all the desire in the child's Oedipus complex, and that the child's desires were a fantasy, not a reality. This reformulation helped to conceal actual child sexual abuse from exposure for another 90 years.

I know of some psychoanalysts and therapists who would wholly embrace the audit of psychology that I propose, firmly focussing on what your parents did to you, if not necessarily agreeing with all the aspects that I consider to be most important, but the majority would not. Where psychoanalysis does score higher than other kinds of therapy is that it acknowledges the complexity of our inner lives. Humans are not straightforward creatures. Only psychoanalysis confronts the fact that an awful lot of what we feel and think simply does not stack up, that all of us are, to some extent, not really playing with a full deck of cards, are several sandwiches short of the proverbial picnic. It is the only theory that can cope with our tendency to have mixed feelings, making us very contrary. For example, my weak conscience with regard to driving means I am liable to queue-jump if in a hurry and when I think I can get away with it, I have been known to nip the wrong way down a short bit of one-way street near my home, to save time. Yet, just a minute later, when I see another person doing something similar, inconveniencing me, I fulminate loudly and judgmentally against them, making no connection between my own selfishness and that which I have just witnessed. I am hypocritical in this respect, and believe that only psychoanalytic theory can cope with this kind of everyday complexity. Equally important, at least most analysts do regard our childhood as an important area for investigation in explaining such psychological design faults, even if they don't take what is actually done to the child as seriously as they should. That is more than can be said for the other main type of therapy, known as the cognitive kind.

Whereas analysts are generally trying to make sense of our thoughts by exploring our feelings with reference to childhood experience, cognitive therapists believe that feelings are caused by wrong thinking and usually maintain that there is no value in examining our past. Rather, they ask us to describe our precise thought patterns and then teach us better ones with which to replace them. This can be helpful for some problems, such as phobias, but equally it can seem very superficial. Telling someone who is feeling useless and loathsome to think of themselves as the opposite can seem like defining the solution without explaining how it is achieved – like saying, 'Just get better.' Depressed people know what not being depressed is; the trouble is, they don't feel that way. What is more, the relentless focus on thought patterns leaves out the potential value of exploring the patient's emotional reactions to the therapist. Whilst some analysts overdo this, it can be enormously revealing to the patient to notice that they are attributing feelings to the therapist, because it is so obvious at such moments that these feelings come from their own past and have nothing to do with what the therapist is actually like. Much insight is achieved in this way.

That is not to say that cognitive treatment is worthless, and it does have the advantage that the therapist is prepared to offer practical suggestions. To date, the nearest I have come to a therapy that would help with your emotional audit combines the best of both worlds. Called Cognitive Analytic Therapy (CAT), it begins with four sessions which look at how your parents cared for you and culminate in the therapist writing you a letter summarizing what you have told him. The remaining twelve sessions use this understanding to focus on a particular problem that you have chosen to tackle. I believe that almost anyone, however mentally healthy, would benefit from this therapy if done well. For in the end, all the research shows that it is not the theory that a therapist subscribes to which is the best indicator of whether they are helpful, but what he or she is like as a person. Predictably enough, whether he is warm and full of insight is more important than whether he is clever or full of theories.

Of course, therapy is not the only kind of help on offer. There are also psychiatric drugs and, as I explained at some length in my book *Britain on the Couch* (see Chapter 7), I am all for them under certain circumstances. For schizophrenia they can be a vital first step back to greater stability. For manic depression, lithium can be life-saving. Antidepressants can transform the lives of depressives. Taking these drugs can get them back into a state of mind where insight becomes possible again. But in all cases therapy should be attempted as well, and only if it fails should drugs be regarded as a permanent fixture. Often, if good therapy is available, the drugs can be phased out completely. Unfortunately some childhood damage is almost irreversible, establishing an electro-chemical pattern in the brain that can only be shifted by drugs; and some rare, major mental illnesses can have a strong genetic component. In either case, no amount of therapy can make a sufficient difference for drugs to be relinquished altogether.

For all their shortcomings, I do believe that psychoanalytic treatments remain the ones most likely to produce enduring and profound change. Although I cannot pretend that either of my own analyses were unqualified successes, I know dozens of people who were transformed for the better by theirs. The two cases that I know best of all are those of my parents.

My mother's childhood and early adulthood were packed with adversity. Her own mother was, by all accounts, a remote, probably depressive, woman who played little part in her care. My mum could only recall one occasion on which she had taken her out on her own in a pram, and, apparently her mother found this so unpleasing that she did not repeat the experiment. So my mum was cared for by a nanny – a woman from Tasmania who was illiterate. Although she used to hit my mum and sometimes threaten her with abandonment, she was primarily loving and loved. That is probably one of the reasons why my mother was able to withstand various subsequent family tragedies. When she was fourteen, her father killed himself during a deep depression. Ten years later her mother had to be admitted to a mental hospital, after my mum had spent some unpleasant months preventing her committing suicide, and she died soon afterwards. A few years after that, my mum's favourite brother became insane and killed himself.

265

The other reason that my mother was able to cope with all this was that she underwent two psychoanalyses. One of her brothers recommended that she attempt the first one when she left school, aged eighteen, in 1932. After three years of this (having previously shown no interest in academic matters) she rapidly passed the exams which enabled her to go on to university, where she trained as a psychiatric social worker. A second analysis with a particularly good practitioner helped my mother to progress further, and she herself trained as an analyst. However, soon afterwards she married my father and eventually gave up her career to look after her four children, with the express intention of breaking the cycle of deprivation – of giving us the experience that she did not have. In this, she largely succeeded.

My father was the fifth of six brothers, with one sister. His own father was a famously domineering and selfish man, his mother a hard-pressed carer. Like my mother, he was absolutely determined not to repeat the neglect and abuse that he had suffered in childhood. His analysis was with Freud's daughter, Anna, who was, so he said, not a very good clinician; but even so, he felt it helped him a lot. Whilst he was an imperfect parent in various respects, as was my mother, he was able to give the love and respect that he had been denied to each of his four children, albeit in different amounts and ways.

This book is a concrete manifestation of their success in overcoming their difficult childhoods. They regarded parental care as the single most important social issue and our home was a talking shop about it; indeed, as a consciousness-raising exercise, in 1960 and 1961 my mother wrote in the *Observer* what was, I believe, the first ever newspaper column answering parents' letters about their problems. Because they loved me and because, for the most part, they practised what they preached, their fascination with the subject endures in me. Whilst I would like to think that I have managed to make their interest my own, my desire to persuade you of the importance of parenting for your own mental health and for that of our society as a whole is the direct consequence of their having been psychoanalysed. It was this that enabled them to parent me in such a loving way that I have pursued what mattered so much to them.

Therapeutic horses for psychiatric courses

To be specific about which kinds of problem are best treated by which therapies is not particularly easy because there is insufficient reliable evidence. However, some observations can be made here (and further detail is in Chapter 8 of my book *Britain on the Couch*).

The good news about Personality Disorder is that many of the symptoms tend to diminish as people get older. Criminality decreases substantially, especially violence, and the more florid forms of narcissism or omnipotence become less pronounced. There is good evidence that some kinds of Personality Disorder are most effectively treated by the psychoanalytic approach, and CAT, too, has been shown to help with this. Insecure relationships (Chapter 3) are also most likely to be helped by analytic therapies and CAT. For depression, it may depend on which of the variety of childhood experiences have caused it.

At a basic level, there are two kinds of depression. One is primarily focused on a sense of the lack of others in your life, with feelings of emptiness, loneliness and desperation very similar to a state of mourning, and is often caused by the parenting described in Chapters 4 and 5. At least half of people with Personality Disorders also have this kind of depression and it is possible that they will respond best to types of treatment that are focused on their relationship with the therapist. The other kind of depression is focused on fear of failure and the person's status relative to others – he or she feels constantly subordinate. This is often the result of the parenting described in Chapters 2 and 3. It may well be that such people will also benefit from relationship-focused treatment, but changing their way of thinking is also important.

The kind of depression that results from unempathic early care, where the infant simply withdraws from the world (Chapter 5), may be hardest to cure. Antidepressants may help, and so may cognitive, 'think positive' therapy, but a cure, as such, may not be possible. The later the origin of the depression, the more treatable it is likely to be.

267

For depressions caused by having an emotionally unresponsive or absent carer when a child is between the ages of six months and three years analytic therapies are probably best, exploring the repetition of the insecurity in the childhood relationship with the therapist; they should perhaps be combined with some cognitive help. The same is likely to be true for depression as a result of a punitive conscience (Chapter 3), like that of my patient Mrs B who had a harsh internal voice running her down, and for the kinds of depression that entail Dominant Goal or perfectionist mindsets (Chapter 2).

Cognitive treatments are probably most effective for specific neuroses, like a phobia, or for challenging an isolated habit that you cannot seem to kick, like falling out with friends or bosses. Cognitive work may also help with childhood delinquency and hyperactivity, although by far the most effective therapy here combines cognitive with analytic approaches in seeking to help the parents change their way of parenting, based on understanding their own childhood maltreatment and the way they are repeating it.

With criminal adults it is not so easy, but in the case of violent men far too little attention is paid by prison psychologists to the childhood causes of their destructiveness, described in Chapter 3. However, no amount of therapy will help those who are so damaged that they have become irrevocably psychopathic.

For addictions, again the root cause is important, so that ones which date back to earliest infancy will prove less easy to cure than ones with later causes. In most cases, whatever the cause, in my experience the approaches which start with a 'Just Say No' programme of total abstinence are the best. This usually entails joining a group of fellow sufferers and following some version of the '12 Steps' programme, originally developed for alcoholics but applicable to almost any addiction. This method has the huge advantage of creating a society of like-minded friends who provide support and offer an alternative to the people with whom you have been pursuing your self-destructive habit. Once weaned off it, you can then proceed to the other kinds of therapy.

The most extreme mental illnesses of schizophrenia, autism and manic depression are very hard to treat, since they sometimes have both a genetic component and an environmental one from early infancy. As Rufus May's story in Chapter 1 illustrated, drugs for schizophrenia can be a mixed blessing, but they can be enormously helpful as a way of getting on to a sufficiently even keel to be able to benefit from talking therapies. Lithium for treatment of manic depression is less controversial than the drugs used to treat schizophrenia, and many sufferers have been completely cured because the right dosage was prescribed. Whilst some psychoanalysts have had success with cases of this kind, and whilst there are therapeutic communities which claim the same, it is difficult work. The only certain evidence is as described in Chapter 5 – if the patient returns home from hospital, the parents should be helped so as to avoid being negative and confusing.

Of course, most people would prefer to avoid both drugs and therapy in the pursuit of greater fulfilment and better mental health. They would rather find an activity into which they can throw themselves and which recycles their problem as creativity. Whether through work, sport or art, many people are able not only to keep unhappiness at bay but to gain insight as well.

Insight through creativity

Apart from being fun, the products of others' creativity often stimulate insights in us, whether it be through reading novels and poetry, listening to opera and pop music, or watching films and television programmes. For anyone who feels that their life lacks meaning, for example, I can strongly recommend Tolstoy's short story 'The Death of Ivan Illytch'. Mothers having trouble with teenage daughters could do worse than watch a video of Mike Leigh's film *Life is Sweet*. Of course, what works for one person will not necessarily work for others but nearly all of us derive solace and greater awareness through art of some form. But it is not only through consuming art that we benefit; creating it ourselves can help as well. Broadening this concept, insight can also come from creativity in non-artistic fields, such as sport or our professional lives. However, just as self-revelatory talk with others or writing about ourselves can be a way of avoiding the truth, so it is with other activities.

According to Freud, most creative activity contains an element of turning away from reality, a substitution of fantasy for unpalatable truths. At the deepest level, he regards the symptoms of a neurotic, the play of children and artistic work as indistinguishable. The content of novels, poems or plays can often be traced back to the childhood experiences of the author. A person whose parents were excessively punitive about potty training and cleanliness in childhood may express this in adulthood through a hand-washing ritual. Alternatively, they might respond to that sort of childhood, as the eighteenth-century satirist Jonathan Swift did, by writing *Gulliver's Travels*, a book replete with anxieties about defecation and urination. The phobic and Swift are the same in the technical sense that both achieve the goal of giving expression to a repressed, unconscious anxiety.

There are some difficulties with this view. Far from seeing artistic creation as a pathological symptom, it is also possible within Freud's own schema to see it as a thoroughly healthy way of dealing with problems, a healthy 'sublimation' of inner conflicts into some socially valued activity. For example, the newspapers report that the author Alan Bennett is suffering from writer's block. 'I've been unable to write for about four or five months,' he tells a journalist. 'It's something that is depressing and concerning me.' Bennett declares himself fearful of analysis of his motivations. 'I know there have been a couple of books written about me by critics, but I don't read them because I don't want to know what my work is about. I'm worried that if I think about how I work too much I will lose the ability to write altogether.' Without the opportunity to sublimate his anxieties into fictions, Bennett feels depressed: does that mean he would be better off being psychoanalysed? In the same way a comedienne who is filled with self-hatred, like Ruby Wax, whom I interviewed in 1988, may be depressed but she can sublimate that depression by making us laugh. She gains our approbation, serves a useful social purpose and does wonders for her self-esteem (as well as her bank balance) in the process.

Freud's account of art as neurosis also neglected the fact that it can be a way of enhancing our grasp on reality, a royal road to insight. As the literary critic Lionel Trilling put it, 'The illusions of art are made to serve the purpose of a closer and truer relation with reality.' Being false, in the sense of pretence, occupies a central place in being real in all our lives. Just as art is achieved through the illusions of artifice, so a healthy personality playfully invents and embellishes. An angst theory of symbolism which regards this as purely defensive, is missing much of the point. For it has nothing to say about aesthetics, as Anthony Storr points out: 'Psychoanalysis neither distinguishes between good and bad art; nor . . . between a work of art and a neurotic symptom.' Any theory of the causes of artistic creation must surely address the fact that its goals are often beauty and emotional resonance. Not to be able to distinguish a phobic hand-washing ritual from *Gulliver's Travels* seems a grave shortcoming of any theory.

Above all, in terms of gaining insight, it does not matter in the least whether you are good or bad at art, nor whether your golf handicap rises or falls. What matters is the process of creation.

The importance of play

One theme that may have emerged from your emotional audit is a sense that your life is not as joyful as you might wish – that some of the exuberance and freshness which you can remember as a child has been lost. In his book *Playing and Reality*, Donald Winnicott shows that children's play is not just a method for exorcizing fears and conflicts; it also exists as a joyful activity, in and of itself, vital to emotional wellbeing in adulthood. As adults, we have a strong tendency to become excessively concerned with our internal states, such as our instinct-driven desires for sex or food. Equally, we are very liable to become swamped with the external, such as the need to meet deadlines or to conform with social pressures. In Winnicott's account of play we are combining inner and outer – but not to achieve some preordained goal, whether instinctive or external.

The loss of our capacity to be playful at work or at home is a dreadful impoverishment, yet finding a way to combine the demands of a busy adult life with personal needs in a playful manner is not easy. Some aspects of modern life, like our freedom from strict morality or greater opportunity to choose careers, should make it more possible, but there are a great many countervailing pressures. For example, being able to work as a freelance, as the name suggests, offers liberty; yet that very freedom can become oppressive if the freelance is overburdened by a sense that he is not doing as well as his contemporaries. The insecurity of the freelance about getting work is not a bad metaphor for the Modern Man or Woman – someone who is always worried about how to succeed in their personal as well as professional life, obsessively and enviously comparing themselves with others, never satisfied with what they have got. If we live our lives beyond hierarchies, there is a danger of creating new ones that are even more totalitarian. As a result, for too many of us the simple enjoyment of a playful life is not possible for much of the time.

A simple test of our aliveness is provided by the ways we go about responding to everyday requests for information about our state of mind and recent experience. Asked 'How are you?' by a friend, most people do not give an honest reply. 'Fine,' we say, omitting to mention the blazing row we just had with our partner or the pall of gloom that descended when reading our bank statement over breakfast. Of course, when people ask after your wellbeing they rarely want an honest answer, and rarely would it be appropriate. But there are different ways of going about representing our mental state, even in this trivial exchange. It is one thing to offer a lively 'Fine' with enough spin to indicate to the observant listener that you mean nothing of the kind, quite another simply to go through the motions. Even more significant is the way you go about accounting for your life to strangers when you are introduced. Asked 'What do you do?', you can engage in an empty ritual, flatly describing the facts of your professional life, or you can attempt to bring it alive each time, connecting each narrative to the particular person with whom you are communicating. Indeed, half the battle of living a playful life lies in committing yourself to authentic communication. The fun of engagement with others soon follows.

One playful application of our emotional audit entails making sense of other people by understanding the past in their present. This is a whole lot easier than analyzing ourselves – and who does not enjoy analyzing someone else? Apart from being fun, it can be very practical. For example, if we have a boss who is so pernickety that he drives all his staff mad, we can ease the burden of dealing with him by paying close attention to how he makes us feel. Very often, what he is doing is using us as a place to deposit his own unwanted feelings, so if *we* feel unaccountably humiliated or angry or fearful, it may well be that this is really how *he* is feeling. Merely grasping that this is what is happening will be surprisingly soothing. Should we happen to get to know him better and he tells us something of his childhood we will soon see what he is re-creating in his dealings with us.

I witnessed a particularly creative use of this technique at the hands of a consultant who was advising a team of health professionals who had an unusually disturbed and disturbing boss. The consultant asked the team to tell him how they felt after spending time with their boss. One member said she felt stupid, another inferior, and yet another reported a sense of empty futility. When they had finished, the consultant asked them to picture the daily experience of being the person who was their boss. After envisaging a tremendously insecure, lonely man, they were able to see that the boss himself felt what he was making them feel. After that, although they continued to have to cope with his destructiveness, at least they were able to avoid confusing his problems with their own. They became adept at trying to work out what was wrong with him on a particular day from examining how he made them feel.

Playfulness is the cornerstone of emotional fulfilment, a key to insight and volition. It can be activated by any kind of everyday task from gardening to washing up. Although we need to earn a living, and although there are many practical tasks we have to perform, living as playfully as possible makes our existence infinitely more worthwhile.

Sanity, creativity and schizophrenia

The idea that there are sharp divisions between sanity and madness is doubtful. Rather, there is a spectrum, ranging from super-realism at one extreme, via average realism, minor illness and Personality Disorder, to major delusion at the other extreme.

At one end is the most realistic kind of adult, whose perceptions of the world are as undistorted as is possible for a human being.

Not many of us are like that. Next is where the average person lives, with a perception largely unclouded by depression or neurosis. Then come people suffering from minor mental illnesses and after them, veering at times towards the delusional, are people with Personality Disorders or the eating disorder anorexia. At the delusional end of the spectrum are those in the grip of manic depression, major depression and schizophrenia. There are large grey areas, such as that between many Personality Disordered people, with their subpersonalities and dissociation, and the fully fledged delusions of the schizophrenic. A very startling parallel can be drawn between schizophrenia and the normal thinking of small children.

At the age of four we engage in plentiful magical thinking, making all manner of playful connections between things that are actually not connected. I was recently playing with a lively girl of this age, my niece Lydia. Speaking of her older brother, she said, 'Jack has gone away for a year and he told me he will only return when the weather changes.' When quizzed in more detail as to the likely reality of this, she simply elaborated the fantasy to accommodate any inconvenient facts. Since I knew that Jack was actually at home with Lydia's parents, and since I knew that she knew that I knew this, I asked what sort of weather would get Jack back. Pointing at the blue sky, she replied, 'Jack said he won't come back while the sky is blue.'

Many of Lydia's comments were surreal non sequiturs, such as 'Jack is sixteen and he has six fishes and eight videos.' She adores puns and wordplay, so that her current favourite joke goes, 'What's the favourite food of a monkey in space? An astronut.' She likes to carry a doll around with her called Fizz, and gave me regular bulletins about Fizz's feelings which, on many occasions, you did not need to be a psychoanalyst to realize were very thinly veiled accounts of herself. When it was time for Lydia to go to bed, as if apropos of nothing, she reported that Fizz was not feeling tired. She was possessive of what happened to Fizz, resisting suggestions that I knew anything about her. She had invented the idea that it was Fizz's eleventh birthday the next day, and when I said that I could make her a birthday cake she immediately said, 'Oh no, it's not Fizz's birthday.' Fizz was clearly a useful character for feeling in control.

There are interesting parallels between Lydia's play and the

274

work of creating literary or other artistic fictions, sheer pleasure in creation being only one. The novelist knows his characters are not real, but at the moment of creation he believes in their reality, just as the reader does when reading the eventual book. Lydia knows at one level that Fizz is not a real person, and any attempt to expose the connection between what Lydia and Fizz feel is not pleasing to Lydia. Anyone who has tried suggesting to a novelist that their fiction is 'really' about them usually gets equally short shrift. It spoils the pleasure and seeks to undermine a satisfying defensive activity. Whether it be child or novelist, their fictions are often a way of working out uncomfortable realities, recasting them in more tolerable form or living them out secondhand. Of course, neither children's play nor artistic work should be reduced purely to these functions, but they do show the proximity of normality to schizophrenia.

There is a thin line between a child believing in its imaginary friends, a novelist locked into the fictional lives of characters and a schizophrenic conviction (like that of Rufus May, the recovered schizophrenic turned psychologist described in Chapter 1) that the plot of your life has developed in such a way that you have become a spy. Every bit as much as the child or the novelist, the schizophrenic is inventing a character and believing in its reality; but the character is himself rather than someone else. So, in people with a weak sense of self, schizophrenia can be viewed as a regression to the childish stage of development as an escape from unbearable adult realities.

Be your own scriptwriter: an exercise

Whether you think of yourself as a good or bad writer, as creative or uncreative, I propose that you attempt the writing of a story, as long or as short as you wish, based on your emotional audit. By far the most important aim of this exercise is that it should be enjoyable, so don't feel constrained by any of my ideas. I shall offer a basic structure to help you get started, but feel free at any point to deviate.

The idea is to create a fiction which enables you to capitalize on what you have discovered through the audit. There are two main kinds of insight which I imagine you should be seeking through writing this story, although you may be able to think of others.

The first insight is into an aspect of your present self or life with which you are unhappy. It could be something relatively insignificant, like a tendency to eat too much chocolate or someone at work getting on your nerves. Or it could be something manifestly major, like severe depression or an abusive partner. The crucial thing is that it intrigues you, that you have a strong desire to understand it better.

The second insight is into what you can do about it and what is preventing you from doing this. It does not matter whether or not you have any ideas regarding the solution now; writing the story should provide one. Even if the conclusion is that only one tiny change in your life or in who you are is desired and possible, that will be sufficient.

BEGINNING YOUR STORY

The first step is to settle on a subject: something reasonably specific that you wish could be different. You may already have one, but if not reread the notes from your emotional audit. It is in any case desirable that your subject, the question you are posing, is one to which your emotional audit has suggested some answers.

The key point is to identify something that really fascinates or mystifies or troubles you. Here are some examples:

- What is the matter with me?
- Why aren't I more successful at work?
- Why can't I find the right partner?
- Should I stay with my present partner?
- How can I get on better with my younger son?
- Did my mother love my father?
- Why do I always fall out with bosses?
- Why am I depressed and irritable so much of the time?
- Why do I spend all my spare time watching TV?

276

DEFINE THE CENTRAL CHARACTER: YOU

Whatever question you have chosen you need a central character, based on yourself, who is exploring it. Let us suppose that you have chosen the question 'Why am I depressed and irritable so much of the time?' Going back to the emotional audit, consider what it has revealed about the causes of your tendency to be depressed and irritable, taking each of the four elements analyzed in Chapters 2–5 and making a brief note of each. For example, suppose you are a thirty-five-year-old full-time mother of two, the youngest in your family, with one sister and one brother. You might write the following notes about yourself:

FAMILY DRAMA: Least pretty sister, always looked worse by comparison; pressure to be 'nice' and compliant one; didn't feel Mum thought I had much to offer – don't value my own intellect.

CONSCIENCE: Weak overall (benign sexual, weak attitude to authority and conscientiousness). Parents depressingly disorganized, never felt sure what would be punished; no pressure to perform at school from Mum, Dad didn't care.

ATTACHMENT PATTERN: Wobbler. Mum was depressing if not depressed, which I think she was; don't think she was very responsive (Aunt Joan says Mum was too busy looking after us to spend much time playing).

SENSE OF SELF: Some symptoms of weak self/Disorder (e.g. overeating). Mum not very empathic nowadays, if well-meaning. How empathic was she when I was three months old? Watching her with my kids, she doesn't seem to 'get' children.

Bearing all this in mind, you can now invent a central character for your story, supplying a name and a history. If this is your first attempt at fiction, then, as most first novelists find, the least difficult way to proceed is as follows. Continuing with the example of a thirty-five-year-old mother, make your main character a woman in her mid-thirties with two siblings and parents very similar to yours – in other words you, but represented as a fiction.

Of course, even if you are a thirty-five-year-old woman and so on, the story does not have to be written like that. You could decide to turn yourself into a man with all the above characteristics, but it will be harder to do. Remember, this is not a test or a competition; you are doing it for your own satisfaction.

THE PLOT

A story usually requires some sort of narrative. Choose something that will enable you to grapple with your question.

Continuing to assume that you are the woman described above (whose fictional name, let us say, you have given as Jane), once again the simplest approach is to base your story loosely on actual events: something significant that you feel relates to the causes of Jane's depression. The timescale of your story can be extremely short, such as an account of one evening in the pub, or it can cover the whole of her life in snapshots.

For example, you might describe a particular Christmas dinner when Jane was ten and her sister got preferential treatment. You might then cut forward twenty-five years to Jane's Christmas with her own two children and husband, and the connections between the feelings of irritability and depression at the two events.

Or you might tell the story from Jane's husband's standpoint, still focusing on Jane as the central character and basing it on your recollection of real events. You could describe a Sunday in Jane's life through his eyes – waking up and seeing her lying there, her reaction when the children come into their parents' bed, lunch with Jane's parents, a walk in the park, supper in front of the TV, ending with some sex after the children have gone to bed. You might portray how Jane's feeling of being the less pretty sister shows up in a lack of confidence that she is attractive to her husband. The 'depressing' lack of organization in her original childhood home might be expressed by the chaotic Sunday lunch at her parents and be analyzed in the way that she overcompensates with her own children, constantly organizing them, especially her daughter.

FOCUSING ON YOUR PRESENT LIFE

The ideas offered so far give you a pretty free rein, but there is also a more specific approach which requires slightly more rigour.

First, on a separate piece of paper for each, write down some names:

- The name of your lover (or lovers); or, if you have none at the moment, the last one; or, if you have never had one, the one you would like to have had
- The name of the friend who matters most to you
- The name of the friend who most annoys you.

Taking each character in turn, summarize their personality and how they relate to you.

For example, you might write the following about your annoying friend:

> PERSONALITY: Loud and aggressive. Only thinks of herself. Lonely yet sexually promiscuous. Can be funny. Gets jealous. Boring a lot of the time. Why do I bother?
> RELATION TO ME: When she is feeling down she makes my life a misery by always managing to call at an inconvenient time. When I try to get a word in edgeways she always shouts me down. She wants me to look after her or to be someone she can rely on.

When you have profiled each one in this way, use the audit of your past life to analyze the role each person is playing in your present life. Taking each character in turn, write down who they are in terms of your family script, your conscience, your attachment pattern and your sense of self.

Begin with the heading 'FAMILY DRAMA' and consider the ways in which your role in it may be evident in this friendship. Ask yourself if any of the things you have written about your friend remind you of the personality, or relation to you, of someone from your family drama. She could be that brother or sister who always treated you in the same way, or who had that sort of personality. Or perhaps she reminds you of just one aspect of the way one of your parents or siblings treated you. Perhaps your father was traumatically overbearing, especially to you rather than your siblings, when he had had a lot to drink. Perhaps one of your siblings was jealous of you.

Now write down the heading 'CONSCIENCE'. Consider how yours is evident in this relationship, and how this reflects what caused it. Perhaps you are punitive about sex and feel critical of your friend's promiscuity, believing it gets her into difficulties and causes problems to your mutual male friends. But on reflection, knowing what you now do about the punitive conscience, you realize that your condemnation of her is fiercer than her behaviour actually merits – that your response is really more about your inhibitions. Your mother never encouraged your relationships with boys, and was actually hostile to manifestations of your sexuality and your aggression. You feel that these are bad, so you are highly critical of them in your friend.

Next, write down 'ATTACHMENT'. Consider how your relationship with this friend reflects your pattern and, in turn, how this relates to what caused it in your childhood. Perhaps you are a Clinger and this is your bullying, Avoidant friend – Clingers often team up with Avoidants in later life. You may then be able to relate this need for an Avoidant in your life to the specific experiences which you believe caused your Clinging. Perhaps your mother was unresponsive, or a substitute carer was, and you can detect this in the way that you cling to this friend – like the feeling of being abandoned that you sometimes had in your past. This feeling need not be a memory from when you were a toddler, but can be one from later in your childhood.

Finally, write down 'SELF'. Think about how your more or less strong or weak self is evident in this friendship. Perhaps you have identified some features of Personality Disorder in yourself and you can see that, despite your friend's obnoxiousness, you actually quite enjoy it because you share it in some respect. For example she 'only thinks of herself' but maybe sometimes you can be or want to be like that too, so you feel comfortable being with someone who possesses this trait. Perhaps it is manifest in the relationship with your friend when you act selfishly with her, or perhaps she does it on your behalf – she is free to be the unpleasant person you dare not openly be. Using the information from your audit, you can then consider what specific aspect of your relationships with your parents is being relived here. Perhaps you were abused or humiliated or felt very neglected in some respect. Your friend may be doing what they did to you, re-enacting the maltreatment. But this time, you hope, it will turn out right. Alternatively, you may be doing to your friend what your parents did to you – a form of indirect revenge. The more you think about it, you may come to realize it is you as well as her who is unpleasant in this relationship.

When you have completed your remarks, your piece of paper might look like this:

TRISH: ANNOYING FRIEND

PERSONALITY: Loud and aggressive. Only thinks of herself. Lonely yet sexually promiscuous. Can be funny. Gets jealous. Boring a lot of the time. Why do I bother?

RELATION TO ME: When she is feeling down she makes my life a misery by always managing to call at an inconvenient time. When I try to get a word in edgeways she always shouts me down. She wants me to look after her or to be someone she can rely on.

CONSCIENCE: Being punitive, I take a dim view of Trish's philandering. But this is just my bloody mother in me. Why shouldn't Trish sleep around? Actually, am I a bit jealous of her?

ATTACHMENT: As a Clinger, I want Avoidant Trish to like me. Is that why I can't tell her I don't want to talk at length on the phone at inconvenient times? Mum never tuned into me at weekends when she wasn't working. The idea of Trish not liking me feels similar.

SELF: I know I can be selfish and angry. At least Trish doesn't mind when I am, because she is too. I'm not like this with other people. Am I angry because Mum was unresponsive? Does Dad's drunken abusiveness have anything to do with it?

Just by doing this analysis for each of the three people, you should have discovered quite a lot about your relationship with them and its origins. The fictional opportunities the material presents are numerous but two stand out as being particularly likely to give you insight.

The first is to write a story involving these people, with you as the central character, drawing on actual events. This could be as short as the tale of an evening spent together, when you went to see a film and out to supper, describing to the best of your recollection what actually happened. That could be a good way to apply your perceptions of the role they play in your life and its childhood origins.

The second piece of fiction could be the same story, only this time inventing ways in which you would like the characters, including your own, to be different from what they actually are. This might provide clues about what blocks you from behaving differently, and suggest limits to what is possible – even that you may need to find different friends.

This exercise can be applied to other aspects of your life, too. For example, you could do it with your boss at work, and your favourite and least favourite colleagues, as the characters.

David Astor, the late editor of the *Observer* newspaper, once said that 'None of us are responsible for our temperaments – only for our efforts to master them.' If by 'temperament' he meant the electro-chemical brain patterns which our childhood care creates and with which we interpret the world as adults, then I agree with him. But mastering our temperaments is no easy task, nor is it ever complete. I don't pretend to have definitively solved the problem of how to achieve greater volition in this chapter. But I do hope I have pointed the way to help you make your past work for, rather than against, your present. I shall be fascinated to read your stories, should you feel inclined to send them to me.

CONCLUSION

Man hands on misery to man.
It deepens like a coastal shelf . . .

There are three main versions of how nature and nurture cause differences between you and your siblings. The first goes like this:

> We are born with very distinct, genetically inherited tendencies which largely determine who we are. In most respects, parents do not have much effect on how we turn out, certainly no greater than our siblings or friends. The only exception is where children are raised in very extreme environments, such as an orphanage, or are the object of persistent sexual abuse from an early age. Since these events only happen to a small minority of us, overall, genes are the key. Early experiences have no greater influence than later ones.

The second view is the commonsensical 'bit of both':

> Yes, the child is born with a genetic trajectory which affects the way the parents behave towards him or her. What is more, how parents care for the child is itself partly genetic. But that is not the whole story. How parents go about reacting to the child can influence how his genes are expressed to some extent, so that an inhibited one who is actively encouraged to come out of his shell can be helped to be less shy, or a born extrovert can be toned down by parents who restrain him. Both types will always have a tendency, respectively, towards introversion and extroversion, but their environment will modify the degree of its expression. Early experiences may be more important than later ones in some cases, but then again not necessarily; and very often what is happening in their contemporary life is more vital.

283

The third view is almost the diametrical opposite of the first:

Differences in most people's psychology, in most respects, are not much influenced by genes. Whilst they can have a strong effect on extreme and rare mental illnesses, like schizophrenia, even these can also be caused largely or completely by upbringing. In general, parental care is critical, especially during the first six years. The patterns of brain electro-chemistry created then are brought to bear in choosing friends, lovers and professions, and in constantly re-creating the patterns of the past. The earlier a pattern was established, the harder it is to change. Although later experiences can modify what happens in the early years they need to be major changes, such as undergoing therapy in the case of changes for the better and severe abuse for the worse. All this is influenced by a wider context. Social trends, such as greater freedom to engage in sexual promiscuity or drug-taking, or greater pressure to compete at school, can increase the likelihood of fulfilling the potential to have problems created by early care.

The reader will recognize the third view as my main conclusion. All that remains is to pull together the various strands of which it is composed.

As you worked your way through this book you will probably have noticed that the same adult outcomes, like depression or high achievement, can have more than one cause. The four aspects of us that I have focused on – family role, conscience, attachment pattern and sense of self, each of them linked to particular age groups and kinds of parenting – probably all play some role in all adult outcomes. In the case of depression, for example, I have suggested a wide variety of kinds of parental care at different ages that can cause it, singly or in combination. I shall conclude by applying the various explanations to the three main aspects of human psychology: the causes of levels of achievement, the causes of adult relationship patterns and personality, and the causes of different mental illnesses.

The causes of levels of achievement

It is widely claimed that intelligence is the most heritable general

284

trait, accounting for about half of differences between adults in twin studies. However, it is not just adoption studies which call this finding into question.

Children from low-income homes who are adopted into affluent ones in infancy tend to do as well on IQ tests or at school as the biological children of their adopting parents. That people from the bottom social classes or from Afro-Caribbean families have, on average, lower scores on IQ tests (created, let us never forget, by white, middle-class researchers) could, in theory, be due to those groups' genes; but it is more likely that this proves the crucial role of good education and early training in fostering intelligence. In any case, intelligence is vastly over-rated as an explanation for levels of career success. Motivation is far more significant.

Interestingly, both low and high achievement are often the result of childhood maltreatment, such as being over-pressurized, abused or neglected. To put this the other way around, few exceptionally high or low achievers have a strong sense of self, a secure pattern of attachment or a benign conscience. In most cases, they suffered a mixture of unempathic early care in infancy, unresponsive care as toddlers and domineering, chaotic or neglectful care aged 3 to 6 years. Hence, extreme success and failure can both derive from early childhood adversity.

HIGH ACHIEVERS

Many parents are fixated on the educational performance of their children and, even before that, on accelerating the achievement of 'developmental milestones' such as walking or talking. Whilst this may result in a high IQ score and good or even prodigious exam results, it is no guarantee of career success: children with massive IQs are no more likely to be exceptional adults than those with average IQs for someone of their social background. In a significant proportion of cases, hothousing merely produces a lack of creativity and a degree of self-criticism that can develop into full-scale depression.

Far more important than a gargantuan IQ score are Machiavellianism and being prepared to work hard. Few people are motivated to do so if they are satisfied with themselves or have partners and children to whom they are deeply attached. The fact that one in three exceptional achievers (defined by their presence in the top six hundred entries in the *British and American Encyclopaedia*) lost a parent before the age of fifteen proves the causal role of adversity in creating fanatical commitment to success.

Someone with the weak sense of self that results from unempathic care in infancy is liable to encompass any or all of the following: impatience, novelty-seeking, insatiable greed, emptiness, fluidity, attention-seeking, compulsiveness and controlling. In most cases these traits do not advance a career, but it is easy to see how accidents of upbringing can focus the weak-selfed upon achievement as their salvation and in a few cases result in success. Some scientists escape from the pain of emotional contact with others by single-minded manipulation of concepts and objects in experimental laboratories. Equally, many performers lose themself in their public roles, whether that of actor, television presenter or politician, and, desperate for a love with which they are unable to cope in their personal lives, become addicted to the affection they evoke from their audiences. The weak-selfed person's lack of certainty about anything can become originality, a capacity to think beyond convention. Egotism, compensating for low self-regard, can provide the confidence and ruthlessness that often go with success. Omnipotence, covering up extreme vulnerability, can result in breathtaking ambition and imagination coupled with certainty of a providential outcome.

Insecurity in relationships, resulting from the kind of care received between the ages of six months and three years, can make the pursuit of power or status or wealth a very attractive alternative to intimacy. It may be that many high achievers are Wobblers in their pattern of relationships, burying themselves in their work, sometimes using sexual promiscuity as a way of keeping intimacy at bay. The fear of dependence and the abiding assumption that others cannot be trusted may translate into pragmatic, ruthless professionalism. The more lonely and despairing the person becomes, the harder she may work, using her career as a way of gaining social contact and glittering prizes as a means of bolstering self-esteem and countering the feeling of unlovability. At least money or powerful roles or the status of public recognition can be relied upon; where others are relating to us purely on a personal basis, not coerced by financial reliance or a professional hierarchy, there is a danger of them rejecting or abandoning us without our having any means of forcing them to comply with our wishes.

A small proportion of high achievers have managed to convert into sporting or showbusiness success the impulsivity or novelty-seeking that result from erratic, coercive care between the ages of three and six. However, it is much more common for the high achiever's parents to have been perfectionist, demanding and overcontrolling in later childhood. Love is conditional on performance. The child becomes fixated upon fulfilling parental goals, a quest made easier by a weak or weakish sense of self, so that there is less conflict between the child's and the parents' wishes. Backed by the parental taskmaster, teachers and bosses become the new goal-setters in later life; but, increasingly, the setting of incredibly high standards is done by oneself. Once a pattern is developed of only being able to feel good if achievement is exceptional, an addiction to ever greater success is set up. The weak self and insecurity become less visible to others, concealed beneath a carapace of confident proficiency. In many cases, high achievers are not so much exceptionally talented as exceptionally skilled manipulators of colleagues, singleminded in their determination to advance themselves at almost any cost. When I asked the industrialist Sir John Harvey-Jones what made the difference between those who become the chairman of main boards of major corporations and those who are merely directors, he replied, 'They want it more.'

In the backgrounds of most high achievers is a sense that they were special to one or other of their parents. Mothers are particularly likely to have been intelligent, disciplined women who insisted on their child being the same. Whilst the love was liable to have been conditional on performance, it created a feeling that the child was marked out for greater things in later life than were his siblings. It provides an inner confidence, a more or less well-concealed sense that elevation to high office is inevitable and deserved.

There are a small proportion of high achievers, of whom Sir John is probably an example, who are both strong-selfed and secure in relationships. They were loved by one or both parents, with whom they have identified, and their success results from a pleasure in creation within their given field. As toddlers they were encouraged to take pleasure in their deeds and words, but they were also provided with practical and effective training if they sought to improve their performance in a chosen sphere – one chosen by them. They are motivated not so much by money or power as by enjoyment of what they do and, although they work hard, they have rich personal lives as well. There are many such people who could have been equally successful but were unwilling to sacrifice their intimate relationships to the long hours needed for high achievement. Alternatively, they may be hard workers but lack the ruthlessness and Machiavellianism needed to beat less talented competitors in an increasingly dog-eat-dog world that places profit, superficial appeal and commercial pragmatism above substance.

LOW ACHIEVERS

At the simplest level, people unable to fulfil their career potential (allowing for their social class, itself by far the strongest predictor of how successful an adult is likely to be) may be so because they are incompetent, easily distracted, or unmotivated by power, status or wealth; they may dislike working long hours, and lack confidence, diplomacy or cunning in office politics; or perhaps they can find no profession that inspires them. Few people are unsuccessful purely because they are slow-witted. There are people who are not quick of thought or are incapable of sophisticated mental operations, but who are nevertheless very successful because they are good manipulators or have clear, simple ideas that they pursue doggedly. Rather, at a deeper level, people underperform because they are depressed, aggressive, abuse alcohol or drugs, hate any kind of authority, are obsessive or overanxious, have Personality Disorders or symptoms of them, are criminal, or fear that if they conform to any conventions they will be compromised.

Most people with the weak sense of self that comes from unempathic care in infancy are too fragile to sustain a challenging, frustrating or otherwise demanding job. If it means they have symptoms of Personality Disorder, at the first sign of difficulty they fall back on their career-stopping ways of coping which they may have managed to conceal during their interview. Narcissism puts off employers, as does unrealistic, omnipotent claims of what people can achieve. Such people are liable to blame others for what is their fault, and their moods yo-yo between grandiosity and despair. Alternatively, if their weak self has made them prone to depression, they have trouble getting up for work in the morning and are lethargic, irritable and depressingly negative about themselves and their work. Having discovered that office life does not suit them, they may go home and decide they are better off being 'self-starters'. But this is no easier for them. If depressed, they cannot sustain a project that they have begun, soon despairing that it is worthless. If Disordered and self-employed they are liable to flit from idea to idea, unable to see anything through, prone to wildly exaggerated claims about a project's value one day, perplexingly focused on a completely different one the next.

If a weak self is compounded by insecure relationship patterns, created by unresponsive care between six months and three years, career success is even less likely. Those with a strong self combined with insecurity are able to get and keep a job, although their insecurity hampers them. Avoidant types are easily annoyed by colleagues and prefer working alone. Their abrasiveness makes them unclubbable and unpopular. Although they may work very hard, finding it preferable to the hazards of rejection posed by a social life, their uncongeniality is liable to hold them back. Pessimism and inappropriate aggression may colour their workplace decisions, also counting against them when promotion is a possibility.

Clingers do even worse, on average, earning one third less than people with other relationship patterns. They crave constant encouragement and anticipate criticism when none is likely. Their dependence on others makes them easily bested in office politics, and anyway they rarely see themselves as leaders. They are happiest in jobs with predictable tasks and very clear responsibilities – not a prescription for high achievement.

Wobblers with a strong self are probably the kind of insecure person who is most capable of high achievement, but more commonly their Wobbling causes the opposite. Although a few are able to convert it into flexibility, a trait that has become advantageous in a fast-changing employment market, most suffer a mixture of the career disadvantages of Avoidance and Clinging.

In later childhood, pre-school input from parents plays a crucial role in educational performance – lack of basic tuition in the 'three Rs' by parents sets the child off on a trajectory of low achievement, especially if he or she attends poor-quality schools. However, such deficits can be reversed in later life by a determined effort in the late teens or early twenties. It is far more damaging if the care is erratic and coercive or if neither parent ever shows genuine interest in the child's educational or other progress. Even if the child is well tutored, such care is liable to set in motion a lifelong lack of motivation and focus. Self-esteem is battered, for the absence of praise for effort is undermining. Being unwanted, or witnessing favouritism towards a sibling, further diminishes self-belief. On top of all this, if the parents do not get on well, or separate or divorce, their offspring are very liable to underperform, especially if they are girls.

The causes of adult relationship patterns and personality

How we relate to someone depends on how we perceive them. If we imagine a person is getting at us by veiled insults, or that deep down they dislike us, we will react to them very differently from the way we will react if we assume they are benign or find us congenial.

Genes play very little role in how we perceive others, and therefore have virtually no effect on how we relate to other people. Rather, we base our impressions of how others regard us, and our expectations of their future behaviour, on our past relationships, especially the accumulation of our early childhood ones. This has now been confirmed by the experiments which demonstrate that, confronted by a stranger, we impose on them identities from our original family script. What is more, having invented them as ghosts from our past, if they do not behave in the scripted manner we try to manipulate them to do so.

Receiving unempathic care in infancy inevitably gets us off on the wrong foot in relationships. It creates a set of negative fundamental assumptions that we cannot affect the behaviour of others and are unlovable. If, as tends to be the case, this care is followed by the kind that creates fears of abandonment or rejection in toddlers, a template of mistrust and insecurity is established which will take a lot of changing. If this is compounded in later childhood by overcontrolling or coercive, erratic care, we will be even shakier in adult relationships. On top of that, we may be scripted as unreliable or inadequate or unlikeable in our family role, an expectation that we will bring to love and work. This is best illustrated by a little-noticed body of evidence regarding the causes of divorce.

DIVORCE, SEPARATION AND PERSONALITY

There is no doubt that the huge increase in the divorce rate is caused by a host of broad social trends, such as the greater wealth of working women and the availability of state-sponsored legal aid for those wishing to end their marriage. Much unhappiness has been prevented by making it easier for incompatible couples to divorce. But little attention is paid to the uncomfortable fact that the sort of people who do so are also more likely to have disturbed relationships, caused by their childcare and predating their marital problems: there are people whose personalities would have put them at high risk of divorce whoever they had married.

As one author bluntly put it back in 1935, 'One would hardly expect a man and a woman, both highly neurotic, to achieve a very high order of marital happiness.' Since then, studies which tested the personalities of couples shortly before they married and followed up what happened to them subsequently found that 'neuroticism' (including mild depression as part of its definition) and lack of 'impulse control' (in the male partner) predicted subsequent disharmony and divorce compared with people in whom these traits were absent before marriage. Going even further back, before marriage, a British study found that high neuroticism among sixteen-year-old girls predicted subsequent increased risk of divorce. On top of this, since children of divorced parents are at greater risk of a large variety of psychological problems, this may be one of the reasons why they themselves are at least twice as likely to divorce as children from intact marriages. Taken together, this body of evidence suggests that predisposing personality and emotional problems could cause divorce.

One study repays close inspection. It followed three hundred couples from before they had married, in 1940, until 1980. Those who divorced were significantly more likely to have had personality problems before marriage than those who stayed married. Divorcees of both sexes started out higher in neuroticism, and the men were higher in lack of impulse control. The women were more likely to have said at the outset that they came from unstable families which lacked emotional closeness, a predictor of subsequent emotional problems. Altogether, these factors predicted which couples would be divorced forty years later. Furthermore, among the couples who stayed married the personality factors also predicted which marriages were likely to be the happy ones. The authors concluded that 'The husband's impulsiveness and the neuroticism of both spouses are potent predictors of negative marital outcomes . . . in marital relationships, neuroticism acts to bring about distress, and the other traits of the husband help to determine whether the distress is brought to a head (in divorce) or suffered passively (in a stable but unsatisfactory marriage).'

To some, the idea that depressed, neurotic and impulsive people are more at risk of divorce might seem to be plain old-fashioned common sense. But so deep-rooted is our reluctance to stigmatize individuals that we dare not suggest that one individual has got a problem. Rather, we blame it on 'the relationship'. Others might also feel that it may be true but is best left unsaid, since it seems very negative. Yet many marriages might be greatly helped by this understanding. Where one individual is clearly suffering from depression, rather than encouraging both to agonize over their compatibility it can be extremely helpful for the disturbed person to seek treatment. It is particularly so with depressives, who are liable to be paranoid, to blame everything on their intimates and then to launch a barrage of hostility and aggression on them. The 'personal incompatibility' model is very convenient to such people but they may find the same problems recurring with subsequent partners – divorcees are more likely to divorce if they remarry than first-time unions.

It seems highly probable that the new freedoms and accompanying values will not have affected everyone in the same way. Neurotic and impulsive people may always have had a greater risk of unhappy marriages, but in the past they were protected by social pressures from expressing them in ultimately self-destructive attacks on their marriages. Furthermore, such people may have been made even more neurotic and impulsive by the overheated expectations of the times. This might have caused their marriages to become even unhappier and increased the likelihood of divorce in an era when it is fashionable to believe that the relationship is to blame and that somewhere there is a better partner.

SEX

Our sexual personae, preferences and behaviour are hugely affected by early childhood relationships, but mostly unaffected by genes. Starting with our parents' reactions to our oral sensual needs and to our fascination with excretion, their subsequent responses to early attempts at flirtation with them and others are crucial in determining how inhibited we are. Combined with sexplay with siblings and other children aged three to six, this creates our lovemap. In boys, the pattern of care we receive can affect something as fundamental as which gender you are attracted to. In girls, how close we are to our father not only affects what sort of men you choose in later life, it even influences the age at which you reach puberty.

Neglect or sexual abuse can make us sexually promiscuous and indiscriminate. Confusing sex with love, we may find ourselves lurching from partner to partner in a misguided search for intimacy through intercourse. Alternatively, we may be put right off sex or the relationships which go with it.

If we have an insecure pattern of attachment it prescribes particular problems. Avoidant types are liable to prefer sex that avoids intimate eye-gazing and romantic elements, whereas Clingers cannot get enough of them. Wobblers flit between the two.

Family scripting in later childhood greatly influences our confidence about our looks, our ease when coping with the awkward gropings of adolescence and how promiscuous we are. Being deemed the pretty or ugly one, a Mummy's boy or Daddy's girl, one of four siblings of the same gender – any of these affect who we are sexually.

If our parents were disharmonious or separated, we were at

293

greater risk of youthful promiscuity following early puberty. If their fathers are absent, daughters are liable to see men as unreliable.

PERSONALITY

Experience, in some cases from before birth (musicality, for instance), profoundly affects our interests, how nice or nasty we are, how humorous or humourless. Far more than by genes, our personality is forged by our relationships in early childhood.

Disorders of personality, such as antisocial behaviour or gross narcissism, are traceable back to early infancy. If empathy is absent we defend ourselves against it by denial or omnipotence, establishing a distorted perception of ourselves and others.

Such distortions are exaggerated by subsequent maltreatment. The personality characteristics of the Avoidant type, such as being aggressive, solitary and rejecting, are the result of having received rejecting parental care between the ages of three and six; likewise the Clinger's anxiousness, desire to please and tendency to sabotage well-meaning gestures of intimacy are derived from having felt abandoned at this age. Secure traits, like trustfulness and behaviour that makes for popularity, come from reliable, positive parenting. All these foibles of personality are instituted as electro-chemical patterns in the brain, resulting from thousands and thousands of repeated interactions with parents.

Chaotic, erratic and coercive care later on make for an impulsive and sensation-seeking child. If this is coupled with a weak self and insecure attachments, enduring frustration is hard. There is no stable foundation to fall back on, so we become someone who takes short cuts when faced by obstacles to our wishes. Here, not in genes, lies the origin of the addictive personality.

Family scripting encourages or discourages these traits. If we are the oldest in the family, we may be expected to be the responsible, reliable one; yet our weak self or insecure attachment pattern make this impossible. Perhaps we are expected to achieve impossible goals, or are regarded as talented in a particular skill, yet our personality prevents us being able to please our parents in this way.

The causes of different mental illnesses

In general, the more severe the mental illness, the earlier in childhood it was caused. At the same time, insofar as genes do play a significant role, it is far greater in the major and rarer problems such as schizophrenia, autism or manic depression than in the less debilitating, commoner illnesses. Summarizing my findings from the different chapters, I shall take each mental illness in turn, starting with those that are least severe and least likely to have a significant genetic component, followed by the major ones, which do.

MENTAL ILLNESSES WITH LARGELY ENVIRONMENTAL CAUSES

Minor depression

Unempathic care in infancy and being rejected or abandoned between six months and three years creates a predisposition to dejection in later life: adults with insecure attachment patterns are more at risk of depression. Whether this potentiality is fulfilled may depend on later care as well. In later childhood, those who were belittled, abused, overcontrolled or neglected are more at risk. Having an unwanted status or being made to feel insignificant and inadequate in comparison to siblings or peers creates low self-esteem.

Parents who make their love conditional on performance, such as in exams, are liable to depress the child's mood. Perfectionist parents create highly self-critical offspring. Depression from over-punitive consciences is caused if parents demand impossibly high moral or other standards, where the child's best is never good enough.

Warring couples are more likely to have depressed children, whether or not they stay together. Children who suffer loss and disruption caused by divorcing or separating parents are twice as likely to be depressed in later life; those who had disharmonious parents when they were aged five are four times as likely to suffer the illness in adulthood.

Addictions

The notion that there is an addictive personality, for whom alcohol can be substituted for cocaine or sex or shopaholia, may have some truth to it, but it does not appear to have a strong genetic basis. The impulsivity and sensation-seeking of the addict can be a manifestation of the hyper mode in an infant who is enduring unempathic care. This kind of infancy is extremely common in adoptees and may be a core reason why they form a significant proportion of drug addicts; often, they are trying to dull the depression which frequently coexists with addiction.

Much addiction reflects insecure relationship patterns, created in infancy and between the ages of six months and three years. The compulsion becomes a substitute for the safety and satiety that come from intimacy. Addiction can be a way of structuring an otherwise chaotic existence, created by erratic, coercive care from three to six years old. Children who were abused are far more likely to become addicts, sometimes as an expression of self-hatred, nearly always because they feel their wellbeing does not matter.

Drug abuse and out-of-control behaviour are no longer curbed by traditional, collectivist restraints in today's more individualist society. People made vulnerable to addictive behaviour by their childhood care are put at greater risk by advertisements designed to stimulate demand for consumer goods.

Personality Disorder

Adults whose care before the age of two was unempathic are significantly more likely to have symptoms. If subsequent care is of the kind that creates a Wobbler pattern of attachment, Disorder is even more probable. The weak sense of self, combined with an incapacity to cope consistently with frustrations in relationships, puts the child at risk, in adulthood, of developing Disordered defences against depression (half suffer from this as well) or to evade psychotic fragmentation (Disorder is one step away from schizophrenia).

Where there is physical or sexual abuse in later childhood, the vulnerable are liable to dissociate themselves and develop sub-personalities. Adults who had empathic early care are protected by it from developing Disorder if they are subsequently abused.

Cultural trends, like increased tolerance of exhibitionism, rudeness and selfishness, help Disordered potential to flourish. The many Disordered role models who populate television programmes, and admired figures such as pop stars, legitimize such behaviour.

Violence and criminality

Lack of empathic care from birth to three years old creates angry, belligerent boys and predicts violence in later life. Subsequently, erratic, coercive care from three to six also strongly predicts violence, especially when combined with parental disharmony and an irritable – because depressed – mother.

Early violence breeds later violence. Being physically abused creates depression, which men often 'medicate' with alcohol; this removes the man's inhibitions so that he attacks outwards. At the same time, the abusive childcare he received offers a model of violence as an acceptable means of expressing frustration and anger. Implicit encouragement of violence by society, such as through films, TV drama or permissive gun laws, increases the risk of vulnerable men using violence to deal with paranoid and depressed emotion.

People who steal may never have been taught that it is wrong – or even that it is admirable. They are likely to have been emotionally deprived – one third of the prison population spent time in an orphanage at some point in their childhood, mostly put there because their parents had been maltreating them. Shoplifters may confuse money and the goods that they steal with love, unconsciously pilfering from strangers the care that they did not receive as children. Robberies or muggings may serve this purpose as well, but with displaced aggressive revenge on parents as an additional attraction.

Advertisements exist to stimulate aspirations for consumer goods, which are unaffordable by many young people, and encourage those who are predisposed to steal by their upbringing to do so. When TV advertising was introduced in the USA, there was a 6 per cent increase in larceny.

Eating disorders

Unempathic care in infancy includes not being fed when we are hungry and having food forced upon us when we are not. This can become a template for not eating at all, or for overeating. Self-starving or vomiting toddlers tend to have the insecure relationship patterns that are caused by unresponsive care from six months to three years.

In later life, having a perfectionist mother predicts a greater likelihood of bulimia if the daughter has low self-esteem. Mothers of bulimics are liable to be belittling and censorious, and since these are both significant causes of depression it is not surprising that bulimics are liable to suffer from it. By contrast, anorexics tend to have confusing mothers who give mixed messages, similar to the mystifying 'double-binds' in which parents of schizophrenics may place their children. Indeed, compared with bulimics anorexics are more liable to have delusions, which are a defining symptom of schizophrenia – most notably hallucinating an image of their body far larger than it actually is. Many anorexics probably have a weak sense of self and use not eating as a way of trying to be in control in a family where they feel they have none.

Especially in teenage girls, the vulnerability to eating disorders created by parenting is more likely to be fulfilled in societies that present abnormal standards of thinness in advertising and showbusiness models.

Neuroses

Insecurely attached children and adults are more prone to neuroses such as phobias and obsessions. General anxiety is much commoner in children whose parents are strict or overcontrolling. Frightening parents make for frightened children. Obsessive Compulsive Disorder (OCD) – irrational, repetitive thoughts or behaviour – is far commoner in children whose parents were aggressive, chaotic or intrusive. It can develop as a way of creating reliable order, the repetition of a ritual becoming the one safe feature of an otherwise scary world. OCD is also often a way of regaining control in the face of overcontrolling, perfectionist parents.

Hyperactivity

Intrusive and overstimulating care measured at six months predicts hyperactivity at the ages of both three and eleven. Having an insecure attachment pattern at one year old predicts subsequent hyperactivity. Adopted children who suffered maltreatment prior to adoption are more at risk.

Children of single parents, often stressed by low income and lack of support, are three times more likely to suffer the illness than children with two parents. Children who have either had erratic discipline, between three and six, or were too heavily controlled and received punitive care, are also at risk.

The style of some modern media, such as frantic computer games and manic TV programmes, may increase the risk of children with hyperactive potential expressing it.

MENTAL ILLNESSES WITH A SIGNIFICANT GENETIC COMPONENT

Autism

This may be the most genetic of illnesses, according to one twin study. However, there are also grounds for supposing that nurture can play an important role.

For one thing, if parents are helped to provide specialized care early in life, the child has less severe symptoms than if the care starts later. For another, a study of children who had been put in an orphanage early in life found remarkably high rates of autism and autistic symptoms: 6 per cent had a full diagnosis (compared with 0.2 per cent among children in the general population) and a further 6 per cent had many symptoms. This strongly suggests that the autistic child's turning away from social contact towards the relative safety of self-generated compulsive rituals, like interminably running up and down a room, can be a response to severe emotional deprivation (a theory advanced long

299

ago by psychoanalysts and scorned into obscurity by psychiatrists).

Manic depression

Where this, the second most genetic illness, has an environmental origin it probably dates back to very early infancy. An infant who suffers unempathic care responds in one of two main ways. He can develop a hyper mode, becoming permanently vigilant, unable to sleep, physically tense and frantically oversensitive to his environment, like a cornered animal. Or he can withdraw into a depressive, lethargic inactivity, focusing on his own body as the sole source of gratification, having given up hope that his carer will satisfy him. In later childhood he may alternate between these modes; this state is the precursor of manic depression. Adult manic depressives who suffered childhood sexual or physical abuse develop the illness at a younger age, swing more rapidly between the two moods, and are more prone to suicide and to added problems like addictions, than adults who did not suffer in this way in childhood. That suggests that childcare can play a significant role, even if genes are also important.

If the parenting is intrusive and badgering, the mania may be exaggerated; if the child is put down and belittled, the depression predominates. When the manic depression emerges as a recognizable mental illness, with days of frenetic delusions followed by weeks of total passivity, the weak sense of self set up in early infancy leaves the person at the mercy of what is possibly the most horrific mental state a human can endure.

Schizophrenia

If half of cases are caused partly or completely by genetics, the other half are not. Unempathic early care may create the potential for the illness. Dissociation and the development of subpersonalities are more common in adults who suffered disrupted care before the age of two than afterwards. The illness is more common if there is less than two years between siblings, the small gap increasing pressures on mothers and the likelihood of unempathic care. If a parent is schizophrenic, a child is twice as likely to have the illness if it is the mother (usually the main carer) and a schizophrenic mother is liable to be unempathic. Nearly all schizophrenics smoke, and smoking is linked to orality (a predilection for oral stimulation), which is linked to infantile deprivation.

Stressful subsequent environments greatly increase the likelihood of schizophrenic potential being expressed, whether it exists for genetic reasons or because of the kind of care provided in infancy. Britons of West Indian heritage are many times more likely to develop the illness than relatives who did not emigrate. The severity and duration of sufferers is greater in developed, rather than developing, nations. Whereas over half of sufferers relapse if they are returned from mental hospital to parents who are negative, only 16 per cent do so if the home environment is more positive. The illness is more common among adults who suffered abuse as a child. Above all, where adopted children have a biological parent with the illness they are only at greater risk of developing it themselves if the care they receive in their adopted home is negative or confusing.

Major depression

Again, genes can sometimes be the sole cause, since half of the siblings of depressives who have an identical twin also have the illness. But this also proves that the illness can be solely caused by parenting – if one of a pair with identical genes has the illness but not the other, only the environment can explain it.

Unempathic care in early infancy can result in depressive withdrawal from the world. This may be the foundation of the lonely, bereft kind of depression, in which there is a constant feeling of being let down and unloved, or it may create the potential for this, which may then be exacerbated by subsequent care. If this kind of care gives toddlers insecure attachment patterns that increases the risk, as does parental disharmony.

Unempathic care in infancy can also lay the foundation of the other main kind of major depression, in which self-blame and a pervasive sense of failure are prominent. The infant never develops the conviction that her actions have an effect. Infantile inefficacy probably explains why some children but not others become severely depressed by coercive, erratic care between the ages of three and six, or by belittlement, abuse or neglect. Likewise, it may be the decisive factor in those children of perfectionist, hypercritical parents who subsequently succumb to Dominant Goal major depression.

The bigger picture: the tyranny of geneticism

If the practical consequences of knowing about the past in our present are profound, they are equally significant for our society as a whole. One of the most fundamental implications of the fact that parental care, not genes, is the main cause of how we turn out is that it destroys many of the most cherished political tenets of right-wing social policy. They rely heavily on the assumption that the rich man is in his castle because he has superior genes, whilst the poor man is in a housing estate because his are inferior. Just how much 'no such thing as society' Thatcherism was founded on this idea was proved to me in 1990 when I attended a seminar on 'Crime Culture' held by the right-wing think tank, the Centre for Policy Studies (CPS). Two of the most influential right-wing American thinkers on the causes and prevention of crime presented their theories to an audience of like-minded opinion-formers. This was an occasion on which the true beliefs of Margaret Thatcher's Conservative party were freely expressed, behind closed doors.

The keynote speaker, party chairman Kenneth Baker, addressed an aside to one of the Americans. 'I am just at the moment reading your latest book with the greatest of interest,' he said to Charles Murray, author of *Losing Ground*, the bible of Conservative social policy-makers and subsequent co-author, with Richard Herrnstein, also present, of *The Bell Curve*, a book which in 1995 was to put the case for the genetic inferiority of the poor in general and Afro-Caribbeans in particular. That the Conservative government took these Americans very seriously was demonstrated when they were given a tour of the Home Office, Scotland Yard and the Downing Street Policy Unit. What they believed was not simply music to the ears of anyone seeking justification for letting the poor be, strengthening the police and stiffening sentences. Murray and Herrnstein also maintained that educational and social programmes to change the poorest and most criminal classes were bound to fail because such people were largely born that way. State schemes for preventing crime were futile, because it was no more possible to change genetically determined low IQs and innate criminality than it was to alter the colour of people's eyes by talking to them.

That such eugenic theories formed a crucial building block of Margaret Thatcher's policies is not often realized. At the seminar David Willetts, director of the CPS, dismissed British social research as 'at worst Marxist, at best Fabian'. Echoing Thatcher's pronouncements on crime ('Blow away the fog of excuses . . . the main person to blame is the criminal himself,' she had said), Willetts rejected social explanations. Instead he commended us to the genetic determinism of Professor Herrnstein. When Herrnstein's turn to speak came he set the ball rolling by saying that 'the main advance has come from an area of research known as quantitative genetics'. He told the conference that in Scandinavia '60 to 70 per cent of criminal behaviour is genetic' and that, although the percentage was less in the USA and UK, it was still 'not negligible'.

When I asked him to confirm this he reiterated that what he meant was that 'there are different genes for different classes. That is what I am saying.' But how different? If a thousand boys from poor homes were swapped with a thousand rich ones at birth, would the formerly poor ones still be crime-prone? 'Good question . . . The parents of the swapped boys in your experiment would start to realize they were different from soon after birth.' Herrnstein placed great significance on intelligence. 'A low IQ score is the single best predictor of criminal behaviour,' he claimed, and consequently we should no longer 'support the claims of the fifties and sixties that crime is due to poverty. Boys are born with genetic propensities to have low IQs and become criminal.' He went on to state that 'I cannot believe our ability to anticipate crime from a person's parents, early in life or even before their life had begun, will be ignored. Somewhere down the line this knowledge . . . will be applied.' One of these applications could be eugenics.

Certainly, Herrnstein and Murray should have known all there was to know at that time about how to create a crime epidemic, since theirs was by far the most violent developed nation on earth. Both were supremely confident that Britain was headed the same

303

way, but it seemed strange that the government were turning to Americans for a solution. Surely, with five times fewer murders and three times fewer assaults, it should have been us advising them on how to prevent crime. But as the American guests made free with their advice on what we should be doing, a massive contradiction arose. They maintained that crime is a rational choice and therefore that, faced with severe penalties, criminals would desist. But how was this to be reconciled with their belief, equally earnestly held, that crime is primarily caused by genes? If true, surely criminals are no more responsible for their crimes than they are for their height or shoe size?

It was Murray who best exemplified the confusion between free will and determinism. Since his criminals were rational, 'to say the threat of punishment does not work is to say criminals are utterly different from the rest of us'. Yet this was exactly what he had been trying to persuade us to believe just five minutes earlier, with an excursion into the evidence that criminals have criminal genes and different brains from normal members of society. Likewise, Kenneth Baker was muddled when he argued that we should make the parents of criminals legally liable for the behaviour of their children. How could either be blamed if neither were responsible for their genes?

Since 1990, Charles Murray has been even more explicit about the underpinning of right-wing thinking by genetic explanation. In an article in 2000 he claimed that science was proving the right-wing case:

> The story of human nature, as revealed by genetics and neuroscience, will be conservative in its shape. We will learn, for example, that women innately make better nurturers of small children than men do and that men make better soldiers than women do. Regarding these and many other human characteristics impinging on marriage, the upbringing of children and the enforcement of social order, I predict that adages of the right will usually prove to be closer to the mark than the adages of the left and that many of the causes of the left will be revealed to be incompatible with the way human beings are wired ... when we know the complete genetic story, it will become evident that the population below

the poverty line in the US has a configuration of the relevant genetic makeup significantly different from the configuration of the population above the poverty line. This is not unimaginable. It is almost certainly true.

In fact, it is 'almost certainly' nothing of the kind. As I described at the end of Chapter 2, for example, there are strong grounds for doubting that genes play any role in causing most violence. What is more, other evidence proves conclusively that, far from being rational, criminals mostly veer towards being mad, not bad, and that this is mostly caused by nurture, not nature.

In 1997, the results of an official survey of the mental health and histories of a representative sample of three thousand British prisoners was published. All the prisoners had undergone a full psychiatric examination and a staggering 90 per cent of them were found to have a mental illness, the majority having two simultaneously and one fifth having four or five. Two-thirds had a Personality Disorder, or suffered from depression, or drank heavily, or had used drugs, or had a neurotic disorder; 7 per cent were suffering from schizophrenia or some other delusional psychotic illness. These remarkable facts, replicated in similar US surveys, prove that madness goes hand in hand with badness, which might lead some people to suggest that the prisoners needed treatment, not punishment. That would not be the view of the political right. Since they believe that this panoply of disorder is largely caused by genes, they argue that therapy would be pointless. However, it is possible to prove almost beyond doubt that genes play a minimal role in causing either the criminality or the mental illness of most prisoners.

One third of the British prison sample had spent some time in a local authority care institution at some point during their childhood. In half of cases, the reason children are taken into care is that there is sexual or physical abuse going on, or a high risk of it, and the vast majority of children in care have suffered early emotional deprivation. The combination of abusive, deprived early experiences with the subsequent damage caused by having constantly changing carers after being taken into care could cause both the mental illness and criminality. The geneticist's retort is that it could be the inheritance of the genes of the children's parents which is causing these problems. The geneticist might accept that the adverse childcare does not help, but would still argue that the genes are the key. However, this has been incontrovertibly disproven.

As described in Chapter 3, a large number of studies have compared what happens to children who are adopted when young after being taken into care with children who endure a series of different foster homes or live permanently in institutions or are returned to their biological parents. These are critical comparisons because all the children were born to parents who were very liable to be mentally ill or criminal, so if there is any genetic risk attached the kind of care received should not make much difference. The results leave little doubt that it is the kind of care that is decisive, not the genetic stock. Children adopted in early infancy are no more at risk of mental illness or criminality than children raised by middle-class, law-abiding, married, biological parents. By contrast, children who are fostered rather than adopted do worse. However, very tellingly, when fostered children and those who are permanently institutionalized are compared, the fostered ones do better. Many children who are returned to their (usually disturbed and disturbing) biological parents do worst of all. So the amount of damage correlates precisely with the quality of care received, not with the genetic stock.

For example, when a group of children who were fostered before the age of one were compared with a group of permanently institutionalized ones at the same age, the latter were considerably more likely to be hyperactive. Whilst both groups were more prone to this problem than children raised by their biological parents, suggesting (like the studies described in Chapters 2, 3 and 4) that hyperactivity has a strong environmental component, this also proved that the particular form of childcare is vital. That genes were not important was proven by the fact that both groups had parents with high rates of mental illness and criminality: it was the differences in care that explained the different outcomes, not genetic inheritance.

In general, permanently institutionalized children are many times more likely to be depressed, delinquent and to fail at school than children raised by two biological parents. They are far more likely to be insecure and to be shallow in their relationships, showing a lack of reserve with strangers and a superficial friendliness that is impersonal in its lack of discrimination. In middle childhood and adolescence their teachers describe them as irritable, liable to become angry and aggressive if corrected, prone to arguing and fighting with other children. Interestingly, if the child does manage to become the favourite of one of the carers he or she is less at risk of these problems.

One in seven girls who have been raised in an institution are actually pregnant when they leave, 70 per cent of institution-leavers have passed no exams at all and 40 per cent are already mentally ill. They become the 30 per cent of homeless who have been in care and 40 per cent of clients of drug and alcohol recovery programmes. That permanently institutionalized children suffer these problems so much more than fostered or adopted ones proves that it is not genes that are causing them.

It proves that, in the vast majority of cases, if children receive adequate care in childhood they will be much less likely to suffer mental illness or to become criminal. It also proves that genes inherited from biological parents play little part in determining outcomes. The kind of problems that a child's biological parents had are far less predictive of how he will turn out than whether he was adopted, fostered, returned to his biological parents or permanently institutionalized. Low-quality care and lack of continuity in early childhood are the core causes of crime and mental illness. To believe, as so many British and American leaders seem to, that genes are the main reason that the poor are poor, the bad are bad and the mad are mad is simply incorrect. As Carol Bellamy, executive director of UNICEF, recently put it, 'The greatest tragedy is that many decision-makers simply don't know how crucial those first three years of life are.'

What is more, and perhaps most destructive of all, if parents believe they have little control over their children's psychology – as those who perceive genes to be a major cause of behaviour are liable to – they are more likely to maltreat the child. Parents who see themselves as having played little or no part in their child behaving badly are more likely to parent abusively. Even when these parents are in the company of others' offspring, they become more irritable and rapidly annoyed if the child is difficult. This suggests that the widespread conviction that genes are critical is ultimately poisonous to the development of millions of children.

Genes probably do confer universal emotions like humour, anger and sadness, and uniquely human mental equipment like speech, thought and self-consciousness. These characteristics are not the result of 'a bit of both', but are totally caused by our genes, found in almost all humans regardless of family upbringing, class, gender and society. But the role of genes in determining differences in our individual psychology, like those between us and our siblings, is another matter entirely. Even when it comes to the most genetic of mental illnesses, such as schizophrenia, the environment remains important; and, thankfully, such extreme problems rarely afflict us. In the much commoner difficulties, such as neurosis or mild depression, genes play a negligible role and for particular traits, like my love of Lou Reed's music or my cat Zigzag, they probably play none at all.

It is now clear that the electrical and chemical brain patterns which are our selves, and even the actual size of the different parts of the brain, are heavily influenced by the kind of early care we receive. In most cases, this care determines the content of our brain far more than does our particular genetic inheritance. Since the quality of such care varies considerably according to how much stress the parents are under and what sort of childhoods they had themselves, having a low income makes parents far more likely to care for children in ways that promote electro-chemical dysfunction. People from low-income families are much more likely to have excessively low or high levels of cortisol as a result of disturbing early experiences, as well as of the stress of living on a low income as an adult.

It has also recently been proven for the first time that the functioning of their serotonin differs (this is the brain chemical which increases confidence, dominance and successfulness). In a large sample of healthy middle-aged men and women, serotonin levels were found to be more depleted the lower the social class. Interestingly, lower serotonin levels are also found in people who overeat, who are fat, and who smoke and drink – all of which are much more common in people from lower social classes. The low serotonin levels are almost certainly linked to differing early childhood experiences. Rats deprived of maternal attention during the first seven days of their life show decreased serotonin in the hippocampus of the brain; likewise depleted levels are found in monkeys deprived of normal maternal care for their first six months. It is probable that early deprivation and abuse have the same effect in humans. In short, the scientific evidence is piling up that social hierarchies, not genes, cause different brain and hormonal processes in different classes largely through different patterns of early childcare. It is also becoming undeniable that the solution to this problem is not simply to dole out serotonin-raising drugs like Prozac to individuals from lower social levels.

The implications are considerable. On a personal level, they tell us a lot about how to have a more enjoyable life; and on a wider scale, they suggest some fundamental ways in which our society should be organized. It is time to consider the usefulness of knowing how our parents and society formed us.

Some practical suggestions

If the evidence presented in this book is correct, the key to who we are is how we were cared for in early childhood – and, in turn, that how our parents did so is heavily influenced by the kind of society of which they were members. On this basis, there are four specific proposals that I believe should be seriously considered by governments.

AFFLUENCE IS NOT ENOUGH

Studies of rates of mental illness and wellbeing in developed nations prove that, once we are in the top two-thirds of the population, getting richer does not affect what we are like. Whether we are a multi-millionaire or just have an average income, it makes no difference to the likelihood of our being satisfied with our lot or to our sanity. Having the low income of the bottom third does affect these things, but for the upper two-thirds of citizens in developed nations what matters is the kind of childhood care we received. Given this fact, it is extraordinary that economic growth is the principal plank of all mainstream political parties. It should be replaced by a raft of policies designed to improve the quality of early childhood experience, such as paid leave for parents who wish to care for their children when they are small, and good-quality nursery care or subsidies for paid childminders for parents who want to work.

REGULAR AUDITS OF THE NATION'S MENTAL HEALTH

The obsession with economic performance indicators should be replaced with much greater measurement of the effect of government policy on our mental health. Every two years there should be a nationally representative audit of it, by which the government should be judged. This audit would include an evaluation of how parenting is faring in the light of government policies.

EMOTIONAL AUDITING OF TEENAGERS

All children should undergo an emotional audit during their sixteenth year. The grotesque overemphasis on exam performance should be replaced by a version of Cognitive Analytic Therapy (CAT), in which every child is helped to evaluate the impact of her upbringing on her psychology.

A STUDY OF THE IMPACT OF EARLY EXPERIENCE ON ADULT PSYCHOLOGY

The government should commission a large-scale study of a representative sample of the population, following them from before birth to death, to provide a better understanding of the effect of early childhood experiences on adult traits. This study needs to be unprecedentedly detailed, observing the interactions of whole families frequently enough to be able to state with confidence what are the main determinants of family scripts, sense of self, pattern of attachment and strength of conscience.

I shall not hold my breath waiting for these ideas to be implemented. But we can no longer afford to ignore the messages. We can continue to follow in the footsteps of the most pathological developed nation on earth, the USA, with its disastrous disregard for the needs of parents. A recent World Health Organization study of rates of mental illness in 14 nations, including several developing as well as developed ones, proves just how much more disturbed Americans are compared to the rest of the world. 26 per cent of Americans had suffered a mental illness in the previous 12 months. This was six times more than the inhabitants of Nigeria or Shanghai, and three times higher than the European mainland average. Combined with America's addiction to consumerism and its high levels of inequality, its failure to support parents is a major reason. Alternatively, we can emulate the example set by so many of our European neighbours, like Denmark and France. That way, our current unprecedented affluence will work for, rather than against, our emotional wellbeing.

Having descended into earnest socio-political analysis for the last few pages, I was intending to conclude on Philip Larkin's lighter note, his final tip for contentment:

> *Get out as early as you can,*
> *And don't have any kids yourself.*

However, during the course of writing this book the death of my mother has made me an orphan, I have got married (for the first and, I hope, only time), moved house and now I am to be a father (for the first time, as far as I know). There is a danger that, soon after our baby's birth, in order to adjust the complexion of my particular bubble of positive illusions to the necessary tint of rose, I shall become convinced that the way children turn out is largely genetic, to let my wife and myself off the hook. I can only say that if you should come across newspaper articles or new books by me denouncing everything that is in this one, ignore them. This book is the unvarnished truth, told as only someone who is not yet a parent can.

311

Appendix 1: The Dubiousness of the Minnesota Twins Reared-Apart Study

References appear at the end of this appendix.

Since 1979, Professor Thomas Bouchard and colleagues have sought out twins from across the USA who were separated in childhood. On being identified, the twins undergo fifty hours of assessment, answering over fifteen thousand questions at Bouchard's unit in the University of Minneapolis, Minnesota.

Through newspaper articles, books and television documentaries the Minnesota study has hugely influenced popular conceptions of the impact of genes upon behaviour. In most cases, these report seemingly astonishing similarities between twins who have never met each other. For example, the journalist Lawrence Wright (p. 45, Wright, 1997) writes of two of Bouchard's twins who were separated at birth that they both liked their coffee black and cold, had both fallen down the stairs at the age of fifteen and had weak ankles as a result, at sixteen had both met the man they were to marry at a local dance, laughed more than anyone else they knew, and so on. Even Bouchard admits that such similarities are purely coincidental, but that has not stopped them from being endlessly purveyed as spooky evidence of genetics.

For a full critique of the scientific problems with the studies, it is worth reading the paper by Jay Joseph (2001). A fundamental concern is that Professor Bouchard told Lawrence Wright (p. 60, Wright, 1997) that he is not willing to release or allow for inspection by independent assessors the raw data on which his scientific papers are based. Given the history of fradulence in this field of research it is a particularly unfortunate refusal, and some scientists reject the validity of any of his findings until such an assessment is permitted. Suspicions about Bouchard's genuine open-mindedness

312

are compounded by the fact that in order to carry it out he has received grants to the value of $1.3 million from the Pioneer Fund of New York, which has its roots in the eugenics movement and which backs projects that advocate racial segregation (p. 50, Wright, 1997).

The published results contain several worrying and puzzling omissions in the data, serious absences of basic, vital information in the scientific reports from Bouchard's group. Ideally for such a study, the twins would be separated at birth and given for adoption to parents who provide divergent child-rearing experiences. Yet average ages at which the twins were separated are often not provided. The amount of time that they have been in contact and the form that this has taken are not reported. In many cases the twins seem to have had extensive telephone contact before they visited Bouchard in Minnesota, which could have led them to convince themselves of their similarity.

Another issue is the parameters that Bouchard has studied. He has focused on ones most likely to provide support for genetic theories, such as IQ, but there is very little about the huge swathe of psychology which appears to have little or no heritability, such as choice of mate, violence and attachment patterns.

A fundamental problem is that the twins in his sample are nearly all self-selected. Following massive, nationwide publicity of the study in the media, in which similarities of twins were the focus, the twins usually contacted the researchers. This could easily mean that the twins in the study are motivated to identify themselves as similar to each other. There could be hundreds of other pairs who have not contacted him because they are convinced of their differences. What is more, it is possible that some of the twins have simply invented similarities in order to achieve media prominence and financial gain – several of the pairs have signed book and film deals.

Bouchard rejects calls for detailed case histories on the grounds that it would be a breach of confidentiality. Yet there is a long tradition of publication of full reports without any such problem being raised (Farber, 1981), and doubtless he could easily obtain permission if he felt this were necessary. The media need to adopt a more responsible attitude to the reporting of this study.

REFERENCES

Farber, S.L., 1981, *Identical Twins Reared Apart,* New York: Basic Books.

Joseph, J., 2001, 'Separated twins and the genetics of personality: a critique', *American J. of Psychology,* 114, 1–30.

Wright, L., 1997, *Twins,* London: Weidenfeld and Nicolson.

Appendix 2: Twin Studies – A Warning

References appear at the end of this appendix.

INTRODUCTION

Twin and adoption studies are the foundations upon which the behavioural genetic edifice is built. Until molecular genetics is able to provide direct evidence of genetic determinants, the contention that genes are a significant cause of individual differences in human psychology relies wholly on the findings of twin and adoption studies.

In the case of twin studies, they have a chequered past. The psychologist Cyril Burt was demonstrated to have simply invented results of phantom studies of identical twins in his zeal to prove the role of genes. Serious doubts still remain about the validity of the methods employed by the influential schizophrenia researcher Franz Kallman (Marshall, 1984). But actual fraud is not the main objection to such studies. It is the numerous problems which have been raised by highly respected scientists in this field concerning the validity of the scientific methods and assumptions made in such studies.

Behavioural geneticists tend either to ignore or to make the most fleeting of reference to the problems that this body of literature raises (see, for example, Plomin, 1997, a key textbook that hardly refers to them). If they were more commonly considered, the overconfident assertions about the findings of behavioural genetics would be modified and greater caution emphasized regarding the overall validity and generalizability of both twin and adoption studies.

PROBLEM 1: THE GENERALIZABILITY OF FINDINGS FROM STUDIES USING TWIN SAMPLES (SCHACTER, 1982; HAY, 1987)

A fundamental query about studies based on comparisons of samples of identical and non-identical (fraternal) twins is whether the unique experience of being twins and their unusual genetic origins make them a population upon which generalization to other populations can legitimately be based. The evidence is equivocal. For example, studies of twins have suggested that their parents are forced to devote less time to each individual, that their language may be delayed, and that there is a process of 'de-identification' in which parents make abnormal (compared with parents of singletons) attempts to create difference out of twin offspring. These and other similar findings call into doubt the generalizability of estimates of heritability based on twin samples to singletons, who make up the vast majority of all humans.

PROBLEM 2: THE ASSUMPTION OF EQUIVALENT ENVIRONMENTS (BAUMRIND, 1993; JOSEPH, 1998)

In estimating heritabilities from comparisons between samples of identical and fraternal twins, behavioural geneticists make a critical assumption: that the environments of the two kinds of twins do not differ systematically, and therefore that any differences found between the two groups are due not to environmental differences, such as the kind of parenting, reactions from peers or schoolteachers and so on, but solely to the differences in genetic similarity.

There is, however, evidence to suggest that this assumption is false and that identical and fraternal twins are not likely to be treated in the same way. A simple example is appearance. A large body of research demonstrates that reactions to individuals are substantially influenced by physical attractiveness. Attractive people are more likely to be judged pleasant and to be successful than unattractive ones (Etcoff, 1999). Since identical twins look exactly the same their attractiveness will affect responses to

316

them in the same way, whereas fraternal twins (who do not look identical) will receive different reactions. It follows that some, at least, of the higher concordances in psychology between identical compared with fraternal twins may reflect the impact of the greater similarity of appearance of identical twins on the way they are treated and not genetically inherited differences in psychology.

That the assumption of equivalent environments may be invalid makes estimates of heritability unreliable. Whilst it does not invalidate twin studies altogether, it creates an unknown degree of effect on the concordance rates which lie at the heart of the twin method.

PROBLEM 3: MATHEMATICAL ASSUMPTIONS MADE IN ANALYZING TWIN DATA (WAHLSTEIN, 1990, 1994)

Behavioural geneticists subdivide the effects of environment into shared and nonshared effects. Shared effects are those which are experienced by all offspring, whereas nonshared ones are those which are unique to the individual. Hence, if a parent is depressed and expresses this behaviour equally to all offspring it is a shared effect whereas if the depression is especially directed towards one offspring and not the others, it is a nonshared effect.

Behavioural genetic analyses of twin data make assumptions in calculating the partitioning of heritability between shared and nonshared environmental effects and between additive and nonadditive effects. These assumptions create an insensivity to interactive effects between heredity and environment.

PROBLEM 4: THE IMPACT OF PRENATAL FACTORS ON IDENTICAL TWINS (DEVLIN, 1997)

In reanalyzing 212 twin studies, Devlin demonstrated that maternal womb effects may account for substantial amounts of the concordance between twins – concordances that are currently assumed to be due to genetics.

PROBLEM 5: THE LOW ESTIMATES OF SHARED ENVIRONMENT FROM TWIN AND ADOPTION STUDIES (STOOLMILLER, 1999)

It is frequently contended by behavioural geneticists that, overall, twin studies demonstrate that shared environments have minimal effects. This is more debatable than is often claimed on two grounds. First, there are many soundly conducted twin studies which do not confirm this claim, finding significant shared effects. Secondly, there are numerous studies of the impact of families and socioeconomic status on development and psychopathology which find shared effects.

PROBLEM 6: MEASURES OF THE ENVIRONMENT IN BEHAVIOURAL GENETIC STUDIES (BAUMRIND, 1993)

Nearly all twin and adoption studies either make no attempt to measure the possible impact of environmental factors, or use measures which environmentalists find inappropriate or inadequate. It is widely believed that if the environment were to be properly measured in twin studies, its effect would be shown to be much greater than geneticists currently claim, creating a contradiction between the results from comparing concordance rates in samples of identical and non-identical twins and environmental measures.

PROBLEM 7: CAUTIONS IN THE INTEPRETATION OF ADOPTION STUDIES (STOOLMILLER, 1999)

When twins are placed for adoption, a considerable effort is often made to find families of similar socioeconomic status and race as the biological family of origin. This is known as 'selective placement'. Both class and race of family have very well-established, consistent effects on what children are like. Since adoption studies test the similarity of biological parents to their adopted offspring, some of the similarities found between them could be caused by selective placement.

A further worry about adoption studies is the crudity, from an environmental standpoint, of the comparisons made. It is assumed by geneticists that all similarities between biological parent and adopted child are caused by genes, whereas differences between them are environmental. It is further assumed that similarities between adopted child and non-biological parents are environmental. But none of these assumptions can be safely made for a great many traits. For example, some children react against what their non-biological parents are like – an environmental effect that would not be treated as such by geneticists.

A final and fundamental problem is that geneticists make no allowance for the well-established fact that being adopted has a profound effect on psychology in its own right. Children who are taken into care have different outcomes from adoptees, and adoptees are more at risk of a host of problems than children raised by biological parents.

REFERENCES

Baumrind, D, 1993, 'The average expectable environment is not enough: a response to Scarr', *Child Development*, 64, 1299–1317.

Devlin, B. et al, 1997, 'The heritability of IQ', *Nature*, 388, 468–71.

Etcoff, N., 1999, *The Survival of the Prettiest*, London: Little, Brown.

Hay, D.A. and O'Brien, P.J., 1987, 'Early influences on the school social adjustment of twins', *Acta Genetica Medica Gemellol*, 36, 239–48.

Joseph, J., 1998, 'The equal environments assumption of the classical twin method: a critical analysis', *J. of Mind and Behaviour*, 19, 325–58.

Marshall, J.R., 1984, 'The genetics of schizophrenia revisited', *Bulletin of the British Psychological Society*, 37, 177–81.

Plomin, R. et al, 1997, *Behavioural Genetics*, San Francisco: Freeman.

Schacter, F.F. et al, 1982, 'Sibling de-identification and split-parent identification: a family tetrad', in Lamb, M.E. and Sutton-Smith, B., *Sibling Relationships: Their Nature and Significance Across the Lifecycle*, NJ: LEA.

Stoolmiller, M., 1999, 'Implications of the restricted range of family environments of heritability and nonshared environment in behaviour-genetic adoption studies', *Psychological Bulletin*, 125, 392–409.

Wahlstein, D., 1990, 'Insensitivity of variance to heredity-environment interaction', *Behaviour and Brain Sciences*, 13, 109–61.

Wahlstein, D., 1994, 'The intelligence of heritability', *Canadian Psychology*, 37, 244–58.

Appendix 3: Estimates of the Environality of Human Psychology from Twin Studies

References appear at the end of this appendix.

Normally, when results of twin studies are presented they are tabulated in terms of how much heritability has been identified. In my table of various estimates I have reversed this, by presenting the degree of environmental influence that the studies have found, the 'environality' of behaviour. These estimates show that, in the great majority of individual differences in human psychology, environality is considerably greater than heritability. Unless otherwise stated, all the estimates presented here are taken from the review by Robert Plomin (Chapter 4, Plomin, 1990).

ENVIRONALITY OF INTELLIGENCE

(i.e. the percentage of individual differences in intelligence that are caused by environment)

In childhood

65 per cent (but much higher environality in adoption studies: see Turkheimer and Waldron, in press)

In adulthood

50 per cent (for higher adoption studies environality see Turkheimer and Waldron, in press)

ENVIRONALITY OF SPECIFIC MENTAL ABILITIES

Memory: 68 per cent

Spatial ability: 54 per cent
Success in exams at school: 62 per cent
Quick-wittedness (mental processing speed): 62 per cent
Verbal fluency: 70 per cent
Creativity: 75 per cent

ENVIRONALITY OF PERSONALITY

Personality traits

Overall personality: 60 per cent (but much higher environality [78
per cent] in adoption studies: see p.189, Hoffman, 1991)
Extroversion (outgoingness): 60–70 per cent
Neuroticism: 70 per cent
Emotionality: 60 per cent
Activity level: 75 per cent
Sociability: 75 per cent
Masculinity–femininity: little or no genetic influence
Tolerance of ambiguity: little or no genetic influence

Attitudes and beliefs

Conservatism: 70 per cent
Religiosity: 100 per cent
Racism: 100 per cent
Vocational interests: 50 per cent

Relationships

Male homosexuality: estimates vary from 0 to 50 per cent but most
studies are flawed. The most recent and comprehensive study finds
very little genetic effect (Bailey, 1999)
Female homosexuality: 75–100 per cent
Heterosexual romantic love style: 100 per cent (Waller and Shaver,
1994)
Heterosexual mate preferences: 100 per cent (Lykken and Tellegen,
1993)
Attachment security: 100 per cent (Ricciutti, 1993)

Divorce: 50 per cent (McGue and Lykken, 1992)

ENVIRONALITY OF MENTAL ILLNESS

Schizophrenia: 52 per cent (adoption studies have similar findings: see p. 101, Plomin, 1990)
Manic depression: 35–50 per cent (there are large variations in the findings and adoption studies show much higher enironality: see p. 104, Plomin, 1990)
Major depression: 40–60 per cent (there are large variations in the studies, and adoption estimates of environality are much higher: see p. 104, Plomin, 1990)
Mild depression: estimates vary: 80–100 per cent
Anxiety neurosis: estimates vary: 80–100 per cent
Alcoholism in males: 70 per cent (for a review see pp. 56–61, Welleman and Orford, 1999)
Alcoholism in females: 90–100 per cent
Hyperactivity: 50–70 per cent (Goodman and Stevenson, 1989)
Borderline Personality Disorder: 90–100 per cent (Torgersen, 1984)
Autism: 20–30 per cent

ENVIRONALITY OF ANTISOCIAL BEHAVIOUR

Antisocial Personality Disorder: 40–60 per cent (but much higher environality in adoption studies: see Collins, 2000)
Adult stealing: 40–60 per cent (higher environality in adoption studies: see p. 109, Plomin, 1990)
Juvenile delinquency: 52 per cent (much higher environality in adoption studies: see p. 394, Stoolmiller, 1999)
Violence: little or no genetic influence (Carey, 1989)

REFERENCES

Bailey, J.M. et al, 2000, 'Genetic and environmental influences on sexual orientation and its correlates in an Australian twin sample', *J. of Personality and Social Psychology*, 78, 524–36.
Carey, G. 1989, 'Genetics and violence: human studies', USA: National Academy of Sciences.

Collins, W.A. et al, 2000, 'Contemporary research on parenting – the case for nature and nurture', *American Psychologist*, 55, 218–32.

Goodman, R. and Stevenson, J., 1989, 'A twin study of hyperactivity II: the aetiological role of genes, family relationships and perinatal adversity', *J. of Child and Clinical Psychology*, 30, 691–709.

Hoffman, L.W., 1991, 'The influence of the family environment on personality: accounting for sibling differences', *Psychological Bulletin*, 110, 187–203.

Lykken, D.T. and Tellegen, A., 1993, 'Is human mating adventitious or the result of lawful choice? A twin study of mate selection', *J. of Personality and Social Psychology*, 65, 56–68.

Mcgue, M. and Lykken, D.T., 1992, 'Genetic influence on risk of divorce', *Psychological Science*, 3, 368–73.

Plomin, R., 1990, *Nature and Nurture – an Introduction to Behavioural Genetics*, Pacific Grove: Brooks/Cole.

Ricciuti, A.E., 1993, 'Child–mother attachment: a twin study', *Dissertation Abstracts International*, 54, 3364.

Stoolmiller, M., 1999, 'Implications of the restricted range of family environments for estimates of heritability and nonshared environment in behaviour-genetic adoption studies', *Psychological Bulletin*, 125, 392–409.

Torgersen, S, 1984, 'Genetic and nosological aspects of schizotypal and borderline personality disorders', *Archives of General Psychiatry*, 41, 546–54.

Turkheimer, E. and Waldron, M.C., in press, 'Nonshared environment: theoretical, methodological and quantitative review', *Psychological Review*.

Waller, N.G. and Shaver, P.R., 1994, 'The importance of nongenetic influences on romantic love styles: a twin-family study', *Psychological Science*, 5, 268–74.

Welleman, R. and Orford, J., 1999, *Risk and Resilience – Adults Who Were the Children of Problem Drinkers*, Amsterdam: Harwood.

NOTES AND REFERENCES

Each source note below is introduced by a short phrase indicating the passage to which it refers on the page stated.

Those who are unused to seeking out scientific papers from journals or academic texts, may like to know that it is possible to have them sent by post by joining the British Library at Boston Spa (tel. 01937 546464). There is no charge for joining, but a charge of £3.91 is made for each photocopied scientific paper sent to you. Alternatively, you may be able to gain access to your local university library which will contain many of the references cited here.

Introduction

4   The actress Mia Farrow ... Farrow, 1997.
4   Offspring of families ... see p. 425, Belksy, 1993; for evidence that children from families with five or more children are twice as likely to suffer abuse or neglect see p. 1226, Widom, 1999.
5   It has been shown ... For a brief review see Anderson and Miranda, 2000.
5   The pattern of electricity ... see Schore, 1997.
6   And only show up if the mother ... see Dawson et al, 1999.
6   Dysfunctions in the right brain ... see Cutting, 1992.
6   If we were living ... see Bremner and Vermetten, 2001.
6   Adults who suffered childhood maltreatment ... see p. 482, Bremner and Vermetten, 2001
6   Studies of the volume ... see Teicher, 2000 and 2002; Bremner et al, 2000.
7   The earlier and more severe the maltreatment ... see note for p. 145 [beginning 'In general the earlier that a negative pattern'] for detailed references supporting this claim.
7   For instance, in a study of 800 children ... See Manly et al, 2001.
7   Furthermore, their cortisol levels ... see Cicchetti and Rogosch, 2001a; Ciccetti and Rogosch, 2001b.
7   For example, if maltreated ... see Nishith et al, 2000.
8   90 per cent of people in prisons ... this statistic comes from a very thorough survey of a representative sample of 3,000 prisoners in England and Wales: Singeleton et al, 1998.
9   Alice Miller ... this quote is from her Miller, 1991.
9   Bubble of positive illusions ... for a review see Taylor et al, 1994.
9   When university students ... see Bahrich et al, 1996.
9   Parents of high achieving children ... see Himmelstein et al, 1991.
10 Men who had tested positive for HIV ... see Taylor et al, 1992.
10 One fifth of us are afflicted ... see Robins et al, 1992.
10 Between 20 and 40 per cent more ... see Burvill, 1995; Lewinsohn, 2004.

Chapter 1: Our Genes

12 In 1985 his parents published . . . see Hinckley and Hinckley, 1985.
13 For example, when mothers . . . see Hock et al, 1988; DeMeis et al, 1986.
13 For example, when the publics . . . see Ernulf et al, 1989.
14 As long ago as 1949 . . . see Pastore, 1949.
14 And this appears to be true today . . . see Murray, 2000.
14 As a recent review of the matter . . . Jost et al, 2003.
14 Far more logical, in evolutionary terms . . . for discussion of the competing claims of plasticity and genetic structures in determining individuality, see Greenough et al, 1987; Gottlieb, 1991; Johnson, 1999.
14 The child must attract the interest . . . for a readable account of the evolution of family niches see pp. 86–9, Sulloway, 1996.
15 In his words, his work proves . . . reported on p. 1, *The Observer* newspaper, 11 February, 2001; for the scientific report in which this claim is substantiated see Venter et al, 2001.
15 Taken overall, the results do not support . . . e.g. see Pinker, 1997.
16 That is the view of Robert Plomin . . . see Plomin and Daniels, 1987.
19 In recent years it has become fashionable . . . for the grossest misrepresentation of the evidence see Harris, 1995, which manages largely to ignore the 2000-odd scientific studies showing that early childcare heavily influences later patterns in relationships; for an equally deceptive account see Scarr, 1997.
21 This is because the commoner that an illness is . . . for the estimates in the next paragraph, see McGuffin et al, 1994.
21 The next most heritable mental illness . . . see Bryson, 1996.
21 For example, some (but not all) twin studies . . . for a review see pp. 161–2, Gilbert, 1995.
22 One twin study found it to be about 50 per cent heritable . . . see Mcgue and Lykken, 1992.
22 The number of divorces . . . for details of the increase in divorce see pp. 150–7, James, 1997.
22 Also very telling . . . James, 2007a.
22 I have argued elsewhere . . . James, 2007b
22 A New Zealand study . . . Caspi, 2002, 2003, 2005.
23 Four studies have partially replicated . . . Eley et al, 2004; Kendler et al, 2005; Zalsman et al, in press; Wilhelm et al, 2006.
23 In two of them . . . Eley et al, 2004; Grabe et al, 2005.
23 Three other small studies . . . Grabe et al, 2005; Fox, in press, Kaufman et al, 2004.
23 On the other hand, three have not . . . Gillespie et al, 2005; Willis-Owen et al, 2005; Surtees et al, 2005.
23 One of these had a much larger sample . . . Surtees et al, 2006.
23 Furthermore, in large studies of samples of depressed patients . . . Mendlewicz et al, 2004; Lasky-Su et al, 2005.
23 Two of the strongest predictors . . . James, 2007b.
23 Two studies have found no greater . . . Manuck et al, 2004, 2005.

24   **Similar studies** . . . Timimi, 2005.
24   **Whilst I was working** . . . The finest account of the schizophrenic's inner world can be found in Laing, 1960.
25   **Conventional psychiatric wisdom** . . . see Gottesman, 1992.
25   **Emil Kraepelin** . . . see Bentall, 2003 for references to the studies of symptomatology in succeeding paragraphs.
27   **Studies of twins are the cornerstone** . . . see Gottesman, 1992.
27   **But as we shall see in the next chapter** . . . for studies of discordant identical twins, one of whom has schizophrenia but not the other, see Belmaker et al, 1974; DiLalla and Gottesman, 1995; Mosher et al, 1971; Stabenau et al, 1967; Stabenau, 1973.
27   **Apart from twin studies, the fact that** . . . see Jablensky et al, 1992; for an interesting critique of what this study reveals, see the review of the book reporting it, by Marshall, 1991.
27   **A survey of 2 million** . . . Pedersen, 2001.
28   **That its rate varies considerably** . . . for evidence regarding class, race and nations, see pp. 76–9, Gottesman, 1992.
28   **This latter is almost certainly nothing to do with genes** . . . see Hutchinson et al, 1996; see also McKenzie and Murray, 1988.
28   **Non-white people** . . . Boydell, 2001.
28   **Of 38 Afro-Caribbeans** . . . Bhugra, 1997.
28   **In fact, if you go mad** . . . Susser, 1994.
28   **Psychiatrists also play down the uncomfortable fact** . . . see Jablensky et al, 1992.
28   **Whereas only 1 per cent of the general population** . . . see the table on p. 96, Gottesman, 1992.
28   **That this could partly be explained** . . . for a telling analysis of why psychiatrists do not tend to consider this possibility see pp. 260–6, Johnstone, 1992.
29   **About one quarter** . . . for the statistics in this sentence and the next, see pp. 118–19, Horrobin, 2001.
29   **Nurses and family members are very actively** . . . for an example of instructions written by psychiatrists for nurses and family members (e.g. 'stop all hostile exchanges immediately. Focus on structure of expression, not content of problems') see Falloon et al, 1988.
29   **At least 20 per cent of schizophrenics completely recover** . . . see p. 118, Horrobin, 2001.
29   **Now mentally healthy** . . . *Taking a Stand*, produced by Jane Beresford, broadcast in Spring 2001.

Chapter 2: Scripting Our Family Drama

36   **As R.D. Laing** . . . see p. 78, Laing, 1971.
39   **In general** . . . for a comprehensive account of gender differences see Hoyenga et al, 1993.
40   **For example, when adults are told** . . . see Condry and Condry, 1976.
40   **Hence, even today** . . . see Parsons et al, 1982.
40   **Since parents' beliefs** . . . see chapters 1 and 7 in Sigel, 1985.

44–52 **At least as big a determinant** ... For references during these pages on birth order and the development of family niches, unless I have otherwise provided a specific one, see Sulloway, 1996. This book includes his analysis of the birth order of politicians and scientists and a review of the literature.

47 **It affects which of us suffer from schizophrenia** ... see Mackenzie and Wright, 1996.

47 **Children who had been exceptionally gifted** ... see Jarvelin et al, 1999.

47 **Large families of five or more** ... p. 425. Belsky, 1993; Widom, 1999.

48 **Tommy Lascelles, a leading courtier** ... see p. 115, Bradford, 1996.

50 **Physical beauty** ... for the beauty studies cited here, see Etcoff, 1999.

51 **Interestingly, both parents** ... see Baker and Oram, 2000.

52 **Described in his sister, La Toyah's autobiography** ... Jackson, 1991.

52 **Although it is often assumed** ... for a review see Sloboda et al, 1994.

53 **Lineker's grandfather** ... see Connolly, R., 1990, 'Football was the only thing that I ever wanted to do as a kid', London: *The Times*.

54 **If we are prodigiously good** ... see chapter 6, Howe, 1999.

55 **A famous study** ... Terman et al, 1983.

55 **Overall, our intelligence** ... see Gould, 1991.

55 **If Afro-Caribbeans** ... see Herrnstein and Murray, 1994.

55 **This is almost certainly more to do with** ... see Fraser, 1995.

55 **This is proven by the fact that children from poor homes** ... see Capron and Duyme, 1989. Equally telling, when children adopted from poor homes into middle-class ones are compared with siblings who had stayed with their biological parents, the adopted have an average of 14 more IQ points: see Schiff et al, 1982.

55 **One third of the six hundred** ... see Albert, 1983.

55 **In contemporary surveys** ... see Cox and Jennings, 1988; Jennings et al, 1994.

55 **In the world of the arts** ... see Brown, 1968.

56 **In science think of** ... see Roe, 1952.

57 **Many of the most famous** ... see Eisenstadt et al, 1989.

57 **The long list of such ruthless** ... see Eisenstadt et al, 1989.

58 **In his autobiographical work** ... see Storr, 1972.

59 **Writers tend to be depressives** ... see Jamison, 1993.

59 **A high incidence of alcoholism** ... see Dardis, 1989.

59 **Charles Darwin** ... see Storr, 1972.

59 **Adult criminals** ... see Albert, 1983

60 **They refuse to conform** ... see Storr, 1972.

61 **There are many examples of this** ... see Eisenstadt et al, 1989.

61 **Several other environmental factors** ... see Cox and Jennings, 1988.

61 **In general, when adults** ... for a very thorough review of the scientific evidence regarding the childhood causes of depression, see Blatt and Holman, 1992.

61 **They do so regardless** ... see Gotlib et al, 1988.

62 **A study of 800 women** ... see Bifulco and Moran, 1998.

62 **A recent British study** ... Sadowski et al, 1999.

62 **Another study** ... Koestner et al, 1991.

62 This depression is known as Dominant Goal ... for a full description including the case history quoted in the following pages, see Arieti and Bemporad, 1980.

66 Interestingly, Fry might have been able ... see Kasser and Ryan, 1993; Kasser and Ryan, 1996; Carver and Baird, 1998; Srivastava et al, 2001.

66 A remarkable series of studies ... Luthar, 2002, 2003, 2005.

66 Many studies have proven ... for a review, see Blatt and Holman, 1992. See also pp. 63–9 of my book, James, 1997.

67 At least one in ten of all children ... this was found in a survey of 2689 18–24-year-olds: NSPCC, 2001.

67 The pressure on the daughter makes her prone to Dominant Goal depression ... for example, see Frost et al, 1991. This shows that hypercriticism from mothers, but not fathers, predicts pathology. For other evidence of new pressures on young women see also chapter 3, James, 1997.

67 Comparison to inappropriate models ... see pp. 51–68, James, 1997.

67 Children of parents who used healthy ... see Frey, 1987.

67 In America, over one fifth ... Luthar, 2002.

67 In Britain ... West et al, 2003.

67 The pressures ... WHO, 1999.

67 An in-depth British study ... see Walkerdine, 1995.

69 This is typical of the perfectionist's childhood ... for a good review of this literature see Blatt, 1995.

69 In general ... see Frost et al, 1991.

69 When children are followed over time ... see p. 1012, Blatt, 1995.

70 Academic pressure to succeed ... see pp. 112–22, James, 1997.

70 Perfectionist girls are at greater risk ... see Vohs et al, 1999.

70 More generally ... see the remarkable study by Humphrey, 1989.

71 That there are large fluctuations ... see pp. 201–3, James, 1997.

71 It is hardly surprising that in a recent British survey ... 901 18–24-year-old women were interviewed face to face, for a report, Bread for Life, 1988.

71 New Zealand has the highest ... based on a United Nations report discussed by Shenon, 1995, in the *New York Times*, 15 July, p. 3, 'New Zealand seeks causes of suicides by young'.

72 As might be expected, children with a positive relationship to both parents ... see Matthijssen et al, 1998.

72 Favouritism is found in all families ... see Daniels et al, 1985; see also pp. 52–4, Bifulco and Moran, 1998.

72 Two-thirds of children ... see Dunn and Plomin, 1990.

72 A Czechoslovakian study ... published in two papers: Matejcek et al, 1978; and Matejcek et al, 1980.

72 They attribute the preferential treatment ... see McConville, 1985.

73 Adults who report ... see the excellent Gilbert and Gerlsma, 1999.

74 Shaming is one of the most ... see Gilbert and Andrews, 1998; see also Gilbert et al, 1996.

74 Shaming in early childhood ... see Schore, 1998.

74 At least one in ten ... see NSPCC, 2001.

74 This is not borne out by the scientific evidence ... for a brief yet lethal

demolition of the idea that peers are highly influential, see pp. 227–8, Collins et al, 2000.

75   **The commonest emotions** ... see chapter 5 of my monograph, James, 1995.

75   **Two-year-old children** ... see Cole et al, 1992.

75   **Precisely which child** ... for a review of the differential impact of paternal support and course and timing of mother's depression, see Goodman and Gotlib, 1999.

78   **Children of separated parents** ... see Rodgers and Pryor, 1998.

78   **Having a positive relationship** ... see Matthijssen et al, 1998.

79   **These studies are very rarely referred to nowadays** ... for an example of the hostility of the psychiatric community see the issue devoted to schizophrenia of *J. of Mental Health*, Vol. 2, no. 3; see also Leff and Vaughn, 1994; Lam and Kuipers, 1993.

79   **There is now a large body of evidence** ... for a summary see Johnstone, 1993.

79   **In about half of the families** ... see p. 259, Johnstone, 1993.

79   **Whereas 55 per cent of patients who return** ... see Brown et al, 1972.

80   **If negative parents are trained to be less so** ... see Kuipers et al, 1992.

80   **Whether or not patients keep on taking their prescribed drugs** ... see Johnstone, 1993.

80   **Compared to families with no schizophrenic members** ... see Liem, 1980.

80   **So the critical question becomes** ... for reviews, see Lidz, 1985; Asarnow and Goldstein, 1986; Goldstein, 1990.

80   **In one, families with problem boys** ... see Goldstein, 1985.

81   **Fifty-six children adopted** ... Wahlberg et al, 1997.

82   **The cornerstone of Read's evidence** ... Read et al, 2005

82   **A review of 13 studies** ... Read et al, p. 227, 2004.

82   **Crucially, psychiatric patients or schizophrenics** ... Read et al, 2005.

82   **Studies already show** ... Bentall, 2003.

82   **This has been shown many times over** ... see chapter 5.

82   **About 15 per cent of schizophrenics** ... Bentall, 2003.

83   **Among the sexually abused** ... Read et al, 2005.

83   **For example, the mildly abused are twice** ... Jansen et al, 2004.

83   **In a review of the 33,648 studies** ... Read et al, 2005.

83   **From sociological studies** ... for facts in this paragraph, see Read et al, 2004b.

83   **Rates of schizophrenia** ... for facts in this paragraph see Kleinman et al, 1997, and Ross et al, 2004.

83   **They have been shown** ... Ross et al, 2004.

83   **There is a close relationship** ... Mosher et al, 2004.

84   **Surveys find that the majority** ... Read et al, 2006.

85   **There are several other studies** ... listed in the chapter by Lidz, in Asarnow and Goldstein, 1985.

85   **There are also at least seven** ... cited on p. 261, Read et al, 2004c.

85   **In these cases** ... see Belmaker et al, 1974; DiLalla and Gottesman, 1995.

85   **They show that from a young age** ... see Mosher et al, 1971; Stabenau et al, 1967; Stabenau, 1973.

86   The controversial theory of R.D. Laing . . . see Laing, 1960; Laing et al, 1964.

86   In a famous example, a mother visited her schizophrenic son . . . this was originally described, along with the theory of Double Bind, in Bateson, 1973.

87   Sexual abuse often leads to a symptom known as . . . see pp. 92–5, Bifulco and Moran, 1998.

88   Children whose mothers are schizophrenic . . . see p. 99, Gottesman, 1992.

88   It has become clear . . . the assertions in this paragraph are supported by a review of this evidence in an edition of *The Psychologist* devoted to it, 2000, vol. 13, and the comprehensive review by Westen, 1998.

89   Participants in one experiment were asked to describe . . . see Chen and Andersen, 1999.

89   We tend to feel more favourably . . . see Andersen et al, 1996.

89   For example, if a person resembles our brother or sister . . . see Hinkley and Nadersen, 1996.

89   Most dramatic of all . . . see Berk and Anderson, 2001.

Chapter 3: Scripting Our Conscience, Aged Three to Six

95   The starting point . . . for those who have never read anything by Freud, there is the highly entertaining *Introducing Freud* and *Introducing Psychoanalysis* (Icon Books). For a more detailed idea of his theories try Anthony Storr's biography of Freud.

97   Alcoholics or chocolate addicts . . . for a review of this evidence and that relating to smoking, see pp. 127–30, Fisher and Greenberg, 1985.

97   (Incidentally, smoking is twice as common . . . see pp. 149–61, Gilbert, 1995.

97   Mothers issue a command . . . see Forehand et al, 1975.

98   In experiments, adults who had previously . . . see Fisher and Greenberg, 1985.

98   Anal personality . . . see pp. 137–62, Fisher and Greenberg, 1985.

98   Even in the far more sexually inhibited . . . see Kinsey et al, 1948; Kinsey et al, 1953.

98   More recent surveys . . . see Friedrich et al, 1991; see also Wyatt et al, 1988.

99   But these days of sybaritic living . . . best described in Friedrich et al, 1991.

99   Far from it deriving from fear . . . this is one of the most important findings of Fisher and Greenberg, 1985, see pp. 207–12.

102   Genetics have been shown . . . for examples, see: Waller and Shaver, 1994; Lykken and Tellegen, 1993; Hershberger et al, 1997; and most telling of all in regard to the alleged genetics of homosexuality, Bailey et al, 2000.

102   According to John Money . . . Money, J., 1986, *Lovemaps*, New York: Irvington.

102   When couples are asked . . . see Collins and Read, 1990.

103   When teenage girls . . . see Wilson and Barret, 1987; see also the unpublished study conducted by students at Davis, Laing and Dick, 10 Pembridge Square, London, W2.

103   **The group studied consisted of 980** . . . see Jedlicka, 1980.

103   **Divorcees who remarry** . . . see *Marriage and Divorce Statistics*, 2000.

103   **When feelings are buried** . . . there is a large literature showing that sexually prohibitive, repressive parenting, often allied to religious beliefs, causes sexual inhibition or feelings of inadequacy and that children tend to have similar sexual attitudes to parents if they are close to them. See Lewis, 1963; Thornton and Camburn, 1987; Moore et al, 1986; Weinstein and Thornton, 1989; Cado and Leitenberg, 1990.

103   **Fifteen per cent of us** . . . see Finkelhor, 1980.

104   **Despite the huge publicity** . . . see Bailey and Hershberger, 1998.

104   **Gay men are twice as likely** . . . Since it is such a controversial matter I shall provide detailed references for my observations about the causes of homosexuality. A further review of this literature by me is accessible on www.Bloomsbury.com. The picture painted in this paragraph comes from twenty-four studies of 1,944 gay men: Bieber et al, 1962; Saghir and Robins, 1973; Bell et al, 1981; Whitam and Zent, 1984; Terman and Miles, 1936; Jonas, 1944; Miller, 1958; West, 1959; Brown, 1963; O'Connor, 1964; Van Den Aardweg, 1984; Roberston, 1972; Westwood, 1960; Symonds, 1969; Schofield, 1965; Evans, 1969; Apperson and McAdoo, 1969; Snortum et al, 1969; Braatan and Darling, 1965; Bene, 1965; Thompson et al, 1973; Stephans, 1973; Siegelman, 1981; Buhrich and McConaghy, 1973.

104   **Gay men are much more likely to report having been 'feminine' boys** . . . If pooling the results of the four main studies, 65–70 per cent of 954 gay men compared with 5–10 per cent of comparable heterosexuals reported themselves as having been labelled 'sissies' in childhood. Gay boys were much less likely than heterosexuals to have enjoyed traditionally male sports, especially team games. They more often liked playing with girls and with dolls. At the extreme, about one third liked to dress in girls' clothing and were over-ridingly effeminate compared with 0–5 per cent of the heterosexual men. The four studies were by Bieber et al (1962), Saghir and Robins (1973), Bell et al (1981) and Whitam and Zent (1984).

105   **About two-thirds of feminine boys** . . . see Bell et al, 1981, and Green, 1987, reports his study in which 75 per cent of the 44 feminine boys followed from childhood turned out gay. Green also reviews a number of other studies with similar findings, including that of Zuger (1970) which followed 48 effeminate boys into adulthood and found that 73 per cent became gay.

105   **Gay men tend to have had negative relationships** . . . at least 2750 gay men in 24 published studies have been quizzed about their relationship with their father. In 20 of them, 60–70 per cent of gays reported having had a negative relationship compared with 20–30 per cent of heterosexuals. The 20 studies are listed in the previous note. The four studies which found no difference in relations with father are: Robertson, 1972; Zuger, 1970; Siegelman, 1974; and Buhrich and McConaghy, 1975.

105   **At most, only 3 per cent are exclusively** . . . for a brief review of prevalence surveys and a report of the results of by far the most authoritative British one, see pp. 188–90, Wellings et al, 1994.

106   **The kind of parenting that creates a benign conscience** . . . see Baumrind,

1966; Baumrind, 1967; Baumrind, 1971; and the chapters by Power and by Jannsen in Jannsen and Gerris, 1992.

106 **But, interestingly, not for intelligence** . . . see Mueller and Dweck, 1998.

106 **Eventually, moral behaviour** . . . for free access to many important papers demonstrating this, go to http://www.selfdeterminationtheory.org, and take the 'publications' option, then look in the 'internalization and self-regulation' and the 'development and parenting' sections.

106 **The resultant benign conscience meant that** . . . see Kochanska and Aksan, 1995.

107 **Being totally dependent on parents** . . . see Zahn-Wexler et al, 1992.

107 **Toddlers with depressed** . . . see Cole et al, 1992.

107 **This is particularly true** . . . see Duggal et al, 2001; Zahn-Wexler, 2002; Klimes-Dougan and Bolger, 1998.

107 **In later life** . . . see Steinberg et al, 1989; see also Baumrind, 1991.

107 **Of course, both our parents were not authoritative** . . . see Watson and Getz, 1990; see also McGuire et al, 1995.

108 **For example, if daughters were close to their fathers** . . . see Lewis and Janda, 1988.

108 **If they felt able to talk openly** . . . Ellis et al, 1999.

108 **Children of parents who teach them** . . . see Patterson, 1990.

109 **A study of junior doctors** . . . see Firth-Cozens, 1992.

109 **Paul Feyerabend** . . . Feyerabend, 1995.

109 **The punitive conscience** . . . see Baumrind, 1967 and 1971.

111 **The definitive British sexual survey** . . . see Wellings et al, 1994.

114 **In simple terms, as several studies** . . . see Bushman et al, 2001.

115 **Alice Miller** . . . see Miller, 1980.

115 **The Authoritarian Personality** . . . for a comprehensive account see Stone et al, 1992.

116 **The American Christian fundamentalists** . . . see Milburn, 1996. This remarkable book provides a detailed analysis of fundamentalism in America.

117 **Bush Junior** . . . the main source for this profile of Bush is *Minutaglio*, 1999, a biography written with the full cooperation of many of the Bush family and friends. Quotations by David Frum come from Frum, 2003.

121 **But it is not only right-wingers** . . . Stone, 1993.

121 **A particularly striking example is Germany** . . . see Milburn, 1996.

123 **One of the most sophisticated accounts** . . . see Patterson, 1982.

124 **Just one of numerous clinicians** . . . for a review of studies that have intervened in parenting seeking to create authoritative care, see Mcmahon and Wells, 1998.

124 **Increases in crime** . . . see Rutter and Smith, 1995.

124 **There are also many other causes** . . . is the subject of my last book, James, 1997.

127 **Since depression is nearly always** . . . see Tangney et al, 1992.

129 **If our parents** . . . see Steinberg, 1986; Vandell and Corasaniti, 1988; Vandell and Ramanan, 1991; Greenberger and Goldberg, 1989; and Moorehouse, 1991.

129 **We were twice as likely** . . . see Richardson et al, 1989.

129   For instance those raised in an institution . . . see Quinton and Rutter, 1988.
129   The absence of a father . . . see pp. 71–97, Surbey, 1990; Bereczkei and Csanaky, 1996; Moffit et al, 1992; Wierson et al, 1993; Graber et al, 1995.
129   This pubertal precocity . . . see Graber et al, 1997; Barber, 1998.
129   Fatherless girls are also liable to have negative attitudes to men . . . see Bereczkei and Csanaky, 1995.
129   In intact families . . . see Ellis et al, 1999.
130   In summary, the host of studies . . . for a review of studies pre-1992, see Beitchman et al, 1992; see also Bushnell et al, 1992; Rowan et al, 1994; Peters et al, 1995; Mullen et al, 1994; Mullen et al, 1996; Fergusson et al, 1996; and McCauley et al, 1997.
130   The tendency for abused to become abuser . . . see Widom, 1989; see also Coid, J. 2001, on the website of the Institute of Psychiatry: www.mds.gmw.ac.uk/psychiatry/forensic.html.
130   Abused women are particularly prone . . . see Coid, previous note.
130   For abused men it is more common . . . see chapter 5, James, 1995.
137   Almost invariably . . . see chapter 1, James, 1995.
137   There is a close relationship . . . see chapter 5, James, 1995.
137   Six times more likely to have self-inflicted scars . . . see Bach-Y-Rita and Veno, 1974.
137   A study of British murderers . . . see West, 1965.
138   Freud writes . . . see p. 130, Freud, 1930.
139   Some studies of identical twins . . . for an introduction see pp. 188–90 in Plomin et al, 1997.
139   Adopted children with biological parents who have been convicted of crimes . . . see Cloninger et al, 1982; Mednick et al, 1984.
139   Where the mother is five times . . . see Horn et al, 1975.
139   On top of all this . . . Hyun Rhee, S. et al, 2002.
139   The number of murders . . . US Department of Justice, Federal Bureau of Investigation, 2004.
139   In 1950, in England . . . see chapter 7, James, 1995.
140   There is also a major contradiction . . . see chapter 6, James, 1995.
140   Demonstrated by Jerry Patterson . . . see Patterson, 1982, 1990.
141   There is little or no connection . . . see Vaughan and Bost, 1999.
141   A lot of evidence that the parental care is critical . . . see James, 1995; see also, Wakschlag and Hans, 1999.
141   But when these mothers are asked to care . . . see Halverson and Waldrop, 1970.
141   The more unequal the society . . . see Braithwaite and Braithwaite, 1980; Messner, 1982.
141   The states of the USA . . . see Currie, 1985.
141   Some toddlers are fearless . . . the succeeding two paragraphs refer to the study by Kochanska, 1997.

Chapter 4: Scripting Our Relationship Patterns in
Our First Three Years

151 **John Bowlby** ... his attachment theory is presented in *Attachment and Loss*, Vols 1, 2 and 3, London: Penguin; see Karen, 1998 for a very readable account of its subsequent development.

153 **Toddlers with eating disorders** ... Chatoor et al, 1998; for a brief review see pp. 508–9, Cassidy and Shaver, 1999.

153 **The insecure child has difficulties** ... see pp. 76–82, in Cassidy and Shaver, 1999.

153 **Aggressive, depressive and antisocial** ... see Greenberg et al, 1993; Speltz et al, 1999.

153 **Their brains and bodies** ... for a review see pp. 230–7 in Cassidy and Shaver, 1999; see also the review by Dawson et al, 2000.

153 **As adults** ... for a summary of correlations of insecurity with adult pathologies see pp. 411–16 in Cassidy and Shaver, 1999.

154 **In both animals and humans** ... for a review of human evidence see Kolb et al, 1998; for a wider review, see Schore, 1996; for increased numbers of synapses in humans resulting from education, see Jacobs et al, 1993; for a popular account see Kotulak, 1996. For animal studies see Greenough, 1976; Greenough et al, 1987.

154 **The greater the frequency and intensity of stimulation** ... see Turner and Greenough, 1985.

154 **The brain is growing most rapidly** ... see Reiner, 1997.

154 **Rats that were deprived of maternal care** ... see Ladd et al, 1996.

154 **Rats raised in early environments** ... see Greenough, 1976.

155 **The kind of early care a monkey receives** ... A review of all the monkey studies reported in succeeding paragraphs is in pp. 182–97, in Cassidy and Shaver, 1999; see also Suomi, 1997.

156 **In general, the earlier a negative pattern** ... for examples of age in relation to consequences of sexual abuse, see Gibb et al, 2001; Fossati et al, 1999; Meiselman, 1978; Cortois, 1979; Bagley, 1986; Kirby et al, 1993; McLellan et al, 1996; Nash et al, 1993; and p. 874, Ogawa et al, 1997.

For age of physical abuse, see Manly, 2001; Keiley et al, 2001; Famularo et al, 1994; Bolger et al, 1998; Keiley and Martin, in preparation.

For age of neglect, see Verhulst et al, 1992; Sroufe et al, 1999; Appleyard et al, 2005; see also Sroufe et al, 1990.

For age at infantile feeding deprivation, see Skuse et al, 1994: this shows that infants who were malnourished during the first six months are three times more damaged than those malnourished during the second six.

For age at parental divorce or separation, see Rodgers and Pryor, 1998; Wallerstein and Kelly, 1979; Kalter and Rembar, 1981; Allison and Furstenberg, 1989.

For age at maternal depression, see p. 473, Goodman and Gotlib, 1999; Alpern and Lyons-Ruth, 1993; Coghill et al, 1986; Wolkind et al, 1980.

For age at parental financial misfortune, see Elder, 1974.

156   In the case of sexual abuse . . . see Bremner et al, 1999; De Bellis, 2001, Teicher, 2002.
156   In the case of physical abuse . . . see Keiley et al, 2001.
156   In the case of divorce . . . see Rodgers and Pryor, 1998.
156   Stress in infancy . . . Lyons et al, 2000; Dettling et al, 2002.
156   Whatever electro-chemical . . . for a thorough yet readable account of the scientific evidence demonstrating the impact of parental care on the infantile and early childhood brain, and its adult outcomes, see Gerbner, 2004.
156   If as children we secrete large quantities of cortisol . . . see Johnson et al, 1992; for evidence of long-term damage to cortisol levels in rats, see Ladd et al, 1996.
157   Infants of depressed mothers . . . for reviews of all neurological effects, see Dawson et al, 2000; Graham et al, 1999.
157   The precise kind of disturbed behaviour . . . see Caspi et al, 1996.
157   Such adults have abnormal brain chemicals . . . see Koob et al, 1993; Kendler et al, 1993; Gabbard, 1994; Schore, 1997; pp. 346–58, James, 1997.
157   In some bird species . . . see Konishi, 1995.
157   Thus, among a group of immigrants . . . see Johnson and Newport, 1989.
158   Rats that have been distressed . . . see Greenough, 1987.
158   Children who were insecure in their relationships . . . see p. 701, Dawson et al, 2000.
158   Romanian children . . . see pp. 701–2, Dawson et al, 2000.
158   The pattern a child will have . . . see pp. 655–6, Cassidy and Shaver, 1999.
158   For every three months spent . . . Martorell et al, 1997.
158   These deficits can sometimes . . . see Rutter et al, 1988; Castle et al, 1999; O'Connor et al, 2000; Verhulst, 1992.
158   Many of these children suffer long-term damage . . . for aggression and delinquency as outcomes, see Quinton and Rutter, 1988; Vorria et al, 1998.
         For hyperactivity, see Roy et al, 2000; O'Connor et al, 1999.
         For emotional insecurity, see Chisholm et al, 1995; Chisholm, 1998; O'Connor, 1999.
         For autism, see Rutter et al, 1999.
         For indiscriminate friendliness, see Wolf and Fesseha, 1999; Chisholm, 1995 and 1998.
159   First, if the child . . . see Dumaret et al, 1997; Vorria et al, 1998; Verhulst et al, 1992.
159   Secondly . . . see Marvin and Britner, 1999; Chisholm, 1995 and 1998, for differences between more than eight months versus less than four months; O'Connor, 1999 and 2003b; Verhulst et al, 1992; Roy et al, 2000.
159   Thirdly . . . see Wolf and Fesseha, 1999; Hodges and Tizard, 1989; Triseliotis, 1989.
159   Fourthly . . . see Verhulst, 1990; O'Connor, 1999; Chisholm, 1998.
159   Finally . . . see Hodges and Tizard, 1989; Verhulst et al, 1992.
160   In general, mothers have a greater effect . . . see Matthijssen et al, 1998.
160   Infants whose mothers are depressed . . . see Field, 1992.

160 Their attention span . . . see Radke-Yarrow et al, 1985; Zahn-Wexler et al, 1984.
160 Language and mental abilities . . . see previous note.
160 As toddlers . . . see pp. 38–42, James, 1995.
160 Such mothers did not respond . . . see Cohn et al, 1986; Cohn et al, 1989.
160 Their voices had . . . see Murray et al, 1993.
160 Eight per cent of children whose mothers . . . see Hammen et al, 1990.
160 The longer and more profoundly . . . see NICHD, 1999(b); Teti et al, 1995; Campbell et al, 1995.
160 If our mother was depressed . . . see Wolkind et al, 1980; Stein et al, 1991; Lee and Gotlib, 1991; Alpern and Lyons-Ruth, 1993.
160 Infants and toddlers with depressed mothers . . . see Dawson et al, 1997.
160 Even more detailed study . . . see Dawson et al, 1999; Dawson et al, (in press).
161 Above all, it would seem that . . . for a summary, see pp. 472–3, Goodman and Gotlib, 1999.
162 To avoid insecurity . . . for evidence regarding the list that follows in this paragraph, see Sroufe et al, 1992.
162 John, for example . . . see Robertson and Robertson, 1989.
163 Studies in the 1970s . . . see Belsky and Steinberg, 1978.
163 However, with the 1980s . . . see Belsky, 1986; Belsky, 1988.
163 Whether measured at four years or ten . . . see Bates et al, 1994; Vandell et al, 1990; Belsky and Eggebean, 1991; Belsky, 2001.
163 One important American study . . . see NICDH, 1998.
163 At its simplest, this is because . . . see Galinsky, 1994.
164 For example, the longer and earlier the day care . . . see NICHD, 1999(a).
164 Those women who are more likely to leave their child . . . see Hock et al, 1988.
164 It is now clear . . . see Belsky, 2002.
164 One that has been is indiscriminate friendliness . . . see Wolf, 1999, and Quinton and Rutter, 1988.
165 Some studies suggest that it can actually promote faster development . . . see Andersson, 1992.
165 Others that it reduces . . . see Baydar and Brooks-Gunn, 1991.
165 The sort of mother who works . . . see Hewlett, 1993. On pp. 9–10, she states that 'Research by Pittman and Brooks has shown that the hard-edged personality cultivated by many successful professionals – control, decisiveness, aggressiveness, efficiency – can be directly at odds with the passive, patient, selfless elements in good nurturing. The last thing a 3-year-old or a 13-year-old needs at 8 o'clock in the evening is a mother – or father, for that matter – who marches into the house in his or her power suit, barking orders, and looking and sounding like a drill sergeant.
    'Compare the ingredients in a recipe for career success with those of a recipe for meeting the needs of a child, beginning with the all-important ingredient of time. To succeed in one's career may require long hours and consume one's best energy, leaving little time to spend together as a family and precious little enthusiasm for the hard tasks of parenting.

Mobility and a prime commitment to oneself are virtues in the job world; stability, selflessness and a commitment to others are virtues in family life. Qualities needed for career success also include efficiency, a controlling attitude, an orientation towards the future and an inclination towards perfectionism, while their virtual opposites – tolerance for mess and disorder, an ability to let go, an appreciation of the moment and an acceptance of difference and failure – are what is needed for successful parenting.'

165 The same is true . . . Hewlett, 1998.

166 There is one study . . . see p. 150, Ainsworth et al, 1978.

166 A surprisingly small . . . Halpern, 2004.

166 Only 11 per cent have one who works . . . see Table 4.8, p. 121, Hakim, 2000.

166 The quality of at least half these set-ups . . . see Moss and Melhuish, 1991.

170 Indeed, one study . . . see Egeland and Hiester, 1995.

170 Another study . . . see Cohn et al, 1991.

171 Some sixty studies show that . . . see p. 584, De Wolff and Van Ijzendoorn, 1997.

171 One study measured responsiveness . . . see Beckwith et al, 1999.

171 In a sample of people followed from infancy to twenty . . . see Waters et al, 1995.

172 In another very stable sample . . . see Hamilton, 1994.

172 For example, in firstborns . . . see Teti et al, 1996.

172 In adulthood, a love affair collapsing or a bereavement . . . see Dozier et al, 1999.

172 In one study of insecure adult patients . . . see Fonagy et al, 1995.

172 Three-quarters of eighteen-month-olds . . . see p. 293, Solomon and George, 1999.

172 The same proportion of adults do not change in any five-year period . . . see pp. 55–9, Feeney and Noller, 1996.

172 The vicissitudes of children of depressed mothers, half of whom . . . see Belsky and Cassidy, 2000.

172 If the mother ceases to be depressed . . . see pp. 256–7, Belsky, 1999.

172 For example, 60–70 per cent of Romanian orphans . . . for a study that found 60 per cent to be insecure, see Chisholm, 1998; for 70 per cent see Marcovitch et al, 1997.

172 The best evidence . . . it comes from a study based in Minneapolis headed by Alan Sroufe. For a very readable, unacademic summary of this groundbreaking study, see pp. 177–90, Karen, 1998. For more scientific reports, see Sroufe et al, 1990; Sroufe et al, 1999; Egeland et al, 1993.

173 In general, part-time mothers . . . see Owen et al, 1988.

173 Part-time work seems to suit about half of all mothers . . . see Hakim, 2000.

173 Where mothers find themselves . . . see pp. 85–90, Mirowsky and Ross, 1989.

176 Scripting the Avoidant pattern . . . For an account of the links between

338

rejecting parental care behaviour and avoidance, see pp. 38–41, James, 1995. A revealing study of Avoidant women, following them over thirty-one years, is Klohnen and Bera, 1998. For summaries of the Avoidant adult, see Feeney, 1999, and Dozier et al, 1999.

178   The Clinger pattern ... For a review of the childhood of the Clinger, see Cassidy and Berlin, 1994. For accounts of the adult Clinger see the chapters by Feeney and by Dozier et al in Cassidy and Shaver, 1999.

181   The Wobbler pattern ... For the Wobbler childhood see Van Ijzendoorn, 1999. See also pp. 36–42 in Fonagy, 2001.

182   2.3 million ... Office of Applied Statistics, 2004.

182   The Secure pattern ... For the secure childhood see Marvin and Britner, 1999. See also Dozier et al, 1999.

184   For example, there are telling differences ... see Elicker et al, 1992.

184   Avoidants and Clingers being especially likely to team up ... see Kirkpatrick and Davis, 1994; Belsky and Cassidy, 1994.

184   That the insecure choose each other partly explains why ... see pp. 368–74, Feeney, 1999.

185   Bowlby espoused a 'branching' view ... see Sroufe et al, 1999 for evidence regarding staying the same and change.

185   Thirty mothers who had been abused ... see Egeland et al, 1996; Egeland et al, 1988.

186   For example, ten-year-olds who show emotional or behavioural disturbance ... see Rutter et al, 1995.

186   In the same way, when antisocial children ... see Rutter et al, 1995.

186   In three cases out of four ... see Van Ijzendoorn, 1995.

187   If a parent's attachment pattern is measured before the birth ... see Fonagy and Target, 1997.

187   The degree of mother–child similarity was compared ... see Sagi et al, 1994.

187   An even more convincing proof ... see van den Boom, 1994.

188   Studies of the attachment patterns of identical twins ... see Ricciuti, 1993; O'Connor et al, 2001; Bokhorst, 2003.

188   In fact, one quarter of children ... see Van Ijzendoorn and De Wolff, 1997.

189   A survey of these studies back in 1987 ... see Goldsmith and Alansky, 1987.

189   More recent surveys ... for the most definitive recent one, see Vaughan and Bost, 1999.

189   Twenty-six per cent of toddlers ... the statistics in this paragraph come from Belsky and Cassidy, 2000.

189   Indeed, for great swathes ... the definitive book on this subject begins memorably with the words, 'The history of childhood is a nightmare from which we are only just awakening': see De Mause, L., 1971, The History of Childhood, London: Souvenir Press.

190   Hunter-gatherers ... see Sahlins, 1974.

190   Under these conditions there is every reason to suppose ... see Whiting and Edwards, 1988; see also Levine et al, 1988.

190 **Humans began to settle** . . . see Durkheim, 1933.
191 **The interests of children** . . . for a thorough explication of the theory in this and the next paragraph, see Belsky et al, 1991.
193 **Studies of mothers** . . . see Bugenthal et al, 1989.

Chapter 5: Scripting Our Sense of Self in Our First Six Months

200 **If 80 per cent of criminals** . . . see Singleton et al, 1998.
200 **13 per cent of the general population do so** . . . see pp. 15–38, de Girolamo and Reich, 1994.
201 **The psychoanalyst Donald Winnicott** . . . a summary of his ideas written for parents is in Winnicott, 1957. Harder to read but rewarding nonetheless are Winnicott, 1971 and 1972. For an interesting biography of the man see Kahr, 1996.
201 **If our mother was an empathic carer** . . . see Brazelton and Als, 1979.
202 **Cross-modal matching** . . . see p. 51, Stern, 1985.
202 **The result is a chronic loss** . . . see Weil, 1992.
203 **Widely touted as being of largely genetic origin** . . . for a critique see Joseph, 2002; see also DeGrandpre, 1999.
203 **Even during pregnancy** . . . this study was carried out at Imperial College, London. See www.tommy.campaign.org/website.html, reported in *Family Today*, 1, 1, London: NFPI.
203 **Raised cortisol levels** . . . O'Connor et al, 2003, 2005; Van Der Bergh et al, 2004.
203 **Intrusive and overstimulating care** . . . see Jacobvitz and Sroufe, 1987; Carlson et al, 1995.
203 **Another study compared** . . . see Roy et al, 2000.
203 **Subsequent progress of the illness** . . . see August et al, 1997.
204 **The deprived infant's lack of pleasure** . . . see Weil, 1992.
204 **A core stratagem** . . . a good general account of what Disordered people are like is to be found on pp. 301–6, Cloninger et al, 1997. A more technical account of the various psychic defences described in this section is in Kernberg, 1967.
205 **I am perfect** . . . this formulation was made in Kohut, 1971.
206 **A study comparing borderline women** . . . see Hurlbert et al, 1992.
207 **Borderlines in particular** . . . see Buss and Shackelford, 1997; Wiederman and Hurd, 1999.
210 **The adoption of multiple personalities is easy** . . . see Gediman, 1985.
214 **Psychopaths are impulsive** . . . for a review see Lykken, 1995.
214 **For, although most Personality Disordered people** . . . see Verycken et al, 2002.
217 **The sort of person Anthony Storr describes** . . . see Storr, 1970.
217 **All of us sometimes** . . . see p. 309, Cloninger et al, 1997.
218 **It is interesting that** . . . see p. 309, Clonginger et al, 1997.
219 **There are 39 well-documented cases** . . . Zingg, 1940.
219 **Systematic study of the impact of infant deprivation** . . . see Spitz, 1946; Bowlby, 1951; Bowlby, 1978.

220 In one study, children were examined . . . see Rutter et al, 1990; Quinton and Rutter, 1988.

220 John Ogawa and his American colleagues . . . see Ogawa et al, 1997.

220 A distinctive sign of Personality Disorder, dissociation . . . for an account of the symptoms of dissociation see ch. 16, Putnam, 1995.

220 An accompanying report from the same study . . . see Carlson, 1998; for a discussion of the association between early care and later dissociation see Liotti, 1992.

221 This is suggested by another study . . . see Grossman et al, 2002.

221 Whether sensitive or insensitive . . . see Perry et al, 1995; Schore, 1997; see also Schore, 2000.

222 No genetic influence was found in one study of borderline personality . . . see Torgersen, 1984.

222 By contrast, strong evidence . . . for a review of these studies see Zanarini et al, 2000.

222 In one large study, 84 per cent . . . see Zanarini et al, 2000.

222 Poor relating could be heritable from both sides . . . see Plomin, 1994.

222 As noted in Chapter 4 . . . see Vaughan and Bost, 1999.

223 Between 10 and 15 per cent develop a full-blown major depression . . . see p. 31, Nicolson, 1998; Zeman, 1997.

223 For the remaining women . . . see p. 55, Nicolson, 1998; Patterson, 1980.

223 In a recent survey . . . see 'Mother and Baby Sleep Survey', 2002.

223 Whilst all women's hormonal levels . . . see Nicolson, 1998

223 Whereas in pre-industrial revolution societies . . . see pp. 122–5, Hess, 1995.

224 In the definitive study . . . see pp. 151–2, Brown and Harris, 1978.

224 Women who breastfeed . . . see Nicolson, 1998; Alder and Cox, 1983; Alder and Bancroft, 1986.

224 Likewise, women who lack strong intimate relationships . . . see Brown and Moran, 1997; see also pp. 145–50 in Bifulco and Moran, 1998.

224 So are those whose own mother died . . . see Brown and Harris, 1978; see also Harris et al, 1987.

224 And, if she is still alive . . . see Brown and Moran, 1994.

224 Having three or more under fourteen . . . see Brown and Harris, 1978.

224 The increased risk of having more than one child . . . see Thorpe et al, 1991.

224 In particular, during the first . . . see Wessel et al, 1954; Barr et al, 1992; Gormally and Barr, 1997.

225 If ever there would seem to be . . . see Gormally and Barr, 1997.

225 On the contrary, the fact that . . . see Barr et al, 1991; St James-Roberts et al, 1994.

225 Nor has the problem been linked to . . . for four references showing this, see p. 602, Rautava et al, 1993.

225 What is more, it is becoming increasingly clear . . . see Gormally and Barr, 1997; St James-Roberts et al, 1998.

226 Picking the baby up . . . see Hunziger and Barr, 1986.

226 Close contact with mothers . . . Anisfield, E. et al, 1990.

226 What is more, babies whose mothers feed on demand . . . see Bell and Ainsworth, 1972; Barr and Elias, 1988; Hubbard and Van Ijzendoorn,

1991; I know of only one study which shows good results from regulation rather than demand-led care in the first eight weeks: Pinilla and Birch, 1993.

226 **Carrying will not necessarily make any difference** ... see Barr et al, 1991.

227 **Three-year-olds who had infant colic** ... for references linking colic to hyperactivity see p. 419, Papousek and von Hofacker, 1998; for links with temper tantrums and sleep disorders see Rautava et al, 1995.

227 **Fully 50 per cent of mothers with extremely colicky babies** ... see Papousek and von Hofacker, 1998.

227 **1200 mothers were interviewed** ... see Rautava et al, 1993.

227 **When asked during pregnancy** ... see Carey, 1968; Raiha et al, 1995; Keller et al, 1998; p. 416, Papousek and von Hofacker, 1998.

228 **The reason has little or nothing to do with genes** ... see Piontelli, 2002.

228 **But the crucial finding is that** ... see Vaughan and Bost, 1999.

228 **Nearly always, a difficult baby will be made less so** ... see Fish et al, 1991.

228 **When mothers who have a 'difficult' infant** ... see Taubman, 1984; Taubman, 1988; Hunziger and Barr, 1986; Van Den Boom, 1994.

228 **Numerous studies of infants born to parents** ... for a review of the impact of parental psychopathology on attachment behaviour see Van Ijzendoorn et al, 1992.

For impact of depression, see pp. 704–5, Dawson et al, 2000.

For impact of abuse, see Morton and Browne, 1998.

For impact of parental alcoholism, see O'Connor et al, 1987; O'Connor et al, 1992.

For impact of parental cocaine use, see Rodning et al, 1989; Rodning et al, 1991.

228 **Easy babies of such parents** ... see Van Ijzendoorn et al, 1999.

228 **It is estimated that at any one time** ... this study was done at Glasgow University by McKeganaey, N., see www.gla.ac.uk/centres/drugmisuse.

229 **In the case of infants of depressed mothers** ... see Dawson et al, 2000.

229 **Thus, it has been shown that if a newborn** ... see Murray, 1993.

229 **Autistic children whose parents are helped to care for them** ... see Folkmar, 1998.

230 **Three-quarters of babies** ... see Leach, 1999.

230 **The infant age group are four times** ... see *Criminal Statistics*, 2000.

230 **When asked in adulthood** ... see Herman et al, 1989; Zanarini et al, 1989; Ogata et al, 1990; Brown and Anderson, 1991; Kirby et al, 1993; Mullen et al, 1996; Lewis et al, 1997; Johnson et al, 1997; Johnson et al, 2000.

230 **People who suffered childhood maltreatment** ... this was found in interviews with 639 young adults and their parents, in Johnson, 1997.

231 **Children abused before the age of six** ... see p. 587 in Putnam, 1995; Kirby et al, 1993; McLellan et al, 1996.

231 **Similarly, the earlier and more severe the abuse suffered** ... see Manly, 2001; Keiley et al, 2001; Famularo et al, 1994; Bolger et al, 1998; Keiley and Martin, in preparation.

231 **The life of the film director Woody Allen** ... see Meade, 2000.

234 **His biographer** ... see Meade, 2000.

238 The number of people suffering from Personality Disorder ... see de Girolamo and Reich, 1994.

238 Fully four times more convicted psychopaths ... see Cooke and Michie, 1999.

238 Individualist American culture, largely through ... see Lasch, 1979.

238 Japan is one of the very few ... see chapter 16, Miyake et al, 1988; Doi, 1973.

238 Average American infant, half of whom are given over ... see Hakim, 2000.

238 Lower rates of Personality Disorder ... see pp. 286–9, Derksen, 1995; Loranger et al, 1997. This international survey shows very low rates of psychopathy in collectivist societies such as Japan and Switzerland.

239 In the collectivist system ... for an account of individualism-collectivism see pp. 321–4, James, 1997.

240 But for the large numbers of children ... for a lively account of the impact of the social changes described in this paragraph see pp. 291–305, Derkson, 1995.

240 The strain of making choices ... see Schwartz, 2000; Iyengar and Lepper, 2000.

240 This has been proven ... see Twenge, 2000.

240 Just as depression has increased ... see Klerman, 1992.

240 One study, which followed ... see Shedler and Block, 1990.

241 Because of its enormous gulf ... see Kennedy et al, 1998.

241 Slightly less than 2 per cent ... see Amnesty International, 1998.

241 Nearly half of American infants ... see Hakim, 2000.

242 Many of the traits that accompany Disorder ... see Keating and Heltman, 1994; Fehr et al, 1992; DeVries and Miller, 1987; Levicki, 2001; for a particularly interesting study of the role of narcissism in high achievement, see Wallis and Baumeister, in press.

242 Being a chameleon ... see Kilduff and Day, 1994.

243 Certainly, there is good evidence ... see Greenwood et al, 1996.

243 Parents, whether responsive and empathic ... see Holden and Miller, 1999.

Chapter 6: Be Your Own Scriptwriter

253 Archer's tendency to lie about his past ... see Crick, 1995.

255 Studies of depressed mothers ... see Brown et al, 1986; see also pp. 59–63, James, 1997.

255 They may reveal far too much about themselves ... see Ruble et al, 1987.

256 As a result, they have trouble making new friends ... see p. 93, Ruble et al, 1991.

256 Another impermeable group ... see Myers, 2000.

256 Detailed studies of differences between the generations ... see pp. 82–3, James, 1997.

256 Sincerity conflated with the more honest truth-seeking ... see Trilling, 1972.

257   Subsequently, she wrote a book . . . see Flett, 1998.
260   Another example is the TV presenter . . . see Robinson, 2001.
261   Studies of people . . . see Main, 1993.
262   This kind of thinking . . . see Beitchman et al, 1992.
262   On 21 April 1896 . . . for a full transcript of the paper Freud gave, see pp. 251–82, Masson, 1984; for Masson's account of the responses to it, see pp. 3–13 of this book.
267   To be specific . . . see Roth and Fonagy, 1996.
267   The good news about Personality Disorder . . . see p. 309, Cloninger et al, 1997.
267   Criminality decreases substantially . . . see pp. 15–21, Tarling, 1993.
267   There is good evidence . . . for psychoanalysis, see Chiesa and Fonagy, 2000; for CAT, see Ryle and Marlowe, 1995.
267   At a basic level, there are two kinds of depression . . . see Blatt and Holmann, 1992.
269   According to Freud . . . see p. 144, Freud, 1959.
270   Jonathan Swift did . . . see p. 123, Storr, 1972.
270   Lionel Trilling . . . see Trilling, 1972.
271   As Anthony Storr points out . . . see p. 18, Storr, 1972.
271   In Winnicott's account of play . . . see Winnicott, 1972.

Conclusion

283   The first goes like this . . . for the most cogent account of this view see Scarr, 1992.
283   The second view . . . a readable account is found in Plomin, 1990.
288   At the simplest level . . . for a summary of a recent analysis, see pp. 13–17, Feinstein, 2001; for the full report, see CEP Discussion paper No 443.
288   Allowing for their social class . . . see MacKay, 1999.
290   This has now been confirmed . . . see Andersen et al, 1996; Hinkley and Nadersen, 1996; Berk and Andersen, in press.
291   But little attention is paid . . . details of the evidence for this section can be found on pp. 172–6, James, 1997.
297   One third of the prison population . . . see Singleton et al, 1998.
297   When TV advertising was introduced . . . see Hennigan et al, 1982.
299   For another, a study . . . see Rutter et al, 1999.
300   Adult manic depressives . . . see Post et al, 2001.
301   Britons of West Indian heritage . . . see Hutchinson et al, 1996.
304   In an article in 2000 . . . see Murray, 2000.
305   In 1997, the results of an official survey . . . see Singleton et al, 1998.
306   For example, when a group of children . . . Roy et al, 2000.
306   In general, permanently institutionalized . . . see Quinton and Rutter, 1988.
307   As Carol Bellamy . . . see Young Minds Magazine, 2001.
307   What is more . . . see Bugenthal et al, 1989.
308   People from low-income families . . . Lupien et al, 2001.
308   It has also recently been proven . . . see Matthews et al, 2000.
311   A recent World Health Organization . . . WHO world mental health survey consortium, 2004.

BIBLIOGRAPHY

Ainsworth, M.D.S. et al, 1978, *Patterns of Attachment*, New Jersey: LEA.

Albert, R.S., 1983, 'Family positions and the attainment of eminence', in Albert, R.S., *Genius and Eminence*, Oxford: Pergamon.

Alder, E. and Bancroft, J. 1986, 'The relationship between breast feeding persistence, sexuality and mood in postpartum women', *Psychological Medicine*, 18, 389–96.

Alder, E. and Cox, J.L., 1983, 'Breast feeding and postnatal depression', *J. of Psychosomatic Research*, 272, 139–44.

Allison, P.D. and Furstenberg, F., 1989, 'How marital dissolution affects children: variations by age and sex', *Developmental Psychology*, 25, 540–9.

Alpern, L. and Lyons-Ruth, K., 1993, 'Preschool children at social risk: chronicity and timing of maternal depressive symptoms and child behaviour at school and at home', *Development and Psychopathology*, 5, 371–87.

Amnesty International, 1998, *United States of America: Rights for All*, London: Amnesty International.

Anisfield, E. et al, 1990, 'Does infant carrying promote attachment? An experimental study of the effects of increased physical contact on the development of attachment', *Child Development*, 61, 1617–27.

Anderson, S.M. and Miranda, R., 2000, 'Transference', *The Psychologist*, 13, no 12, pp 608–9.

Andersen, S.M. et al, 1996, 'Responding to significant others when they are not there', in Sorrentino, R.M. and Higgins, E.T., *Handbook of Motivation and Cognition*, 3, 262–321.

Andersson, B.-E., 1992, 'Effects of day-care on cognitive and socioemotional competence of thirteen-year-old Swedish schoolchildren', *Child Development*, 63, 20–36.

Apperson, L.B., and McAdoo, W.G., 1968, 'Parental factors in the childhoods of homosexuals', *J. Abnorm. Psych.*, 73, 201–6.

Appleyard, K. et al, 2005, 'When more is not better: the role of cumulative risk in child behaviour outcomes', *J of Child Psychology and Psychiatry*, 46, 235–45.

Arieti, S. and Bemporad, M.D., 1980, 'The psychological organization of depression', *American J. of Psychiatry*, 137, 1360–5.

Asarnow, J.R. and Goldstein, M.J., 1986, 'Schizophrenia during adolescence and early adulthood: a developmental perspective on risk research', *Clinical Psychology Review*, 6, 211–35.

August, G.J. et al, 1997, 'Hyperactive and aggressive pathways: effects of demographic, family and child characteristics on children's adaptive functioning', *J. of Clinical Child Psychology*, 25, 341–51.

Bach-Y-Rita, G. and Veno, A., 1974, 'Habitual violence: a profile of 62 men', *American J. of Psychiatry*, 131, 1015–20.

Bagley, C., 1986, 'Sexual abuse in childhood', *J. of Social Work and Human Sexuality*, 44, 33–47.

Bahrich, H. et al, 1996, 'Accuracy and distortion in memory for high school grades', *Psychological Science*, 7, 265–71.

Bailey, J.M. et al, 2000, 'Genetic and environmental influences on sexual orientation and its correlates in an Australian twin sample', *J. of Personality and Social Psychology*, 78, 524–36.

Baker, R. and Oram, E., 2000, *Baby Wars: Parenthood and Family Strife*, New York: Diane.

Balen, M., 1994, *Kenneth Clarke*, London: Fourth Estate.

Barber, N., 1998, 'Sex differences in disposition towards kin, security of adult attachment and sociosexuality as a function of parental divorce', *Evolution and Human Behaviour*, 19, 125–32.

Barr, R.G. et al, 1991, 'Carrying as colic therapy', *Pediatrics*, 87, 623–30.

Barr, R.G. et al, 1991, 'Crying in !Kung San infants', *Developmental Medicine and Child Neurology*, 33, 601–10.

Barr, R.G. et al, 1992, 'The crying of infants with colic: a controlled empirical description', *Pediatrics*, 90, 14–21.

Barr, R.G. and Elias, M.F., 1988, 'Nursing interval and maternal responsivity', *Pediatrics*, 81, 529–36.

Bates, J.E. et al, 1994, 'Child care history and kindergarten adjustment', *Developmental Psychology*, 30, 690–700.

Bateson, G., *Steps to an Ecology of Mind*, 1973, Frogmore, Herts, UK: Paladin.

Baumrind, D., 1966, 'Effects of authoritative parental control on child behaviour', *Child Development*, 37, 887–907.

Baumrind, D., 1967, 'Childcare practices preceding three patterns of preschool behaviour', *Genetic Psychology Monographs*, 4 (1, pt 2).

Baumrind, D., 1971, 'Current patterns of parental authority', *Developmental Psychology Monographs*, 4 (pt 2).

Baumrind, D., 1991, 'The influence of parenting style on adolescent competence and substance abuse', *J. of Early Adolescence*, 11, 56–95.

Baydar, N. and Brooks-Gunn, J., 1991, 'Effects of maternal employment and childcare arrangements on preschoolers' cognitive and behavioural outcomes', *Developmental Psychology*, 27, 932–45.

Beckwith, L. et al, 1999, 'Maternal sensitivity during infancy and subsequent life events relate to attachment representations at early adulthood', *Developmental Psychology*, 35, 693–700.

Beitchman, J.H. et al, 1992, 'A review of the long-term effects of sexual abuse', *Child Abuse and Neglect*, 16, 101–18.

Bell, S.M. and Ainsworth, M.D.A., 1972, 'Infant crying and maternal responsiveness', *Child Development*, 43, 1171–90.

Bell, A.P., Weinberg, M.S., and Hammersmith, S.K., 1981, *Sexual Preference: Its Development in Men and Women*, Bloomington: Indiana University Press.

Bellamy, C., 2001, 'World leaders fail to grasp importance of a child's first years', *Young Minds Magazine*, 50, 9.

Belmaker, R. et al, 1974, 'A follow-up of monozygotic twins discordant for schizophrenia', *Archives of General Psychiatry*, 30, 219–22.

Belsky, J. 1986, 'Infant day care: a cause for concern', *Zero to Three*, September, 1–7.

Belsky, J., 1988, 'The "effects" of day care reconsidered', *Early Childhood Quarterly*, 3, 335–73.

Belsky, J., 1993, 'Etiology of child maltreatment', *Psychological Bulletin*, 114, 413–34.

Belsky, J., 1999, 'Interactional and contextual determinants of attachment security', in Cassidy, J. and Shaver, P.R., *Handbook of Attachment*, New York: Guilford.

Belsky, J., 2002, 'Developmental risks (still) associated with early child care', *J. of Child Psychology and Psychiatry*, 42, 845–60.

Belsky, J., in press, 'Quality of nonmaternal care and boys' problem behaviour/adjustment at 3 and 5', *Child Development*.

Belsky, J. and Cassidy, J. 1994, 'Attachment and close relationships: an individual-difference perspective', *Psychological Inquiry*, 5, 27–30.

Belsky, J. and Cassidy, J., 2000, 'Attachment: theory and evidence', in Rutter, M. et al, *Developmental Principles and Clinical Issues in Psychology and Psychiatry*, Oxford: Blackwell.

Belsky, J. and Eggebean, D., 1991, 'Early and extensive maternal employment and young children's socioemotional development', *J. of Marriage and the Family*, 53, 1083–1110.

Belsky, J. and Steinberg, L.D., 1978, 'The effects of day care: a critical review', *Child Development*, 49, 929–49.

Belsky, J. et al, 1991, 'Childhood experience, interpersonal development and reproductive strategy: an evolutionary theory of socialization', *Child Development*, 62, 647–70.

Bene, E., 1965, 'On the genesis of male homosexuality', *Brit. J. Psychiat.*, 111, 803–13.

Bereczkei, T. and Csanaky, A., 1996, 'Evolutionary pathways of child development', *Human Development*, 7, 257–80.

Bentall, R., 2003, *Madness Explained*, London: Penguin.

Berk, M.S. and Anderson, S.M., 2001, 'The impact of past relationships on interpersonal behaviour', *J. of Personality and Social Psychology*, in press.

Bhugra, D. et al, 1997, 'Incidence and outcome of schizophrenia in Whites, African-Caribbeans and Asians in London', *Psychological Medicine*, 27, 791–8.

Bieber, I. et al, 1962, *Homosexuality: a Psychoanalytic Study of Male Homosexuality*, New York: Basic Books.

Bifulco, A. and Moran, P., 1998, *Wednesday's Child – Research into Women's Experience of Neglect and Abuse in Childhood, and Adult Depression*, London: Routledge.

Blatt, S.J. 1995, 'The destructiveness of perfectionism', *American Psychologist*, 50, 1003–20.

Blatt, S.J. and Holman, E., 1992, 'Parent-child interaction in the etiology of dependent and self-critical depression', *Clinical Psychology Review*, 12, 47–91.

Bokhorst, C.L. et al, 2003, 'The importance of shared environment in mother–infant attachment security: a behaviour-genetic study', *Child Development*, 74, 1769–82.

Bolger, K.E. et al, 1998, 'Peer relationships and self-esteem among children who have been maltreated', *Child Development*, 69, 1171–97.

Bowlby, J., 1951, *Maternal Care and Mental Health*, Geneva: WHO.

Bowlby, J., *Attachment and Loss*, vols I, II and III, London: Penguin.

Boydell, J. et al, 2001, 'Incidence of schizophrenia in ethnic minorities in London', *British Medical Journal*, 323, 1–4.

Braatan, L.J., and Darling, C.D., 1965, 'Overt and covert homosexual problems among male college students', *Genetic Psychology Monographs*, 71, 269–310.

Bradford, S., 1996, *Elizabeth – A Biography of Her Majesty the Queen*, London: Heinemann.

Braithwaite, J. and Braithwaite, V., 1980, 'The effect of income inequality and social democracy on homicide', *British J. of Criminology*, 20.

Brazelton, T.B. and Als, H., 1979, 'Four early stages in the development of mother-infant interaction', *Psychoanalytic Study of the Child*, 34, 349–69.

Bremner, J.D. et al, 1999, 'Neural correlates of memories of childhood sexual abuse in women with and without postraumatic stress disorder', *American J. of Psychiatry*, 156, 1787–95.

Bremner, J.D. et al, 2000, 'Hippocampal volume reduction in major depression', *American J. of Psychiatry*, 157, 115–17.

Bremner, J.D. and Vermetten, E., 2001, 'Stress and development: behavioural and biological consequences', *Development and Psychopathology*, 13, 473–89.

Bread For Life, 1998, *Pressure to Be Perfect*, London: Bread for Life (Flour Advisory Bureau).

Brown, D.G., 1963, 'Homosexuality and family dynamics', *Bull. Mennin. Clin.*, 27, 227–32.

Brown, F., 1968, 'Bereavement and lack of a parent in childhood', in Miller, E., *Foundations of Child Psychiatry*, Oxford: Pergamon Press.

Brown, G.R. and Anderson, M.D., 1991, 'Psychiatric morbidity in adult inpatients with childhood histories of sexual and physical abuse', *American J. of Psychiatry*, 148, 55–61.

Brown, G.W., and Harris, T., 1978, *Social Origins of Depression*, London: Tavistock.

Brown, G.W. and Moran, P., 1994, 'Clinical and psychosocial origins of chronic depressive episodes I: a community survey', *British J. of Psychiatry*, 165, 447–56.

Brown, G.W. and Moran, P., 1997, 'Single mothers, poverty and depression', *Psychological Medicine*, 27, 21–33.

Brown, G.W. et al, 1972, 'Influence of family life on the course of schizophrenic disorders: a replication', *British J. of Psychiatry*, 121, 241–58.

Brown, G.W. et al, 1986, 'Social support, self-esteem and depression', *Psychological Medicine*, 16, 813–31.

Bryson, E., 1996, 'Brief report: epidemiology of autism', *J. of Autism and Developmental Disorder*, 26, 165–7.

Bugenthal, D.B. et al, 1989, 'Perceived control over caregiving outcomes: implications for child abuse', *Developmental Psychology*, 25, 432–9.

Buhrich, N and McConaghy, N., 'Parental relationships during childhood in homosexuality, transvesticism, transexuality', *Aust. New Zeal. J. Psychiat.*, 12, 1103–8.

348

Burvill, P.W., 1995, 'Recent progress in the epidemiology of major depression', *Epidemiologic Reviews*, 17, 1.

Bushman, J., et al 2001, 'Do people aggress to improve their mood? Catharsis beliefs, affect regulation opportunity and aggressive responding', *J. of Personality and Social Psychology*, 81, 17–32.

Bushnell, J.A. et al, 1992, 'Long-term effects of intrafamilial sexual abuse', *Acta Psychiatrica Scandinavia*, 18, 136–42.

Buss, D.M. and Shackelford, T.K., 1997, 'Relation of sex and gender role to love, sexual attitudes and self esteem', *J. of Research in Personality*, 31, 193–221.

Cado, S. and Leitenberg, H., 1990, 'Guilt reactions to sexual fantasies during intercourse', *Archives of Sexual Behaviour*, 19, 49–63.

Campbell, S.B. et al, 1995, 'Depression in first-time mothers: mother-infant interaction and depression chronicity', *Developmental Psychology*, 21, 349–57.

Capron, C. and Duyme, M., 1989, 'Assessment of socio-economic status on IQ in a full cross-fostering study', *Nature*, 340, 552–4.

Carey, W.B., 1968, 'Maternal anxiety and infantile colic: is there a relationship?', *Clinical Pediatrics*, 7, 590–5.

Carlson, D.E. et al, 1995, 'A developmental investigation of inattentiveness and hyperactivity', *Child Development*, 66, 37–54.

Carlson, E.A., 1998, 'A prospective longitudinal study of attachment disorganization/disorientation', *Child Development*, 69, 1007–28.

Carver, C.S. and Baird, E., 1998, 'The American dream revisited: is it what you want or why you want it that matters?', *Psychological Science*, 9, 289–92.

Caspi, A. et al, 1996, 'Behavioural observations at age 3 years predict adult psychiatric disorders', *Archives of General Psychiatry*, 53, 1033–9.

Caspi, A. et al, 2002, 'Role of genotype in the cycle of violence in maltreated children', *Science*, 297, 851–4.

Caspi, A. et al, 2003, 'Influence of life stress on depression: moderation by a polymorphism in the 5-HTT gene', *Science*, 301, 386-9.

Caspi, A. et al, 2005, 'Moderation of the effect of the adolescent-onset cannabis-use by functional polymorphism in COMT gene', *Biological Psychiatry*, 57, 1117–27.

Castle, J., et al, 1999, 'Effects of qualities of early institutional care on cognitive attainment', *American J. of Orthopsychiatry*, 69, 424–37.

Cassidy, J. and Berlin, L.J., 1994, 'The insecure/ambivalent pattern of attachment: theory and research', *Child Development*, 65, 971–91.

Cassidy, J. and Shaver, P.R., 1999, *Handbook of Attachment*, New York: Guilford.

CEP Discussion paper No. 443, 'The relative economic importance of academic, psychological and behavioural attributes developed in childhood'.

Chatoor, L. et al, 1998, 'Attachment and feeding problems', *J. of the American Academy of Child and Adolescent Psychiatry*, 37, 1217–74.

Chen, S. and Andersen, S.M. 1999, 'Relationships from the past in the present', in Zanna, M.P., *Advances in Experimental Psychology*, 31, 123–90.

Chiesa, M. and Fonagy, P., 2000, 'Cassel Personality Disorder study', *British J. of Psychiatry*, 174, 485–91.

Chisholm, K., 1998, 'A three-year follow-up of attachment and indiscriminate

friendliness in children adopted from Romanian orphanages', *Child Development*, 69, 1092–1106.

Chisholm, K. et al, 1995, 'Attachment security and indiscriminately friendly behaviour in children adopted from Romanian orphanages', *Development and Psychopathology*, 7, 283–94.

Cicchetti, D. and Rogosch, F.A., 2001 (a), 'Diverse patterns of neuroendocrine activity in maltreated children', *Development and Psychopathology*, 13, 677–93.

Cicchetti, D. and Rogosch, F.A., 2001 (b), 'The impact of child maltreatment on neuroendocrine functioning', *Development and Psychopathology*, 13, 783–804.

Cloninger, C.R. et al, 1982, 'Predisposition to petty criminality in Swedish adoptees', *Archives of General Psychiatry*, 39, 1242–7.

Cloninger, C.R. et al, 1997, 'Personality Disorders', in Guze, S.B., *Adult Psychiatry*, New York: Mosby.

Coghill, S.R. et al, 1986, 'Impact of maternal postnatal depression on cognitive development of young children', *British Medical Journal*, 292, 1165–7.

Cohn, J. et al, 1991, 'Infant response in the still face paradigm at 6 months predicts avoidant and secure attachment at 12 months', *Developmental Psychology*, 3, 367–76.

Cohn, J.F. et al, 1986, 'Face-to-face interactions of depressed mothers and their infants', in Tronick, E.Z., and Field, T., *Maternal Depression and Infant Disturbance*, San Francisco: Jossey-Bass.

Cohn, J.F. et al, 1989, 'Specificity of infants' responses to mothers' affective behaviour', *J. of the American Academy of Child and Adolescent Psychiatry*, 28, 242–8.

Cole, P.M. et al, 1992, 'Emotion displays in two-year-olds during mishaps', *Child Development*, 63, 314–24.

Collins, N.L. and Read, J., 1990, 'Adult attachment, working models, and relationship quality in dating couples', *J. of Personality and Social Psychology*, 58, 644–63.

Collins, W.A. et al, 2000, 'Contemporary research on parenting: the case for nature and nurture', *American Psychologist*, 55, 218–32.

Condry, J.C. and Condry, S., 1976, 'Sex differences', *Child Development*, 47, 812–19.

Cooke, D.J. and Michie, C., 1999, 'Psychopathy across cultures: North America and Scotland compared', *J. of Abnormal Psychology*, 108, 58–68.

Cortois, C.A., 1979, 'The incest experience and its aftermath', *Victimology*, 4, 337–47.

Cox, C.J. and Jennings, R., 1988, *High Flyers*, Oxford: Basil Blackwell.

Crick, M., 1995, *Jeffrey Archer – Stranger than Fiction*, London: Penguin.

Criminal Statistics, 2000, London: Office of National Statistics.

Currie, E., 1985, *Confronting Crime*, New York: Pantheon.

Cutting, J., 1992, 'The role of right hemisphere dysfunction in psychiatric disorders', *British J. of Psychiatry*, 160, 583–8.

Daniels, D. et al, 1985, 'Environmental differences within the family and adjustment differences within pairs of adolescent siblings', *Child Development*, 56, 764–74.

Dardis, T., 1989, *The Thirsty Muse*, London: Abacus.

Dawson, G. et al, 1997, 'Infants of depressed mothers exhibit atypical brain activity: a replication of extension of previous findings', *J. of Child Psychology and Psychiatry*, 38, 176–86.

Dawson, G. et al, 1999, 'Frontal brain electrical activity in infants of depressed and nondepressed mothers: relation to variations in infant behaviour', *Development and Psychopathology*, 11, 589–605.

Development and Psychopathology, 11, 589–605.

Dawson, G. et al, 2000, 'The role of early experience in shaping behavioural and brain development and its implications for social policy', *Development and Psychopathology*, 12, 695–712.

Dawson, G. et al (in press), 'Infants of depressed mothers exhibit reduced left frontal brain activity during interactions with mother and a familiar nondepressed adult', *Child Development*.

De Bellis, M.D., 2001, 'Developmental traumatology: the psychobiological development of maltreated children and its implications for research, treatment and policy', *Development and Psychopathology*, 13, 539–64.

de Girolamo, G. and Reich, J.H., 1994, *Personality Disorders*, Geneva: World Health Organization.

DeGrandpre, R., 1999, *Ritalin Nation*, New York: Norton.

De Mause, L., 1974, *The History of Childhood*, London: Souvenir Press.

DeMeis, D. et al, 1986, 'The balance of employment and motherhood', *Developmental Psychology*, 122, 627–32.

Derksen, J., 1995, *Personality Disorders: Clinical and Social Perspectives*, Chichester: John Wiley.

DeVries, K. and Miller, D., 1987, *Unstable at the Top*, New York: Signet.

De Wolff, M.S. and Van Ijzendoorn, M.H., 1997, 'Sensitivity and attachment: a meta-analysis on parental antecedents of infant attachment', *Child Development*, 68, 571–91.

Detting, A. et al, 2002, 'Repeated parental deprivation in the infant common marmoset', *Biological Psychiatry*, 52, 1037–46.

DiLalla, D.L. and Gottesman, I.I. 1995, 'Normal personality characteristics in identical twins discordant for schizophrenia', *J. of Abnormal Psychology*, 104, 490–9.

Doi, T., 1973, *The Anatomy of Dependency*, New York: Kodansha International.

Dozier, M. et al, 1999, 'Attachment and psychopathology in adulthood', in Cassidy, J. and Shaver, PR, *Handbook of Attachment*, New York: Guildford.

Duggal, S. et al, 2001, 'Depressive symptomatology in childhood and adolescence', *Development and Psychopathology*, 13, 165.

Dumaret, A.-C. et al, 1997, 'Adult outcome of children reared for long-term periods in foster families, *Child Abuse and Neglect*, 21, 911–27.

Dunn, J. and Plomin, R., 1990, *Separate Lives: Why Siblings are so Different*, New York: Basic Books.

Durkheim, E., 1933, *The Division of Labour in Society*, New York: Free Press.

Egeland, B. et al, 1988, 'Breaking the cycle of abuse', *Child Development*, 59, 1080–8.

Egeland, B. et al, 1993, 'Resilience as process', *Development and Psychopathology*, 5, 517–28.

Egeland, B. et al, 1996, 'Dissociation as a mediator of child abuse across generations', *Child Abuse and Neglect*, 20, 1123–32.

Egeland, B. and Hiester, M., 1995, 'The long-term consequences of infant day-care and mother-infant attachment', *Child Development*, 66, 474–85.

Eisenstadt, M. et al, 1989, *Parental Loss and Achievement*, Madison, Connecticut: International Universities Press. 52.

Elder, G.H., 1974, *Children of the Great Depression*, Chicago: Chicago UP.

Eley, T.C. et al, 2004, 'Gene-environment interaction analysis of serotonin system markers with adolescent depression', *Molecular Psychiatry*, 9, 1–8.

Elicker, J. et al, 1992, 'Predicting peer competence and peer relationships in childhood from early parent-child relationships', in Parke, R. and Ladd, G., *Family-Peer Relationships*, New Jersey: LEA.

Ellis, B.J. et al, 1999, 'Quality of early family relationships and individual differences in the timing of purbertal maturation in girls', *J. of Personality and Social Psychology*, 77, 387–401.

Ernulf, K.E. et al, 1989, 'Biological explanation, psychological explanation, and tolerance of homosexuals: a cross-national analysis of beliefs and attitudes', *Psychological Reports*, 65, 1003–10.

Etcoff, N., 1999, *The Survival of the Prettiest*, London: Little Brown.

Evans, R.B., 1969, 'Childhood parental relationships of homosexual men', *J. Consulting Clin. Psych.*, 33, 129–35.

Falloon, I. et al, 1988, *Behavioural Family Therapy: A Workbook*, Buckingham: Buckingham Mental Health Services.

Famularo, R. et al, 1994, 'Maternal and child post-traumatic stress disorder in cases of child maltreatment', *Child Abuse and Neglect*, 18, 27–36.

Farrow, M., 1997, *What Falls Away*, New York: Doubleday.

Feeney, J. and Noller, P., 1996, *Adult Attachment*, London: Sage.

Feeney, J.A., 1999, 'Adult romantic attachment and couple relationships', in Cassidy, J. and Shaver, P.R., *Handbook of Attachment*, New York: Guilford.

Fehr, B. et al, 1992, 'The construct of Machiavellianism: twenty years later', in Spielberger, C.D. and Butcher, J.D. *Advances in Personality Assessment*, Vol 9, Hillsdale: Erlbaum.

Feinstein, L., 2001, 'Getting the balance right', *CentrePiece*, Autumn.

Fergusson, D.M. et al, 1996, 'Childhood sexual abuse and psychiatric disorders in young adulthood', *J. of American Academy of Child and Adolescent Psychiatry*, 34, 1365–74.

Feyerabend, P., 1995, *Killing Time: The Autobiography*, Chicago: Univ. of Chicago Press.

Field, T., 1992, 'Infants of depressed mothers', *Development and Psychopathology*, 4, 49–66.

Finkelhor, D., 1980, 'Sex among siblings: a survey on prevalence, variety and effects', *Archives of Sexual Behaviour*, 9, 171–94.

Firth-Cozens, J., 1992, 'The role of early family experiences in the perception of organizational stress', *J. of Occupational Psychology*, 64.

Fish, M. et al, 1991, 'Conditions of continuity and discontinuity in infant negative emotionality: newborn to five months', *Child Development*, 62, 1525–37.

Fisher, S. and Greenberg, R.P., 1985, *The Scientific Credibility of Freud's Theories and Therapy*, New York: Basic Books.

Flett, K., 1998, *The Heart-Shaped Bullet*, London: Heinemann.

Folkmar, F., 1998, *Autism and Pervasive Developmental Disorders*, Cambridge: Cambridge University Press.

Fonagy, P. and Target, M., 1997, 'Attachment and reflective function: their role in self-organization', *Development and Psychopathology*, 9, 679–700.

Fonagy, P, 2001, *Attachment*, London: Routledge.

Fonagy, P. et al, 1995, 'Attachment, the reflective self and borderline states', in Goldberg, S. et al, *Attachment Theory*, New Jersey: Analytic Press.

Forehand, R. et al, 1975, 'Mother-child interactions: comparison of a non-compliant clinic group and a nonclinic group', *Behaviour Research and Therapy*, 13, 79–84.

Fossati, A. et al, 1999, 'Borderline personality disorder and childhood sexual abuse: a meta-analytic study', *J. of Personality Disorder*, 13, 268–80.

Fox, N., in press, 'Stress and the serotonin transporter polymorphism interact to predict early childhood inhibited temperament', *Psychological Science*.

Fraser, S., 1995, *The Bell Curve Wars*, New York: Basic Books.

Freud, S., 1930, *Civilization and Its Discontents*, Standard Edition of the Complete Psychological Works of Sigmund Freud, Vol XXI, London: Hogarth Press.

Freud, S., 1959, 'Creative writers and day-dreaming', Standard Edition of Sigmund Freud's Scientific Works, IX, London: Hogarth Press.

Frey, K.S., 1987, 'Coping responses of parents of disabled children', unpublished data.

Friedrich, W.N. et al, 1991, 'Normative sexual behaviour in children', *Pediatrics*, 88, 456–64.

Frost, R.O. et al, 1991, 'The development of perfectionism', *Cognitive Therapy and Research*, 15, 469–89.

Frum, D., 2003, *The Right Man: The Surprise Presidency of George W. Bush*, New York: Random House.

Gabbard, G.O., 1994, 'Mind and brain in psychiatric treatment', *Bulletin of the Menninger Clinic*, 58, 427–46.

Galinsky, E., 1994, *Zero to Three*, April/May, p.33.

Gediman, H.K., 1985, 'Imposture, inauthenticity and feeling fraudulent', *J. of the American Psychoanalytic Association*, 33, 911–36.

Gerbner, S., 2004, *Why Love Matters – How Affection Shapes a Baby's Brain*, East Sussex: Brunner-Routledge.

Gibb, B. E. et al, 2001, 'Emotional, physical and sexual maltreatment in childhood versus adolescence and personality dysfunction in young adulthood', 15, 505–11.

Gielgud, J., 1973, *Distinguished Company*, New York: Doubleday.

Gilbert, D.G., 1995, *Smoking – Individual Differences, Psychopathology and Emotion*, London: Taylor and Francis.

Gilbert, P. et al, 1996, 'Parental representations, shame, interpersonal problems and vulnerability to psychopatholoy', *Clinical Psychology and Psychotherapy*, 3, 23–34.

Gilbert, P. and Andrews, B., 1998, *Shame: Interpersonal Behaviour, Psychopathology and Culture*, Oxford: OUP.

Gilbert, P. and Gerlsma, C., 1999, 'Recall of shame and favouritism in relation to psychopathology', *British J. of Clinical Psychology*, 38, 357–73.

Gillespie, N.A. et al, 2005, 'The relationship between stressful life events, the serotonin transporter (5-HTTLPR) genotype, and major depression', *Psychological Medicine*, 35, 101–11.

Goldsmith, H.H. and Alansky, J.A., 1987, 'Maternal and infant temperamental predictors of attachment: a meta-analytic review', *J. of Consulting and Clinical Psychology*, 55, 805–16.

Goldstein, M.J., 1985, 'Family factors that antedate the onset of schizophrenia and related disorders: the results of a fifteen-year prospective longitudinal study', *Acta Psychiatrica Scandinavia*, 319, 71, 7–18.

Goldstein, M.J., 1990, 'Family relations as risk factors for the onset and course of schizophrenia', in Rolf, J. et al, *Risk and Protective Factors in the Development of Psychopathology*, Cambridge: Cambridge University Press.

Goodman, S.H. and Gotlib, I.H., 1999, 'Risk for psychopathology in the children of depressed mothers: a developmental model for understanding mechanisms of transmission', *Psychological Review*, 106, 458–90.

Gormally, S. and Barr, R.G., 1997, 'Of clinical pies and clinical clues: proposal for a clinical approach to complaints of early crying and colic', *Ambulatory Child Health*, 3, 137–53.

Gotlib, I.H. et al, 1988, 'Depression and perceptions of early parenting: a longitudinal study', *British J. of Psychiatry*, 152, 24–7.

Gottesman, I.I., 1992, *Schizophrenia Genesis – the Origins of Madness*, New York: Freeman.

Gottlieb, G., 1991, 'Experiential canalization of behavioural development', *Development Psychology*, 27, 4–13.

Gould, S.J., 1991, *The Mismeasure of Man*, New York: W.W. Norton.

Grabe, H.J. et al, 2005, 'Mental and physical distress is modulated by a polymorphism of the 5-HT transporter gene interacting with social stressors and chronic disease burden', *Molecular Psychiatry*, 10, 220–4.

Graber, J.A. et al, 1995, 'The antecedents of menarchal age: heredity, family environment and stressful life events', *Child Development*, 66, 346–59.

Graber, J.A. et al, 1997, 'Is psychopathology associated with the timing of pubertal development?', *J. of the American Academy of Child and Adolescent Psychiatry*, 36, 1768–76.

Graham, Y.P. et al, 1999, 'The effects of neonatal stress on brain development: implications for psychopathology', *Development and Psychopathology*, 11, 545–65.

Green, R., 1987, *The Sissy Boy Syndrome*, New Haven: Yale University Press.

Greenberg, M.T. et al, 1993, 'The role of attachment in the early development of disruptive behaviour problems', *Development and Psychopathology*, 5, 191–213.

Greenberger, E. and Goldberg, W.A., 1989, 'Work, parenting and the socialization of children,' *Developmental Psychology*, 25, 22–35.

Greenough, W.T., 1976, 'Enduring brain effects of differential experience and training', in Rosenzweig, M.R., *Neural Mechanisms of Learning and Memory*, Cambridge, MA: MIT Press.

Greenough, W.T. et al, 1987, 'Experience and brain development', *Child Development*, 58, 539–59.

Greenwood, P. et al, 1996, *Diverting Children from a Life of Crime: Measuring Costs and Benefits*, Santa Monica: RAND.

Grossman, K.E. et al, 2002, 'Attachment relationships and appraisals of partnership', in Pulkinnen, L., and Caspi, A., *Paths to Successful Development*, Cambridge: Cambridge University Press.

Gunnar, M.R. et al, 2001, 'Salivary cortisol levels in children adopted from Romanian orphanages', *Development and Psychopathology*, 13, 611–28.

Hakim, C., 2000, *Work–Lifestyle Choices in the 21st Century*, Oxford: OUP.

Halpern, D.F., 2004, *Public Policy, Work and Families*, Washington: American Psychological Association.

Halverson, C.F. and Waldrop, M.F., 1970, 'Maternal behaviour toward own and other preschool children', *Child Development*, 41, 839–45.

Hamilton, C.E., 1994, *Continuity and Discontinuity of Attachment from Infancy Through Adolescence*, PhD dissertation, University of California, Los Angeles.

Hammen, C. et al, 1990, 'Longitudinal study of diagnoses in children of mothers with unipolar and bipolar affective disorder', *Developmental Psychology*, 26, 24–30.

Harris, J.R. 1995, 'Where is the child's environment? A group socialization theory of environment', *Psychological Review*, 102, 485–9.

Harris, T.O. et al, 1987, 'Loss of parent in childhood and adult psychiatric disorder: the role of social class position and premarital pregnancy', *Psychological Medicine*, 17, 163–83.

Hennigan, K.M. et al, 1982, 'Impact of the introduction of television on crime in the United States: empirical findings and theoretical implication', *J. of Personality and Social Psychology*, 42, 461–77.

Herman, J.L. et al, 1989, 'Childhood trauma in Borderline personality disorder', *American J. of Psychiatry*, 146, 490–5.

Herrnstein, R.J. and Murray, C., 1994, *The Bell Curve*, New York: Free Press.

Hershberger, S.L., 1997, 'A twin registry study of male and female sexual orientation', *J. of Sex Research*, 34, 212–22.

Hess, L.E., 1995, 'Changing family patterns in Western Europe', in Rutter, M. and Smith, D.J., *Psychosocial Disorders in Young People*, Chichester: Wiley.

Hewlett, S.A., 1993, *Child Neglect in Rich Nations*, UNICEF.

Hewlett, S.A. et al, 1998, *The War Against Parents*, New York: Houghton Mifflin.

Himmelstein, S. et al, 1991, 'An attributional analysis of maternal beliefs about the importance of child-rearing practices', *Child Development*, 62, 301–20.

Hinckley, J. and Hinckley, J.A., 1985, *Breaking Points*, Michigan: Chosen Books.

Hinkley, K. and Nadersen, S.M., 1996, 'Activating transference without consciousness', *J. of Personality and Social Psychology*, 71, 1279–95.

Hock, E. et al, 1988, 'Maternal separation anxiety – its role in the balance of employment and motherhood in mothers of infants', in Gottfried, A.E. and

Gottfried, A.W., *Maternal Employment and Children's Development*, London: Plenum Press.

Hodges J. and Tizard B., 1989, 'IQ and behavioural adjustments of ex-institutional adolescents', *J. of Child Psychology and Psychiatry*, 30, 53–75.

Holden, G.W. and Miller, P.C., 1999, 'Enduring and different: a meta-analysis of of the similarity in parents' child-rearing', *Psychological Bulletin*, 125, 223–54.

Horn, J.M. et al, 1975, 'Bias against genetic hypotheses in adoption studies', *Archives of General Psychiatry*, 32, 1365–7.

Horrobin, D., 2001, *The Madness of Adam and Eve*, London: Bantam.

Howe, M.J.A., 1999, *Genius Explained*, Cambridge: CUP.

Hoyenga, K.B. et al, 1993, *Gender-Related Differences: Origins and Outcomes*, London: Allyn.

Hubbard, F.O.A. and Van Ijzendoorn, M.H., 1991, 'Maternal unresponsiveness and infant crying across the first 9 months', *Infant Behaviour and Development*, 14, 299–312.

Humphrey, L.L., 1989, 'Observed family interactions among subtypes of eating disorders using structural analysis of social behaviour', *J. of Consulting and Clinical Psychology*, 57, 206–14.

Hunziger, U.A. and Barr, R.G., 1986, 'Increased carrying reduces infant crying: a randomized controlled trial', *Pediatrics*, 77, 641–8.

Hurlbert, D.F. et al, 1992, 'An examination into the sexuality of women with borderline personality disorder', *J. of Sex and Marital Therapy*, 18, 231–42.

Hutchinson, G. et al, 1996, 'Morbid risk for schizophrenia in the parents and siblings of African-Caribbean and White probands', *British J. of Psychiatry*, 169, 776–80.

Hyun Rhee, S. et al, 2002, 'Genetic and environmental influences on antisocial behaviour: a meta-analysis of twin and adoption studies', *Psychological Bulletin*, 128, 490-523.

Iyengar, S.S., and Lepper, M.R., 2000, 'When choice is demotivating: can one desire too much of a good thing?', *Personality Processes and Individual Differences*, 79, 995–1006.

Jablensky, A. et al, 1992, *Schizophrenia: Manifestations, Incidence and Course in Different Cultures. A World Health Organization Ten-Country Study*, Cambridge: CUP.

Jackson, L. 1991, *LaToyah: Growing up in the Jackson Family*, New York: Dutton.

Jacobs, B. et al, 1993, 'A quantitative dendritic analysis of Wernicke's area: gender, hemispheric and environmental factors', *J. of Comparative Neurology*, 387, 167–78.

Jacobvitz, D. and Sroufe, A.L., 1987, 'The early caregiver-child relationship and attention-deficit disorder with hyperactivity in kindergarten: a prospective study', *Child Development*, 58, 488–95.

James, O.W., 1995, *Juvenile Violence in a Winner-Loser Culture*, London: Free Association Books.

James, O.W., 1997, *Britain on the Couch – Why We're Unhappier Compared With 1950, Despite Being Richer*, London: Arrow.

James, O.W., 2007a, *Affluenza – How to be Successful and Stay Sane*, London: Vermilion.

James, O.W., 2007b, *Selfish Capitalist Origins of Mental Illness*, London: Vermilion.

Jamison, K.R., 1993, *Touched by Fire*, New York: Free Press.

Jannsen, J.M.A.M. and Gerris, J.R.M., 1992, *Child Rearing*, Amsterdam: Swets and Zeitlinger.

Jansen, I. et al, 2004, 'Childhood abuse as a risk factor for psychotic experience', *Acta Psychiatrica Scandinavica*, 109, 38–45.

Jarvelin, I.I. et al, 1999, 'Can excellent school performance be a precursor of schizophrenia?', *Acta Psychiatrica Scandinavia*, 100, 17–26.

Jedlicka, D., 1980, 'A test of the psychoanalytic theory of mate selection', *J. of Social Psychology*, 112, 295–9.

Jennings, R. et al, 1994, *Business Elites*, New York: Routledge.

Johnson, E.O. et al, 1992, 'Mechanisms of stress: a dynamic overview of hormonal and behavioural homeostasis', *Neuroscience and Biochemical Review*, 16, 115–30.

Johnson, J.G. et al, 1997, 'Childhood maltreatment increases risk for personality disorders during early adulthood', *Archives of General Psychiatry*, 56, 600–6.

Johnson, J.G. et al, 2000, 'Association between four types of childhood neglect and personality disordered symptoms during adolescence and early adulthood; findings of a community based confidential study', *J. of Personality Disorders*, 14, 177–87.

Johnson, J.S. and Newport, E.L., 1989, 'Critical period effects in second language learning', *Cognitive Psychology*, 21, 60–99.

Johnson, M.H., 1999, 'Cortical plasticity in normal and abnormal cognitive development', *Development and Psychopathology*, 11, 419–37.

Johnstone, L., 1992, 'Family management in schizophrenia: its assumptions and contradictions', *J. of Mental Health*, 2, 255–69.

Jonas, C.H., 1944, 'An objective approach to the personality and environment in homosexuality', *Psychiatric Quarterly*, 184, 626–41.

Joseph, J., 2002, 'Not in their genes: a critique of the genetics of attention-deficit hyperactivity disorder', *Developmental Review*, 20, 539–67.

Jost, J.T. et al., 2003, 'Political conservatism as motivated cognition', *Psychological Bulletin*, 129, 339–75.

Kahr, B., 1996, *D.W. Winnicott*, London: Karnac.

Kalter, N. and Rembar, J., 1981, 'The significance of a child's age at the time of parental divorce', *American J. of Orthopsychiatry*, 51, 85–100.

Karen, R., 1998, *Becoming Attached*, Oxford: OUP.

Kasser, T. and Ryan, R.M., 1993, 'A dark side of the American dream: correlates of financial success as a central life aspiration', *J. of Personality and Social Psychology*, 65, 410–22.

Kasser, T. and Ryan, R.M., 1996, 'Further examining the American dream: differential correlates of intrinsic and extrinsic goals', *Personality and Social Psychology Bulletin*, 22, 280–7.

Kaufman, J. and Charney, D., 2001, 'Effects of early stress on brain structure and function: implications for understanding the relationship between child maltreatment and depression', *Development and Psychopathology*, 13, 451–71.

Kaufman, J. et al, 2004, 'Social support and serotonin transporter gene moderate depression in maltreated children', *Proceedings of the National Academy of Sciences USA*, 101, 17316–21.

Keating, C.F. and Heltman, K.R., 1994, 'Dominance and deception in children and adults: are leaders the best misleaders', *Personality and Social Psychology Bulletin*, 20, 3, 312–21.

Keiley, M.K. et al, 2001, 'The timing of child physical maltreatment: a cross-domain growth analysis of impact on adolescent externalizing and internalizing problems', *Development and Psychopathology*, 13, 891–912.

Keiley, M.K. and Martin, N.C., in preparation, 'Child abuse, neglect and juvenile delinquency: how new statistical approaches can inform our understanding of old questions – a reanalysis of Widom, 1989'.

Keller, H. et al, 1998, 'Relationships between infant crying, birth complications and maternal variables', *Child: Care, Health and Development*, 24, 377–94.

Kendler, K.S. et al, 1993, 'The prediction of major depression in women', *American J. of Psychiatry*, 150, 1139–48.

Kendler, K.S. et al, 2005, 'The interaction of stressful life events and serotonin polymorphism in the prediction of episodes of major depression', *Archives of General Psychiatry*, 62, 529–35.

Kennedy, B.P. et al, 1998, 'Income distribution, socioeconomic status and self rated health in the United States: a multilevel analysis', *British Medical J.*, 317, 917–21.

Kernberg, O., 1967, 'Borderline personality organization', *J. of the American Psychoanalytic Association*, 15, 641–85.

Kierman, G.L., 1992, 'The changing rate of depression', *J. of the American Medical Association*, 268, 3098–3105.

Kilduff, M. and Day, D.V., 1994, 'Do chameleons get ahead? The effects of self-monitoring on managerial careers', *Academy of Management J.*, 37, 1047–60.

Kirby, J.S. et al, 1993, 'Correlates of dissociative symptomatology in patients with physical and sexual abuse histories', *Comprehensive Psychiatry*, 34, 258–63.

Kirkpatrick, L.A. and Davis, K.E. 1994, 'Attachment style, gender and relationship stability: a longitudinal analysis', *J. of Personality and Social Psychology*, 66, 502–12.

Klimes-Dougan, B. and Bolger, A.K., 1998, 'Coping with maternal depressed affect and depression: adolescent children of depressed and well mothers', *J. of Youth and Adolescence*, 27, 1–15.

Kleinman, A. et al, 1997, 'Psychiatry's global challenge: An evolving crisis in the developing world for a better understanding of the links between culture and mental disorders', *Scientific American*, 276(3), 86–9.

Klohnen, E.C. and Bera S., 1998, 'Behavioural and experiential patterns of avoidantly and securely attached women across adulthood: a 31-year longitudinal perspective', *J. of Personality and Social Psychology*, 74, 211–23.

Kochanska, G., 1997, 'Multiple pathways to conscience of children with different temperaments: from toddlerhood to age 5', *Developmental Psychology*, 33, 228–40.

Kochanska, G. and Aksan, N., 1995, 'Mother–child mutually positive affect, the quality of child compliance to requests and prohibitions, and maternal control as correlates of early internalization', *Child Development*, 66, 236–54.

Koestner, R. et al, 1991, 'Family origins of adolescent self-criticism and its continuity into adulthood', *J. of Abnormal Psychology*, 100, 191–7.

Kohut, H., 1971, *The Analysis of the Self*, New York: IUP.

Kolb, B. et al, 1998, 'Age, experience and the changing brain', *Neuroscience and Biobehavioural Review*, 22, 123–59.

Konishi, M., 1995, 'A sensitive period for birdsong learning', in Julesz, B. and Kovacs, I., *Maturational Windows and Adult Cortical Plasticity*, MA: Addison-Wesley.

Koob, G.F. et al, 1993, 'The role of corticotropin-releasing factors in behavioural response to stress', in Chadwick, D.J. et al, *Corticotropin-Releasing Factor, CIBA Foundation Symposium 172*, Chichester: John Wiley.

Kotulak, R., 1996, *Inside the Brain*, Kansas City: Andrews McMeel.

Kuipers, L. et al, 1992, *Family Work for Schizophrenia: a Practical Guide*, London: Gaskell.

Ladd, C.O. et al, 1996, 'Persistent changes in corticotropin-releasing factor neuronal systems induced by maternal deprivation', *Endocrinology*, 137, 1212–18.

Laing, R.D., 1960, *The Divided Self*, London: Penguin.

Laing, R.D. et al, 1964, *Sanity, Madness and the Family*, London: Penguin.

Laing, R.D., 1971, *The Politics of the Family*, London: Penguin.

Lam, D. and Kuipers, L. 1993, 'Being critical is not a critical review', *Clinical Psychology Forum*, July.

Lasch, C., 1979, *The Culture of Narcissism*, London: Sphere.

Lasky-Su, J.A., 2005, 'Meta-analysis of the association between two polymorphisms in the serotonin transporter gene and affective disorders', *American Journal of Medical Genetics*, 133b, 110–5.

Leach, P., 1999, *Physical Punishment of Children in the Home*, London: National Children's Bureau.

Lee, C.M. and Gotlib, I.H., 1991, 'Adjustment of children of depressed mothers: a ten-month follow-up', *J. of Abnormal Psychology*, 100, 473–7.

Leff, J. and Vaughn, C. 1994, 'Critics of family management in schizophrenia: their assumptions and contradictions', *J. of Mental Health*, 3, 115–16.

Levicki, C., 2001, *Developing Leadership Genius*, New York: McGraw Hill.

Levine, R.A. et al, 1988, *Parental Behaviour in Diverse Societies*, San Francisco: Jossey-Bass.

Lewinsohn, P.M. et al, 2004, 'The prevalence and co-morbidity of subthreshold psychiatric conditions', *Psychological Medicine*, 34, 613–22.

Lewis, D.O. et al, 1997, 'Objective documentation of child abuse and dissociation in 12 murderers with dissociative identity disorder', *American J. of Psychiatry*, 154, 1703–10.

Lewis, R.A., 1963, 'Parents and peers: socialization agents in the coital behaviour of young adults', *J. of Sex Research*, 9, 156–70.

Lewis, R.J. and Janda, L.H., 1988, 'The relationship between adult sexual adjustment and childhood experiences regarding exposure to nudity, sleeping in the parental bed and parental attitudes toward sexuality', *Archives of Sexual Behaviour*, 17, 349–62.

Lidz, T., 1985, 'A psychosocial orientation to schizophrenic disorders', *Yale J. of Biology and Medicine*, 58, 209–17.

Liem, J.H., 1980, 'Family studies of schizophrenia: an update and commentary', *Schizophrenia Bulletin*, 6, 429–55.

Liotti, G., 1992, 'Disorganized/disorientated attachment in the etiology of dissociative disorders', *Dissociation*, 4, 196–204.

Loranger, A.W. et al, 1997, *Assessment and Diagnosis of Personality Disorders – The ICD-10 International Personality Disorder Examination*, Cambridge: CUP.

Lupien, S.J. et al, 2001, 'Can poverty get under your skin? Basal cortisol levels and cognitive function in children from low and high socioeconomic status', *Development and Psychopathology*, 13, 653–76.

Luthar, S.S. et al, 2002, 'Privileged but pressured? A study of affluent youth', *Child Development*, 73, 1593–1610.

Luthar, S.S., 2003, 'The culture of affluence: psychological costs of material wealth', *Child Development*, 74, 1581–93.

Luthar, S.S. et al, 2005, 'Comparable "risks" at the socioeconomic status extremes: preadolescents' perceptions of parenting', *Development and Psychopathology*, 17, 207–30.

Lykken, D.T., 1995, *The Antisocial Personalities*, New Jersey: LEA.

Lykken, D.T. and Tellegen, A., 1993, 'Is human mating adventitious or the result of lawful choice? A twin study of mate selection', *Interpersonal Relations and Group Processes*, 65, 56–68.

Lyons, D. et al, 2002, 'Early environmental regulation of glucocorticoid feedback sensitivity in young adult monkeys', *Journal of Neuroendocrinology*, 12, 723–8.

McCauley, J. et al, 1997, 'Clinical characteristics of women with a history of childhood abuse', *JAMA*, 277, 1362–8.

Maccoby, E.E., 1989, chapter 8 in Garmezy, M. and Rutter, M., *Stress, Coping, and Development in Children*, Baltimore: John Hopkins Press.

McConville, B., 1985, *Sisters: Love and Conflict within the Lifelong Bond*, New York: Pan.

Mcgue, M. and Lykken, D.T. 1992, 'Genetic influence on risk of divorce', *Psychological Science*, 3, 368–73.

McGuffin, P. et al, 1994, *Seminars in Psychiatric Genetics*, London: Gaskell.

McGuire, S. et al, 1995, 'Maternal differential treatment of siblings and children's behavioural problems: a longitudinal study', *Development and Psychopathology*, 7, 515–28.

MacKay, T., 1999, 'Education and the disadvantaged: is there any justice?', *The Psychologist*, 12, 344–9.

Mackenzie, C.D. and Wright, L.S., 1996, *Delayed Posttraumatic Stress Disorders from Infancy*, Netherlands: Harwood.

McKenzie, K. and Murray, R.M., 1988, 'Risk factors for psychosis in the UK African-Caribbean population', in *Ethnicity – An Agenda for Mental Health*, London: Gaskell.

McLellan, M.D. et al, 1996, 'Age of onset of sexual abuse: relationship to sexually inappropriate behaviour', *J. of American Academy of Child and Adolescent Psychiatry*, 34, 1375.

Mcmahon, R.J. and Wells, K.C., 1998, 'Conduct problems', in Mash, E.J. and Barkley, R.A., *Treatment of Childhood Disorders*, New York: Guilford Press.

Main, M., 1993, 'Discourse, prediction and recent studies in attachment: implications for psychoanalysis', *J. of the American Psychoanalytic Association*, 41, 209–44.

Manly, J.T. et al, 2001, 'Dimensions of child maltreatment and children's adjustment: contributions of developmental timing and subtype', *Development and Psychopathology*, 13, 759–82.

Manuck, S.B. et al, 2004, 'Socio-economic status covaries with central nervous system serotonergic responsivity as a function of allelic variation in the sertonin transporter gene-linked polymorphic region', *Psychoneuroendocrinology*, 29, 651–68.

Manuck, S.B. et al, 2005, 'The socio-economic status of communities predicts variation in brain serotonergic responsivity', *Psychological Medicine*, 35, 519–28.

Marcovitch, J., 1997, 'Determinants of behavioural problems in Romanian children adopted in Ontario', *International J. of Behavioural Development*, 20, 17–31.

Marriage and Divorce Statistics, 2000, London: HMSO.

Marshall, J.R., 1993, *Clinical Psychology Forum*, 56, 41–2.

Martorell, R. et al, 1997, 'Reversibility of stunting: epidemiological findings in children from developing countries', *European J. of Clinical Nutrition*, 48, 45–57.

Marvin, R.S. and Britner, P.A., 1999, 'Normative development: the ontogeny of attachment', in Cassidy, J. and Shaver, P.R. *Handbook of Attachment*, New York: Guilford.

Marvin, R.S. and Britner, P.A., 1999, in Cassidy, J. and Shaver, P.R., *Handbook of Attachment*, New York: Guilford.

Masson, J.M., 1984, *Freud: The Assault on Truth*, London: Faber.

Matejcek, Z. et al, 1978, 'Children from unwanted pregnancies', *Acta Psychiatrica Scandinavia*, 57, 67–90.

Matejcek, Z. et al, 1980, 'Follow-up study of children born from unwanted pregnancies', *International J. of Behavioural Development*, 3, 243–51.

Matthews, K.A. et al, 2000, 'Does socioeconomic status relate to central serotoninergic responsivity in healthy adults?', *Psychosomatic Medicine*, 62, 231–7.

Matthijssen, J.J.J.P. et al, 1998, 'The relationship between mutual family relations and child psychopathology', *J. of Child Psychology and Psychiatry*, 39, 477–87.

Meade, M., 2000, *The Unruly Life of Woody Allen*, London: Weidenfeld and Nicolson.

Mednick, S.A. et al, 1984, 'Genetic influences in criminal convictions: evidence from an adoption cohort', *Science*, 224, 891–3.

Meiselman, K.C., 1978, *Incest*, San Francisco: Jossey-Bass.

Mendlewicz, J. et al, 2004, 'Serotonin transporter 5-HTTLPR polymorphism and affective disorders: no evidence of association in a large European multicenter study', *European J of Human Genetics*, 12, 377–82.

Messner, S.F., 1982, 'Income inequality and murder rates', *Comparative Social Research*, 3.

Milburn, M.A., 1996, *The Politics of Denial*, London: MIT Press.

Miller, A., 1980, *For Your Own Good*, London: Virago.

Miller, A., 1991, *Banished Knowledge*, London: Virago.

Miller, P.R., 1958, 'The effeminate passive obligatory homosexual', *Arch. Neurol. Psychiat.*, 80(5), 612–18.

Mirowsky, J. and Ross, C.E., 1989, *Social Causes of Psychological Distress*, New York: Aldine De Gruyter.

Miyake, K. et al, 1988, 'Issues in emotional development', in Stevenson, H. et al, *Child Development and Education in Japan*, New York: Freeman.

Moffit, T. et al, 1992, 'Childhood experience and onset of menarche', *Child Development*, 63, 47–58.

Money, J., 1986, *Lovemaps*, New York: Irvington.

Moore, K.A. et al, 1986, 'Parental attitudes and the occurrence of sexual activity', *J. of Marriage and the Family*, 48, 777–82.

Moorehouse, M.J., 1991, 'Linking maternal employment patterns to mother–child activities and children's school competence', *Developmental Psychology*, 27, 295–303.

Morton, N. and Browne, K.D., 1998, 'Theory and observation of attachment and its relation to child maltreatment: a review', *Child Abuse and Neglect*, 22, 1093–1104.

Mosher, L.R. et al, 1971, 'Families with identical twins discordant for schizophrenia: some relationships between identification, thinking styles, psychopathology and dominance-submissiveness', *British J. of Psychiatry*, 118, 29–42.

Mosher, L.R. et al, 'Drug companies and schizophrenia: unbridled capitalism meets madness' in Read, J. et al, *Models of Madness – Psychological, social and biological approaches to schizophrenia*, Hove: Brunner-Routledge.

Moss, P. and Melhuish, E., 1991, *Current Issues in Day Care for Young Children*, London: HMSO.

'Mother and Baby Sleep Survey', 2002, *Mother and Baby Magazine*, April.

Mueller, M.M. and Dweck, C.S., 1998, 'Praise for intelligence can undermine children's motivation and performance', *J. of Personality and Social Psychology*, 75, 33–52.

Mullen, P.E. et al, 1994, 'The effect of child sexual abuse on social, sexual and interpersonal function in adult life', *British J. of Psychiatry*, 165, 35–47.

Mullen, P.E. et al, 1996, 'The long-term impact of the physical, emotional and sexual abuse of children', *Child Abuse and Neglect*, 20, 7–21.

Munson, J.A. et al, 2001, 'Structure and variability in the developmental trajectory of children's externalizing problems: impact of infant attachment, maternal depressive symptomatoloty and child sex', *Development and Psychopathology*, 13, 277–96.

Muntaglio, B., 1999, *First Son: George W. Bush and the Bush Family Dynasty*, New York: Three Rivers Press.

Murray, C., 2000, 'Genetics of the right', *Prospect*, April.

Murray, L., 1993, 'The role of infant irritability in postnatal depression in a Cambridge (UK) community population', in Brazelton, T.B. and Lester, B.M., *The Cultural Context of Infancy, Vol. 3*, New Jersey: Ablex.

Murray, L. et al, 1993, 'Depressed mothers' speech to their infants and its relation to infant gender and cognitive development', *J. of Child Psychology and Psychiatry*, 31, 1083–1101.

Myers, LB, 2000, 'Deceiving others or deceiving themselves', *The Psychologist*, 13, 400–3.

Nash, M.R. et al, 1993, 'Characteristics of sexual abuse associated with greater psychological impairment among children', *Child Abuse and Neglect*, 17, 401–8.

NICHD Early Child Care Research Network, 1998, 'Early child care and self-control, compliance and problem behaviour at twenty-four and thirty-six months', *Child Development*, 69, 1145–70.

NICHD, 1999 (a), 'Child care and the mother-child interaction in the first 3 years of life', *Developmental Psychology*, 35, 1399–1413.

NICHD, 1999 (b), 'Chronicity of maternal depressive symptoms, maternal sensitivity and child functioning at 36 months', *Developmental Psychology*, 35, 1297–1310.

Nicolson, P., 1998, *Post-Natal Depression: Psychology, Science and the Transition to Motherhood*, London: Routledge.

Nishith, P. et al, 2000, 'Prior interpersonal trauma: the contribution to current PTSD symptoms in female rape victims', *J. of Abnormal Psychology*, 109, 20–5.

Norman, P., 2002, *Sir Elton*, London: Pan.

NSPCC, 2001, *Child Maltreatment in the United Kingdom: a Study of the Prevalence of Child Abuse and Neglect*, London: NSPCC.

O'Connor, M.J. et al, 1987, 'Disorganization of attachment in relation to maternal alcohol consumption', *J. of Consulting and Clinical Psychology*, 55, 831–6.

O'Connor, M.J. et al, 1992, 'Attachment behaviour of infants exposed prenatally to alcohol: mediating effects of infant affect and mother-infant interactions', *Development and Psychopathology*, 4, 243–56.

O'Connor, P.J., 1964, 'Aetiological factors in homosexuality as seen in Royal Airforce psychiatric practice', *Brit. J. Psychiat.*, 110, 381–91.

O'Connor, T. et al, 1999, 'Attachment disturbances and disorders in children exposed to early severe deprivation', *Infant Mental Health J.* 20, 10–29.

O'Connor, T. et al, 2000, 'The effects of global severe privation on cognitive competence', *Child Development*, 71, 376–90.

O'Connor, T. et al, 2001, 'A twin study of attachment in preschool children', *Child Development*, 72, 1501–11.

O'Connor, T. G. et al, 2003, 'Maternal anxiety and behavioural/emotional problems in children: A test of a programming hypothesis', *Journal of Child Psychology and Psychiatry*, 44, 1025–36.

O'Connor, T.G. et al, 2003b, 'Child-parent attachment following early institutional deprivation', *Development and Psychopathology*, 15, 19–38.

O'Connor, T.G. et al, 2005, 'Prenatal anxiety predicts individual differences in cortisol in pre-adolescent children', *Biological Psychiatry*, 58, 211–217.

Office of Applied Statistics, 2004, Department of Health and Human Services, Substance Abuse and Mental Health Services Administration, Research Triangle Park, North Carolina: Research Triangle Institute.

Ogawa, J. et al, 1997, 'Development and the fragmented self: longitudinal study of dissociative symptomatology in a nonclinical sample', *Development and Psychopathology*, 9, 855–79.

Ogata, S.N. et al, 1990, 'Childhood sexual and physical abuse in adult patients with borderline personality disorder', *American J. of Psychiatry*, 147, 1008–13.

Papousek, M. and von Hofacker, N., 1998, 'Persistent crying in early infancy: a non-trivial condition of risk for the developing mother–infant relationship', *Child: Care, Health and Development*, 24, 395–424.

Parsons, J.E. et al, 1982, 'Socialization of achievement attitudes and beliefs: parental influences', *Child Development*, 53, 310–21.

Pastore, N., 1949, *The Nature/Nurture Controversy*, New York: Columbia UP.

Patterson, G.R., 1980, 'Mothers the unacknowledged victims', *Monographs of the Society for Research in Child Development*, 85 (3).

Patterson, G.R., 1982, *Coercive Family Processes*, Oregon: Castalia.

Patterson, G.R., 1990, *Antisocial Boys*, Oregon: Castalia.

Pedersen, C.B. et al, 2001, 'Evidence of a dose-response relationship between urbanicity during upbringing and schizophrenia risk', *Archives of General Psychiatry*, 58, 1039–46.

Perry, B.D. et al, 1995, 'Childhood trauma, the neurobiology of adaptation and "use dependent" development of the brain: how "states" become "traits"', *Infant Mental Health J.*, 16, 271–91.

Peters, D.K. et al, 1995, 'Childhood sexual abuse and current suidality in college women and men', *Child Abuse and Neglect*, 19, 335–41.

Pinilla, T. and Birch, L.L., 1993, 'Help me make it through the night: behavioural entrainment of breast-fed infants' sleep patterns', *Pediatrics*, 91, 436–44.

Pinker, S., 1997, *How the Mind Works*, London: Penguin.

Piontelli, A., 2002, *Twins – from Fetus to Child*, London: Routledge.

Plomin, R., 1990, *Nature and Nurture – an Introduction to Behavioural Genetics*, Pacific Grove: Brooks/Cole.

Plomin, R., 1994, *Genetics and Experience – The Interplay Between Nature and Nurture*, London: Sage.

Plomin, R. and Daniels, D., 1987, 'Why are children from the same family so different from each other?', *Behaviour and Brain Science*, 10, 1–16.

Plomin, R. et al, 1997, *Behavioural Genetics* (3rd edition), New York: Freeman.

Post, R.M. et al, 2001, 'Developmental vulnerabilities to the onset and course of bipolar disorder', *Development and Psychopathology*, 13, 581–98.

Putnam, F.W., 1995, 'Development of dissociative disorders', in *Development and Psychopathology, Volume 2: Risk Disorder and Adaptation*, Cicchetti, D. et al, London: John Wiley.

Quinton, D. and Rutter, M., 1988, *Parenting Breakdown*, Aldershot: Avebury.

Radke-Yarrow, M. et al, 1985, 'Patterns of attachment in two- and three-year-old children in normal families and families with depressed mothers', *Child Development*, 56, 884–93.

Raiha, H. et al, 1995, 'Family context of infantile colic', *Infant Mental Health J.*, 16, 206–17.

Rautava, P. et al, 1993, 'Psychosocial predisposing factors for infantile colic', *British Medical J.*, 307, 600–4.

Rautava, P. et al, 1995, 'Infantile colic: child and family three years later', *Pediatrics*, 96, 43–7.

Read, J. et al, 2004a, 'Childhood trauma, loss and stress' in Read, J. et al, *Models of Madness – Psychological, Social and Biological Approaches to Schizophrenia*, Hove: Brunner-Routledge.

Read, J. et al, 2004b, 'Poverty, ethnicity and gender' in Read, J. et al, *Models of Madness – Psychological, Social and Biological Approaches to Schizophrenia*, Hove: Brunner-Routledge.

Read, J. et al, 2004c, 'Unhappy families' in Read, J. et al, *Models of Madness – Psychological, Social and Biological Approaches to Schizophrenia*, Hove: Brunner-Routledge.

Read, J. et al, 2005, 'Childhood trauma, psychosis and schizophrenia: a literature review with theoretical and clinical implications', *Acta Psychiatrica Scandinavica*, 112, 330–50.

Read, J. et al, 2006, 'Prejudice and schizophrenia: a review of the "mental illness is an illness like any other" approach', *Acta Psychiatrica Scandinavica*, 1–16.

Reiner, R., 1997, *Rethinking the Brain: New Insights into Early Development*, New York: Families and Work Institute.

Ricciuti, A.E., 1993, 'Child–mother attachment: a twin study', *Dissertation Abstracts International*, 54, 3364.

Richardson, J.L. et al, 1989, 'Substance use among eighth-grade students who take care of themselves after school', *Pediatrics*, 84, 556–66.

Robertson, G., 1972, 'Parent–child relationships and homosexuality', *Brit. J. Psychiat.*, 121, 525–8.

Robertson, J. and Robertson, J. 1989, *Separation and the Very Young*, London: Free Association.

Robins, L.N. et al, 1992, *Psychiatric Disorders in America*, New York: Free Press.

Robinson, A., 2001, *Memoirs of an Unfit Mother*, London: Little, Brown.

Rodgers, B. and Pryor, J., 1998, *Divorce and Separation: the Outcome for Children*, York: Joseph Rowntree.

Rodning, C. et al, 1989, 'Characteristics of attachment organization and play organization in prenatally drug-exposed toddlers', *Development and Psychopathology*, 1, 277–89.

Rodning, C. et al, 1991, 'Quality of attachment and home environments in children prenatally exposed to PCP and cocaine', *Development and Psychopathology*, 3, 351–66.

Roe, A., 1952, *The Making of a Scientist*, New York: Dodd, Mead.

Ross, C.A. et al, 2004, 'Antipsychotic medication: myths and facts' in Read, J. et al, *Models of Madness – Psychological, Social and Biological Approaches to Schizophrenia*, Hove: Brunner-Routledge.

Roth, A. and Fonagy, P., 1996, *What Works for Whom – a Critical Review of Psychotherapy Research*, London: Guildford.

Rowan, A.B. et al, 1994, 'Posttraumatic stress disorder in a clinical sample of adults sexually abused as children', *Child Abuse and Neglect*, 18, 51–61.

Roy, P. et al, 2000, 'Institutional care: risk from family background or pattern of rearing?', *J. of Child Psychology and Psychiatry*, 41, 139–49.

Ruble, D.N. et al, 1987, 'Social comparison and self-evaluation in the classroom: developmental changes in knowledge and function', in Masters, J.C., *Social Comparison, Social Justice and Relative Deprivation*, New Jersey: LEA.

Ruble, D.N. et al, 1991, 'Changing patterns of comparative behaviour as skills are acquired: a functional model of self-evaluation', in Suls, J. et al, *Social Comparison: Contemporary Theory and Research*, New Jersey: LEA.

Rutter, M. and Smith, D., 1995, *Psychosocial Disorders in Young People*, London: John Wiley.

Rutter, M. et al, 1988, 'Developmental catchup and deficit following adoption after severe global early prevention', *J. of Child Psychology and Psychiatry*, 39, 465–76.

Rutter, M. et al, 1990, 'Adult outcomes of institution-reared children: males and females compared', in Robins, L.N. et al, *Straight and Devious Pathways in Development from Childhood to Adulthood*, Cambridge: CUP.

Rutter, M. et al, 1995, 'Understanding individual differences in environmental-risk exposure', in Moen, P. et al, *Examining Lives In Context*, Washington: American Psychological Association.

Rutter, M. et al, 1999, 'Quasi-autistic patterns following severe early global privation', *J. of Child Psychology and Psychiatry*, 40, 537–49.

Ryle, A. and Marlowe, M., 1995, 'CAT of borderline personality disorder: theory and practice and clinical and research uses of the self states SDR', *International J. of Short-Term Psychotherapy*, 10, 21–34.

Sadowski, H. et al, 1999, 'Early life family disadvantage and major depression in adulthood', *British J. of Psychiatry*, 174, 112–20.

Saghir, M.T. and Robins, E., 1973, *Male and Female Homosexuality: A Comprehensive Investigation*, Baltimore: Williams and Wilkins.

Sagi, A. et al, 1994, 'Sleeping out of home in a Kibbutz communal arrangement: it makes a difference for infant–mother attachment', *Child Development*, 65, 992–1004.

Sahlins, M., 1974, *Stone Age Economics*, London: Tavistock.

Scarr, S., 1992, 'Developmental theories for the 1990s: development and individual differences', *Child Development*, 63, 1–19.

Scarr, S., 1997, 'Why childcare has little impact on most children's development', *Current Directions in Psychological Science*, 6, 143–8.

Schiff, M. et al, 1982, 'How much could we boost scholastic achievement and IQ scores? A direct answer from a French adoption study', *Cognition*, 12, 165–96.

Schofield, M., 1965, *Sociological Aspects of Homosexuality*, London: Longman.

Schore, A.N., 1996, 'The experience-dependent maturation of a regulatory system in the orbital prefrontal cortex and the origin of developmental psychopatholoy', *Development and Psychopathology*, 8, 59–87.

Schore, A.N., 1997, 'Early organization of the nonlinear right brain and development of a predisposition to psychiatric disorders', *Development and Psychopathology*, 9, 595–631.

Schore, A.N., 1998, 'Early shame experiences and infant brain development', pp. 57–77, in Gilbert, P. and Andrews, B., 1998, *Shame: Interpersonal Behaviour, Psychopathology and Culture*, Oxford: OUP.

Schore, A.N., 2000, 'Relational trauma of the developing right brain and to the origin of severe disorders of the self', lecture at the Anna Freud Centre, London.

Schwarz, B., 2000, 'Self-determination: the tyranny of freedom', *American Psychologist*, 55, 79–88.

Shedler, J. and Block, J., 1990, 'Adolescent drug use and psychological health', *American Psychologist*, 45, 612–30.

Shenon, P., 1995, in the *New York Times*, 15 July, p.3, 'New Zealand seeks causes of suicides by young'.

Siegelman, M., 1974, 'Parental backgrounds of male homosexuals and heterosexuals', *Arch. Sex. Behav.*, 3, 3–19.

Siegelman, M., 1981, 'Parental backgrounds of homosexual and heterosexual men: a cross-national replication', *Arch. Sex. Behav.*, 10, 505–12.

Sigel, I.E., 1985, *Parental Belief Systems*, London: Erlbaum.

Singleton, N. et al, 1998, *Psychiatric Morbidity Among Prisoners in England and Wales*, Office of National Statistics.

Skuse, D. et al, 1994, 'Postnatal growth and mental development: evidence for a sensitive period', *J. of Child Psychology and Psychiatry*, 35, 521–45.

Sloboda, J.A. et al, 1994, 'Is everyone musical?, *The Psychologist*, August, 349–54.

Snortum, J.R. et al, 1969, 'Family dynamics and homosexuality', *Psychol. Rep.*, 24, 763–70.

Solomon, J. and George, C., 1999, 'The measurement of attachment security in infancy and childhood', in Cassidy, J. and Shaver, P.R., *Handbook of Attachment*, New York: Guildford.

Speltz, M.L. et al, 1999, 'Attachment in boys with early onset conduct problems', *Development and Psychopathology*, 11, 269–85.

Spitz, R., 1946, 'Hospitalism: a follow-up report', *Psychoanalytic Study of the Child*, 2, 113–17.

Srivastava, A. et al, 2001, 'Money and subjective well-being: it's not the money it's the motives', *J. of Personality and Social Psychology*, 80, 959–71.

Sroufe, L.A. et al, 1990, 'The fate of early experience following developmental change: longitudinal approaches to individual adaptation in childhood', *Child Development*, 61, 1363–73.

Sroufe, L.A. et al, 1992, *Child Development: Its Nature and Course*, New York: McGraw-Hill.

Sroufe, L.A. et al, 1999, 'Implications of attachment theory for developmental psychopathology', *Development and Psychopathology*, 11, 1–13.

Stabenau, J.R. et al, 1967, 'Early characteristics of monozygotic twins discordant for schizophrenia', *Archives of General Psychiatry*, 17, 723–34.

Stabenau, J.R., 1973, 'Schizophrenia: a family's projective identification', *American J. of Psychiatry*, 130, 19–23.

Stein, A. et al, 1991, 'The relationship of between postnatal depression and mother-child interaction', *British J. of Psychiatry*, 158, 46–52.

Stein, M.B. et al, 1997, 'Hippocampal volume in women victimized by childhood sexual abuse', *Psychological Medicine*, 27, 951–9.

Steinberg, L. et al, 1989, 'Authoritative parenting, psychosocial maturity and academic success among adolescents', *Child Development*, 60, 1424–36.

Steinberg, L., 1986, 'Latchkey children and susceptibility to peer pressure: an ecological analysis', *Developmental Psychology*, 22, 433–9.

Stephen, W.G., 1973, 'Parental relationships and early social experiences of activist male homosexuals and male heterosexuals', *J. Abnorm. Psychol.*, 82, 506–13.

Stern, D., 1985, *The Interpersonal World of the Infant*, New York: Basic Books.

St James-Roberts, I. et al, 1994, 'Infant crying patterns in Manali and London', *Child: Care, Health and Development*, 20, 1–15.

St James-Roberts, I. et al, 1998, 'Links between maternal care and persistent infant crying in the early months', *Child: Care, Health and Development*, 24, 353–76.

Stone, W.F. et al, 1993, 'Authoritarianism: Left and Right', in Stone, W.F. et al, *Strength and Weakness – The Authoritarian Personality Today*, New York: Springer-Verlag.

Stone, W.F. et al, 1992, *Strength and Weakness: the Authoritarian Personality Today*, London: Springer Verlag.

Storr, A., 1970, *Human Aggression*, London: Penguin.

Storr, A., 1972, *The Dynamics of Creation*, London: Penguin.

Sulloway, F., 1996, *Born to Rebel*, Abacus: London.

Suomi, S.J., 1997, 'Early determinants of behaviour: evidence from primate studies', *British Medical Bulletin*, 53, 170–84.

Surbey, M., 1990, 'Family composition, stress and human menarche', in Bercovitch, F. and Zeigler, T., *The Socioendocrinology of Primate Reproduction*, New York: Liss.

Surtees, P.G. et al, 2006, 'Social adversity, the serotonin transporter (5-HTTLPR) and major depressive disorder', *Biological Psychiatry*, 59, 224–9.

Susser, E. et al, 1994, 'Epidemiology of nonaffective acute remitting psychosis vs schizophrenia: sex and socio-cultural setting', *Archives of General Psychiatry*, 51, 294–301.

Symonds, M., 1969, 'Homosexuality in adolescence', *Pennsyl. Psychiat. Quart.*, 9, 15–24.

Taubman, B., 1984, 'Clinical trial of the treatment of colic by modification of parent–infant interaction', *Pediatrics*, 74, 995-2–1003.

Taubman, B., 1988, 'Parental counselling compared with elimination of cow's milk or soy milk of the treatment of infant colic syndrome: a randomized trial', *Pediatrics*, 81, 756–61.

Tangney, J.P. et al, 1992, 'Proneness to shame, proneness to guilt and psychopathology', *J. of Abnormal Psychology*, 101, 469–78.

Tarling, R., 1993, *Analysing Offending: Data, Models and Interpretation*, London: HMSO.

Taylor, S.E. et al, 1992, 'Optimism, coping, psychological distress and high risk sexual behaviour among men at risk for acquired immunodeficiency disease', *J. of Personality and Social Psychology*, 63, 460–73.

Taylor, S.E. et al, 1994, 'Positive illusions and well being revisited: separating fact from fiction', *Psychological Bulletin*, 116, 21–7.

Teicher, M.H., 2000, 'Wounds that time won't heal: the neurobiology of child abuse', *The Dana Forum on Brain Science*, 2, 4, 50–67.

Teicher, M.H., 2002, 'Scars that won't heal: the neurobiology of child abuse', *Scientific American*, March, 54–61.

Terman, I.M., and Miles, C.C., 1936, *Sex and Personality*, New York: McGraw-Hill.

Terman, L.M. et al, 1983, *Terman Life-Cycle Study of Children with High Ability, 1922–82*, Ann Arbor: Inter-University Consortium for Political and Social Research.

Teti, D.M. et al, 1995, 'Maternal depression and the quality of early attachment: an examination of infants, preschoolers and their mothers', *Developmental Psychology*, 31, 364–76.

Teti, D.M. et al, 1996, 'And baby makes four: predictors of attachment security among preschool-aged firstborns during the transition to siblinghood', *Child Development*, 67, 579–96.

Thapar, A. et al, 1999, 'Genetic basis of attention deficit and hyperactivity', *British J. of Psychiatry*, 174, 111.

Thompson, N.L., et al, 1973, 'Parent–child relationships and sexual identity in male and female homosexuals and heterosexuals', *J. Consulting Clin. Psychol.*, 41(1), 120–7.

Thornton, A. and Camburn, D., 1987, 'The influence of the family on premarital sexual attitudes and behaviour', *Demography*, 24, 323–40.

Thorpe, K. et al, 1991, 'Comparison of prevalence of depression in mothers of twins and mothers of singletons', *British Medical J.* 302, 875–8.

Timimi, S., 2005, *Naughty Boys*, Basingstoke: Palgrave-Macmillan.

Torgersen, S., 1984, 'Genetic and nosological aspects of schizotypal and borderline personality disorders', *Archives of General Psychiatry*, 41, 546–54.

Trilling, L., 1972, *Sincerity and Authenticity*, London: Secker and Warburg.

Triseliotis, J., 1989, 'Foster care outcomes: a review of key research findings', *Adoption and Fostering*, 13, 5–17.

Turner A.M. and Greenough, W.T., 1985, 'Differential rearing effects on rat visual cortex', *Brain Research*, 329, 357–68.

Twenge, J.M., 2000, 'The age of anxiety? Birth cohort changes in anxiety and neuroticism, 1952–93', *J. of Personality and Social Psychology*, 79, 1007–21.

US Department of Justice, Federal Bureau of Investigation, Crime in the United States, 2002.

Van Den Aardweg, G.J.M., 1984, 'Parents of homosexuals – not guilty?', *Amer. J. Psychoth.*, 38(2), 181–9.

Van den Bergh, B.R.H. et al, 2004, 'High Antenatal Maternal Anxiety is Related to ADHD Symptoms, Externalizing Problems, and Anxiety in 8- and 9-Year-Olds', *Child Development*, 75, 1085–97.

Vandell, D.L. and Corasaniti, M.A., 1988, 'The relation between third graders after-school care and social, academic and emotional functioning', *Child Development*, 59, 868–75.

Vandell, D.L. and Ramanan, J., 1991, 'Children of the national longitudinal survey of youth: choices in after-school care and child development', *Developmental Psychology*, 27, 637–43.

Vandell, D.L. et al, 1990, 'Variations in early childcare: do they predict subsequent social, emotional and cognitive differences?', *Early Childhood Research Quarterly*, 5, 555–72.

Van den Boom, D., 1994, 'The influence of temperament and mothering on attachment and exploration: an experimental manipulation of sensitive responsiveness among lower-class mothers with irritable infants', *Child Development*, 65, 1457–77.

Van Ijzendoorn, M.H. et al, 1992, 'The relative effects of maternal and child problems on the quality of attachment in clinical samples', *Child Development*, 63, 840–58.

Van Ijzendoorn, M.H., 1995, 'Adult attachment representations, parental responsiveness, and infant attachment: a meta-analysis on the predictive validity of the adult attachment interview', *Psychological Bulletin*, 117, 387–403.

Van Ijzendoorn, M.H. et al, 1999, 'Disorganized attachment in early childhood: meta-analysis of precursors, concomitants and sequelae', *Development and Psychopathology*, 11, 225–49.

Van Ijzendoorn, M.H. and De Wolff M.S., 1997, 'In search of the absent father – meta-analysis of infant–father attachment', *Child Development*, 68, 604–9.

Vaughan, B.E. and Bost, K.K., 1999, 'Attachment and temperament: redundant, independent or interacting influences on interpersonal adaptation and personality development?', in Cassidy, J. and Shaver, P.R., *Handbook of Attachment*, New York: Guilford.

Venter, C., 2001, *The Observer*, p.1, 11 February, 2001.

Venter, J.C. et al, 2001, 'The sequence of the human genome', *Science*, 291, 1304–51.

Vereycken, J., et al, 2002, 'Authority conflicts and personality disorders', *J. of Personality Disorders*, 16, 41–51.

Verhulst, F.C., 1990, 'Problem behavior in international adoptees: II. Age at placement', *J. of American Academy of Child and Adolescent Psychiatry*, 29, 104–11.

Verhulst, F.C. et al, 1992, 'Damaging backgrounds: later adjustment of international adoptees', *J. of American Academy of Child and Adolescent Psychiatry*, 31, 518–24.

Vohs, K.D. et al, 1999, 'Perfectionism, perceived weight status and self-esteem interact to predict bulimic symptoms', *J. of Abnormal Psychology*, 108, 695–700.

Vorria, P. et al, 1998, 'A comparative study of Greek children in long-term residential group care and in two-parent families: I. Social, emotional and behavioural differences', *J. of Child Psychology and Psychiatry*, 39, 225–36.

Wahlberg, K.-E. et al, 1997, 'Gene–environment interaction in vulnerability to schizophrenia: findings from the Finnish adoptive family study of schizophrenia', *American J. of Psychiatry*, 154, 355–62.

Wakschlag, L.S. and Hans S.L., 1999, 'Relation of maternal responsiveness during infancy to the development of behaviour problems in high-risk youths', *Developmental Psychology*, 35, 569–79.

Walkerdine, V., 1995 unpublished study, Dept Psychology, Goldsmiths College, London University.

Waller, N.G. and Shaver, P.R., 1994, 'The importance of nongenetic influences on romantic lovestyles: a twin-family study', *Psychological Science*, 5, 168–74.

Wallerstein, J.S. and Kelly, J.B., 1979, 'Children and divorce: a review', *Social Work*, 13, 468–75.

Wallis, J. and Baumeister, R., in press, *J. of Personality and Social Psychology*.

Waters, E. et al, 1995, 'From the strange situation to the Adult Attachment Interview: a 20-year longitudinal study', in Crowell, J.A. and Waters E., *Is the Parent–Child Relationship a Prototype of Later Love Relationships?*, Indianapolis: Society for Research in Child Development.

Watson, D.M. and Getz, K., 1990, 'The relationship between oedipal behaviours and children's family role concepts', *Merril-Palmer Quarterly*, 36, 487–505.

Weil, J.L., 1992, *Early Deprivation of Empathic Care*, Connecticut: IUP.

Weinstein, M. and Thornton, A., 1989, 'Mother-child relations and adolescent sexual attitudes and behaviour', *Demography*, 26, 563–77.

Wellings, K. et al, 1994, *Sexual Behaviour in Britain*, London: Penguin.

Wessel, M.A. et al, 1954, 'Paroxysmal fussing in infancy, sometimes called colic', *Pediatrics*, 14, 421–34.

West, D.J., 1959, 'Parental relationships in male homosexuality', *Int. J. Soc. Psychiat.*, 5, 85–97.

West, D.J., 1965, *Murder Followed by Suicide*, London: Heinemann.

West, P. et al, 2003, 'Fifteen, female and stressed: changing patterns of psychological distress over time', *Journal of Child Psychology and Psychiatry*, 44, 399–411.

Westen, D., 1998, 'The scientific legacy of Sigmund Freud: towards a scientifically informed psychological science', *Psychological Bulletin*, 124, 333–71.

Westwood, G., 1960, *A Minority: A Report on the Life of the Male Homosexual in Great Britain*, London: Longman.

Whitam, F.L. and Zent, 1984, 'A cross-cultural assessment of early cross gender behaviour and familial factors in male homosexuality', *Archives of Sexual Behaviour*, 13(5), 427–41.

Whiting, B.B. and Edwards, C.P., 1988, *Children of Different Worlds*, Massachusetts: Harvard University Press.

WHO World Mental Health Survey Consortium, 2004, 'Prevalence, severity, and unmet need of treatment of mental disorders in the World Health Organization mental health surveys', *JAMA*, 291, 2581–90.

Widom, C.S., 1989, 'The intergenerational transmission of violence', in Weiner, N.A. and Wolfgang, M., *Pathways to Criminal Violence*, California Sage.

Widom, C.S. 1999, 'Posttraumatic stress disorder in abused and neglected children grown up', *American J. of Psychiatry*, 156, 1223–9.

Wiederman, M.W. and Hurd, C., 1999, 'Extradyadic involvement during dating', *J. of Social and Personal Relationships*, 16, 265–74.

Wierson, M. et al, 1993, 'Toward a new understanding of early menarche: the role of environmental stress in pubertal timing', *Adolescence*, 28, 913–24.

Wilhelm, K. et al, 2006, 'Life events, first depression onset and the serotonin transporter gene', *British Journal of Psychiatry*, 188, 210–5.

Williams, V., 1992, *A Different Kind of Life*, London: Bantam.

Willis-Owen, S.A. et al, 2005, 'The serotonin transporter length polymorphism, neuroticism and depression: a comprehensive assessment of association', *Biological Psychiatry*, 58, 451–6.

Wilson, G.D. and Barret, P.T., 1987, 'Parental characteristics: some evidence for Oedipal imprinting', *J. of Biosocial Science*, 19, 157–61.

Winnicott, D.W., 1957, *The Child, the Family and the Outside World*, London: Penguin.

Winnicott, D.W., 1971, *The Family and Individual Development*, London: Tavistock.

Winnicott, D.W., 1972, *Playing and Reality*, London: Penguin.

Wolf, P.H. and Fesseha, G., 1999, 'The orphans of Eritrea: a five-year follow-up study', *J. of Child Psychology and Psychiatry*, 40, 1231–37.

Wolkind, S.N. et al, 1980, 'Continuities in maternal depression', *International J. of Family Psychiatry*, 1, 167–82.

World Health Organization, 1999, *Health and Behaviour Among Young People*, Copenhagen: WHO Regional Office for Europe.

Wyatt, G.E. et al, 1988, 'Kinsey revisited, Part I', *Archives of Sexual Behaviour*, 17, 201–39.

Yates, P., 1995, *The Autobiography*, London: Harper Collins.

Zahn-Wexler, C., 2002, 'The development of empathy, guilt and internalization of distress: implications for gender differences in internalizing and externalizing problems', in Davidson, R., *Wisconsin Symposium on Emotion: Vol. 1 Anxiety, Depression and Emotion*, New York: Oxford University Press.

Zahn-Wexler, C. et al, 1984, 'Problem behaviours and peer interactions of young children with a manic-depressive parent', *American J. of Psychiatry*, 141, 236–40.

Zahn-Wexler, C. et al, 1992, 'Development of concern in others', *Developmental Psychology*, 28, 126–36.

Zalsman, G. et al, in press, 'A triallelic serotonin transporter gene promoter polymorphism (5-HTTLPR), stressful life-events', *American Journal of Psychiatry*.

Zanarini, M.C., 2000, 'Childhood experiences associated with the development of borderline personality disorder', *Psychiatric Clinics of North America*, 23, 89–101.

Zanarini, M.C. et al, 1989, 'Childhood experiences of borderline patients', *Comprehensive Psychiatry*, 30, 18–25.

Zanarini, M.C. et al, 2000, 'Biparental failure in the childhood experiences of borderline patients', *J. of Personality Disorders*, 14, 264–73.

Zeman, S., 1997, *Understanding Depression*, London: Mental Health Foundation.

Zingg, R.M., 1940, 'Feral man and extreme cases of isolation', *American J. of Psychology*, 53, 487–517.

Zuger, B., 1970, 'The role of familial factors in persistent effeminate behaviour in boys', *Amer. J. Psychiat.*, 126, 1167–70.

INDEX

abuse, child: breaking the cycle of 185–6; and depression 6–7, 62, 82–3; effect on cortisol levels 6, 7; enduring effects of 7, 156; Freud on sexual 262–3; generational repetition 130–38; lovemaps 131; parental 100–101; Personality Disorder 230–37; predictor of Personality Disorder 210; projection 131; and schizophrenia xiv, 82–3; sexual xiv, 87, 231; and social class 191; violence 137; weak sexual conscience 130; Wobbler pattern of attachment 181–2

academic performance 70; and day care 165; and gender 40–42; and social background 55

achievement: artistic 58–9; causes of levels of 284–90; childhood adversity 285; depression 61, 62; Dominant Goal type 62–9; family script 52–61; high achievers 285–8; high achieving mothers and child care 165–6; insecure relationships 286–7; intelligence 285; links with child care 285–8; loss of parent 55–61; low achievers 288–90; motivation 55–7, 285; parental relations 287; perfectionism 68–71; Personality Disorder 214, 231–7; punitive conscience 110; qualities needed 337 n.155; self esteem 68; underachievement 110; weak sense of self 200; childhood 'hothousing' 54–5

Acta Psychiatrica Scandinavica 81

Adams, John 54

Adams, Ken 54

addiction: causes of 296; heritability of 21–2; Personality Disorder 207; treatment 268

adoption: long term effects of 158–9, 172; problems with studies of 315–19; studies of 16, 139

adversity: achievement 285; coping with 218; lasting effects of early 154–61; inoculating effects of early security 172

advertising 71, 124; and crime 297

aesthetics 271

affluence 124, 309

Affluenza (James) xiv, 66

ageing: and incidence of Personality Disorder 218

aggression: in coercive family 123; depression 137; impact of day care 164; inhibition of 137; parental projection of 75

Allen, Woody 2, 4, 5, 242; and Personality Disorder 231–7

Amin, Idi 57

Amnesty International 241

amorality 125

anal personality 97–8; stage of childhood 97–8

Anderson, Mabel 170

Andrew, Prince 50

anhedonia 234; *see also* pleasure: loss of capacity for

Animal House 118

Anne, Princess 50

Annie Hall 5

anorexia 70, 153, 274; predictors of 298; schizophrenia 298

antidepressants 265, 267

antisocial behaviour 75; heritability of 139–42; weak conscience 122

Archer, Jeffrey: As If personality 211–14; lack of insight 252–4; father 254

Archer, Mary 213, 253

Archer, William 254

Arieti, Silvano 63
art 58; as neurosis 270; benefits of 269
As If personality 211–13, 233–7; see also Personality Disorder
Astor, David 282
attachment: patterns of 151–3; security of 163–4; see also Attachment Theory
Attachment Theory 151, 161–2, 170, 184–9; Avoidant pattern of attachment 176–8, 184; Clinger pattern of attachment 178–80, 184; Secure pattern of attachment 182–3, 184, 190; Wobbler pattern of attachment 181–2, 189–90
Attention Deficit and Hyperactivity Disorder 24, 203
audit: of conscience 142–50; creating a story from emotional 275–82; emotional 3, 10–11, 263, 310; family role 90–92; mental health 310; patterns of attachment 192–9; sense of self 244–8
Authoritarian Personality: and American Christian fundamentalists 116–17; characteristics of 115–16; George W. Bush 117–21
authoritative parenting 106–8, 171
authority: punitive conscience and fear of 110; weak conscience and battle with 123
autism 21; causes 299–300; treatment of 268–9
Avoidant pattern of attachment 152, 176–8; characteristics 294; low achievement 289; sex 293; see also Attachment Theory

Baker, Kenneth 302, 304
Barrymore, Michael: and As If personality 211
Baudelaire, Charles 56
beauty: influence of 51–2; influence of ideals of 70–71; and the family script 50–52
behaviour: enduring effects of early years on 156–7, 158
Being and Nothingness (Sartre) 58
Bell Curve, The (Herrnstein) 302

Bellamy, Carol 307
Belushi, John 118
Bemporad, Jules 63
benign conscience 93, 94 96; scripting the 106–9, 171; see also conscience
Bennett, Alan 270
bereavement 176; as theme of artistic achievement 58–9
Bergman, Ingrid 213
Biddulph, Steve xv
Big Brother 256
Big Deal at Dodge City 210
birth order: child characteristics 44–50; family script 38, 44–50; schizophrenia 47
Blackburn, Tony: and As If personality 211
Blair, Tony xv, xvi, 120
Blakeney, Gayle 17–20
Blakeney, Gillian 17–20, 42–3
blame: futility of 260–61
Bogart, Humphrey 213
borderline Personality Disorder 205–7; and childhood maltreatment 222; sexual dissatisfaction 206–7; Paula Yates 207–9; see also Personality Disorder
Bouchard, Thomas 8; doubts over twins study 16, 312–13
Bowlby, John 151, 161–2, 170–72, 185, 189
Boyd, William 49
Bradford Sarah 167
brain: childhood and size of 6–7; rapid growth in early years 154
brain patterns: influence of early years in establishing xiii, 5–7, 153–61
Bridget Jones's Diary (Fielding) 71, 257
Britain on the Couch (James) 265, 267
British and American Encyclopaedia 55, 286
Brontë sisters 55
Brown, Helen Gurley 71
bulimia 70; and delusions 298; predictors of 298
Burchill, Julie 259
Burt, Cyril 315
Bush, Barbara 118–19
Bush, George W. 117–21; as

Authoritarian Personality 117, 119–21
Bush, Jeb 118
businessmen: and loss of a parent 55; Personality Disorder 215
Byron, Lord 55

carers: impact of substitute 161–2, 169–70; responsive 171; unresponsive 170–71, 173–5
Casablanca 213
Castro, Fidel 46
catatonia 82
Centre for Policy Studies 302
change: and self-expression 249; *see also* insight
Chapman, Mark 250
Charles, Prince 2, 74, 90; childhood 48–50; early child care 167–8, 169–70
Cheam preparatory school 48, 49
child abuse, *see* abuse, child
child care: absence of 219; accelerated learning 54; adult achievement 52–61, 286; adult addiction 296; adult criminality 297; adult depression 61–71, 295, 300, 301–2; adult divorce 291; adult eating disorders 298; adult neuroses 298–9; adult personality 294; adult Personality Disorder 218, 231–7, 241–2, 296–7; adult relationships 5, 290–91; adult schizophrenia 300–301; adult sexual personae 293–4; adult violence 297; autism 299–300; birth order 44–50; borderlines and childhood maltreatment 222; day care 163–5, 170; development of conscience 94–5, 97–8, 102–9, 122–38; different treatment by parents 37–8, 72–5, 107–8; different types of 306; disrupted parenting 220; empathic 201–2, 219, 228; enduring effects of 2, 5–6, 33–4, 89–90, 156–61, 172–3; erratic 158–9, 287; full-time mothers 170–76; hyperactivity 203, 299; importance of 8–9, 307, 308–9; institutionalized 158–9, 219; intellectual development

and day care 165; Japan 238–9; long-term insecurity 173–5; maternal care 222–30; maternal depression 160–61; New Labour policies xv–xvi; parental projection 113–14; patterns of attachment 152–3, 176–83, 187–9; pleasing parents 107; poverty 308; proposal for national study of 310; reform of 309–11; responsive 171; substitute 162–6, 169–70; unempathic 201, 202, 203, 204, 218, 220, 228–9, 241–2, 286, 290–91, 295; unresponsive 170–71, 173–5; working mothers 162–76, 337 n.155
childhood: establishment of brain patterns 5–6, 153–4; guilt and sexuality 103–4; influence on adult life 2, 5–6, 7, 34, 88–9; repetition of experiences 5; sexual activity 98–9
childhood relationships: history of 189–90
children: anal stage 97–8; attention getting strategies 44–6; body chemistry 6; changes in patterns of attachment 185, 188; effects of an early loss 56; loss of a parent 55–61; uniqueness of upbringing 36–8
China: incidence of mental illness 22, 238
Christian Coalition: and American politics 117
Christian fundamentalists 128; American 116–17
Christmas: and the family script 35–6
Clarke, Kenneth 53–4
Clarke, Michael 53–4
Cleese, John 69–70
Clinger pattern of attachment 152, 178–80; characteristics 294; low achievement 294; sex 293; *see also* Attachment Theory
coercive parents 123–4, 140, 141
Coghlan, Monica 212, 213, 252, 254
Cognitive Analytic Therapy 264, 267, 310; *see also* therapy
colic 224–8
collectivist society: and incidence

of Personality Disorder
239
communication: and a playful life 272
complaint, culture of 9
conscience 33; audit of 142–50;
benign 93, 94, 96, 106–9; child
temperament and parental style
142; developing complexity of 95–6;
genes 138–42; Oedipus complex 99;
parents and development of 95–6;
punitive 93, 96, 109–22; restraint of
desires 96; sexual 96, 102–6, 108;
types of 93–5; weak 93, 94, 96,
122–38
consumerism 311; and addiction 296;
and crime 297
Contented Little Baby Book (Ford)
xiv, 166
Continuum Concept, The (Liedloff)
xv, 190
corporal punishment: and
Authoritarian Personality 116–17
cortisol xiii, 153, 154, 156, 157, 158,
181, 227; childhood levels of 6, 7
counsellors 188, 250
counter-transference 258
Crawford, Marion 168–9
creativity 17, 274–5; and insight
269–71; parental loss 58; release of
repressed unconscious anxiety 270;
story writing 275–82
Crick, Michael 253, 254
crime: childcare and reduction of
8; increase in 139–40; right wing
social policy 303–4
criminality 141; causes of 297;
treatment of 268
criminals: childhood backgrounds
305; early child care 305–7; early
parental loss 59–60; incidence of
mental illness 305; Wobbler pattern
of attachment 192
Critical periods of development 157
cross-modal matching 202

Darwin, Charles 46, 56, 59
daughters: and Dominant Goal
depression 67–71
Dawkins, Richard xiv
day care: advantages of 170;

impact of 163–5, 170; intellectual
development 165
delusions 205; and bulimia 298
denial 231; and Personality Disorder
210
depersonalization: and Personality
Disorder 220–21
depression xiii, 21; and aggression
137; causes of major 301–2; causes
of minor 295; childhood abuse
6–7, 62, 82–3; depressed mothers
and children 75–8; Dominant Goal
type 61–71; drug treatment 265;
early parental loss 60; family scripts
61–71; genes 23; ideals of beauty
70–71; in infants 204; insecure
attachment patterns 295; isolation
255; manic 26, 66, 300; maternal
75–8, 160–61, 170, 194–5, 223–8,
229; parental projection of 75;
perfectionism 68–71; postnatal
223–8; self-revelation 255–6;
smoking 97; treatment 265, 267–8;
writers 59
deprivation: early 218, 219, 230;
emotional 62
De Valera, Eamon 57
development: branching view of
185; Critical periods 157; Sensitive
periods 157, 158, 159
Dickens, Charles 189
dictators: and childhood loss 56–7
Dimbleby, Jonathan 48, 49, 169, 170
discipline xvi; appropriate time to
start 171; consistency of 128; of
weak conscience 123, 125
discordant twins 85–6
displacement 137
dissociation 87–8, 230–31; causes
of 82; and genes 221–2; impact
of early parental care 220–21;
Personality Disorder 210; Wobbly
pattern of attachment 181
divorce 153, 176, 291–3; causes of
291–3; doubts over heritability
of 22; effects of 78; incidence of
homosexuality 105; predictors of
292; roots of 291
Dominant Goal type 62–9
Don't Look Now xiv

Double-Bind 87
drugs: and schizophrenia 83–4,
    265; treatment 265, 269; the
    weak-selfed 241
dysphoria 223

eating disorders 70–71, 153; causes of
    298; *see also* anorexia; bulimia
education: and patterns of attachment
    184; under New Labour xv
Edward, Prince 50
ego: as mediator 95
Eliot, T.S. 9
Elizabeth II, Queen: childhood 168–9;
    and Prince Charles 48, 167–8
emotion: and deveopment of benign
    conscience 109; insight 251–2
emotional audit 3; creating a story
    from 275–82
emotional deprivation: and depression
    62
emotional development: enduring
    effects of early years 154–62
empathy 107; lack of, and Personality
    Disorder 218; maternal 222
environment: and individuality 12–13,
    14–15; mental illness 295–9;
    smoking 22; social conditions of
    divorce 22
eugenics 302–3
exhibitionism: in children 98–9
experience: enduring effects of early
    153–62

failure: and childhood adversity 285;
    low achievement 288–90
false self 202; Machiavellian type 214
families: and beauty of children
    50–52; birth order of children and
    family script 44–50; children 36–9;
    coercive 123–4; effects of large
    4, 47; family scripts and eating
    disorders 70–71; family size and
    family script 47; favouritism 72–4;
    gender of children and family script
    39–43; high achieving children
    52–61; scripted dramas 35–6; sexual
    conscience 102–6; shaming of
    children 74
fantasy 12, 205, 213

Farrow, Mia 4–5, 159, 232, 233
fascism 153; and the Authoritarian
    Personality 115–17; parental 121
favouritism: in families 72–4
femininity: and birth order 47–8; in
    boys 104–5
Feyerabend, Paul 109
fiction: as means of insight 275–82;
    escape into 213
Fielding, Helen 71
Flett, Kathryn 257–9
Ford, Gina xiv
Foster, Jodie 12
Freed, Jack 235
freelance: as metaphor for Modern
    Man or Woman 272
Freud, Anna 266
Freud, Sigmund 1, 10, 60, 88, 117;
    and conscience 95, 96–7, 98, 138–9;
    creativity 269–70; Oedipus complex
    99, 262; sexual abuse 262–3;
    super-ego 138
friendship: and childhood prototypes
    89–90
Frost, Jo xiv
Frum, David 117, 120, 121
Fry, Alan 65
Fry, Stephen: as Dominant Goal
    type 65–6
full-time mothers: and insecurity in
    children 170–76
fundamentalists 128; American
    Christian 116–17, 120

Gandhi, Mahatma 56, 57
Geldof, Bob 208
gender: and Dominant Goal
    depression 67–71; family script 38,
    39–43; parental expectations 40, 42
gender stereotypes 40, 43
genes xiii; and Attention Deficit and
    Hyperactivity Disorder 203; autism
    299–300; conscience 138–42;
    dissociation 221–2; genetic variation
    and vulnerability to depression 23;
    individuality 12–13, 14–15; less
    influential than upbringing 8; major
    depression 301–2; manic depression
    300; mental illness 20–21, 25–7,
    299–302; patterns of attachment

186–9; Personality Disorder 222; proneness to schizophrenia 81; relationship patterns 186–9, 290; role of 308; schizophrenia 27–9, 300–301; sexuality 102; twin studies 15–24; tyranny of geneticism 302–9

Germany: changes in parenting practices 121

Gielgud, John 43, 103

Godley, Sue 207

Gordonstoun 48, 49–50

grandparents: grandmothers 227–8; relationship with grandchildren 53–4

gratification, delayed 95–6

Green, Hughie 208

guilt: and childhood loss 57; childhood sexuality 103–4; punitive conscience 110

Gulliver's Travels (Swift) 270, 271

hallucinations 27, 82

Handel, George Frederick 57–8

happiness 66

Harewood, Lord 169

Harvey-Jones, Sir John 287

Having It All (Brown) 71

Herrnstein, Richard 302, 303

heterosexuality 105

high achievers 17, 285–8; and child care 165; see also achievement

Hinckley, John 12

hippocampus 6–7

Hitler, Adolf 57

Ho Chi Minh 57

Hoffman Process xvi

homicide: and suicide 137

homosexuality 1–14, 43, 332 n.98–99; birth order 47–8; family background 104–5; environmental components of 104–5; incidence of 105

Horizon (BBC 2) 76

'hospitalism' 219

Human Aggression (Storr) 216

Human Genome Project xiii, 15, 22

hunter-gatherers 190

Huntington's Chorea 21

Hutchence, Michael 209

hyperactivity 24, 203; causes of 299

hypnotism: as treatment 250

idealization: and Personality Disorder 209–10

illusions: tenacity of 255

imagination: parental loss and creative 58

incest 87, 102

Inch, Elspeth 68

individualist society: and incidence of Personality Disorder 239; pressures of 239–40

individuality: explanation of 8, 37; parental influences 138; punitive conscience 110

infant deprivation: impact of 219–21

insecurity 152–3, 189; and absence of mother 161–2; effects of 153; effects of early child care 172–3; impact of day care 163–4; in toddlers with full-time mothers 170–76; maternal depression 172; origins of 161; patterns of attachment 176–83, 184; relationships and achievement 286–7; schizophrenia 153; therapy for 172; unresponsive carers 171

insight: 9, 249, 250–54; about what? 260–61; care by our parents 260–61; creativity 269–71; emotion 251–2; key to changing impact of past on present 250; lack of 252–4; play 273; re-experience 251–2; self-revelation and lack of 257–60; self-revelation to intimates 254–7; story writing 275–82; therapy 262–6

instinct: curbing of 95, 96

intellectual development: and day care 165

intelligence: doubts over heritability of 284–5; social background 55

intelligence (IQ) tests 16, 55, 285, 303

intrusive care 170; and hyperactivity 203

isolation: and depression 255

Israel: Clinger pattern of attachment 180

Jackson, La Toyah 52

Jackson, Michael 52, 53, 75
Japan: child care 238–9; incidence of Personality Disorder 238
John, Elton 43, 105; as Dominant Goal type 64–5
Joseph, Jay 312
judgmentalism: and punitive conscience 110, 121
juvenile delinquents: and early parental loss 60

Kallman, Franz 315
Kaunda, Kenneth 57
Keane, Fergal 29
Keaton, Diane 232
Keats, John 55
Kenyatta, Jomo 57
Kinsey, Alfred 98
Klerman, Gerald 25
Knight, Alah 168, 169
Kraeplin, Emil 25

Laing, R.D. 2, 259; and families 36; schizophrenia 81, 86–7
language learning 157
Larkin, Philip 260, 311
Lascelles, Tommy 48
Leigh, Mike 269
Lenin, Vladimir Ilyich 57
Lennon, John 12, 56, 250
Liedloff, Jean xv, 190
Life is Sweet 269
Lightbody, Helen 49, 167, 169–70
Lineker, Gary 53
Lloyd, John 70
Losing Ground (Murray) 302
loss: and achievement 55–61; effects of an early 56
love: and childhood prototypes 89–90; ludic love style 207
lovemaps 102, 103; and sexual abuse 131
low achievement: explanation of 285, 288–90
ludic love style: and Personality Disorder 207

Machiavellianism: and achievement 285, 288; 'false self' type 214

madness 25–6, 273
Madonna 56
major depression 21; causes of 301–2; *see also* depression
manic depression 21, 26, 66, 274; and child care 300; drug treatment 265; genes 300; treatment 268–9; *see also* depression
Marx, Karl 46
masculinity: and birth order 47
maternal care xv, 222–30; empathic 228
maternal depression 75–7, 173, 195, 223–9; and child care 166–7; child insecurity 172; effects of 160–61; *see also* depression
maternal deprivation 154–5, 158; and Frank Williams 217
maternal empathy 222
Maxwell, Robert 55
May, Rufus 87, 268; and schizophrenia 29–33
McCartney, Paul 56
Mein Kampf (Hitler) 57
mental health audit 310
mental illness 274; and birth of child 223–8, 229–30; causes of 295–302; child care 306–7; difficult babies 229; environmental factors 295–9; genes xiii, 299–302; heritability of 20–21; incidence of 10; incidence in prison population 8, 305; national variations in incidence of 22, 311; omnipotence 205; right brain dysfunction 6; schizophrenia 24–33
Messiah, The (Handel) 57–8
Miller, Alice 9, 115
Minnesota Twins Reared-Apart Study: doubts about 312–13
Molière 56
Money, John 102
monkeys 190; effects of early environments 155–6; patterns of mothering 155–6
mothering: and effect of day care 163–5; monkey's patterns of 155–6
mothers: effects of maternal depression 160–61; full-time and insecurity in toddlers 170–76; high achieving and child care 165–6;

interaction with beautiful children
50–51; nature versus nurture debate
13; violent children 140–41
motivation: and achievement
55–7, 285
Mountbatten, Lord 50
multiple personalities: and Personality
Disorder 210–11
Murray, Charles 302, 304–5
musical ability: learned not inherited
52–3

Napoleon Bonaparte 57
narcissism 231; and performing
arts 214; Personality Disorder 66,
205; self-revelation 256; see also
Personality Disorder
Nasser, Gamal 57
nature versus nurture 12–15, 283–4;
and twins studies 16–22
Nausea (Sartre) 58
neglect: as predictor of Personality
Disorder 230; and sexual conscience
129–30, 135–6; Wobbler pattern of
attachment 182
Neumann, Peter 118
neuroses: causes of 298–9
neuroticism: and divorce 291
New Labour xv–xvi
New Zealand 22–3: suicide rates in 71
Newton, Isaac 56
Norman, Philip 43, 65

Obsessive Compulsive Disorder
168, 298–9
obsessiveness: and punitive conscience
110
Oedipus complex 88, 99, 262, 263
Ogawa, John 220–21, 241
omnipotence 214–17, 231, 286; and
Personality Disorder 204–5
only children 46
openness: and self-revelation 256;
misuse of 257
oral fixation 97
originality 286
orphanages: child care in 219

panic attacks 173–5; and punitive
conscience 111–12

paranoia 26, 56; and Personality
Disorder 209
parenting: and antisocial behaviour
in children 140–41; Authoritarian
Personality 116; authoritative 106;
changes in American practices 117,
121; changes in British practices
121; changes in German practices
121; effects of early disrupted 220;
generational transference of style
of 155–6; patterns of attachment
176–83; results of chaotic 124–5
parents: and beauty of children 50–52;
birth order of children 44–50;
blaming 260–61; children's adult
depression 61–71; children's adult
eating disorders 70–71; children's
identification with 100; coercive
families 123–4; development
of children's conscience 94–5,
102–38; different for each child
7–8, 19–20, 36–9, 107–8; effects
of loss of a parent 55–61; effects
of separation of 78; favouritism
72–3; high achieving children
52–61; hypercriticism 69; influence
of own childhood 38–9; influence
on individuality 138; projection
on children 113–14; projections of
emotions on to family members 75;
protective attitude towards 9, 261;
prototypes for adult relationships
102–3; reaction to gender of
children 39–43; shaming of children
74; unempathic 220; unresponsive
170–71, 173–5
patterns of attachment 33–4, 151–2,
176–86; audit of 192–9; child care
187–9; dissociation 220; experiences
likely to change 172; flexibility of
185, 188–9; generational transfer
158, 186; genes 186–9; history
of 189–91; influences on 152–3;
insecure 152–3; low achievement
289; persistence of 171–3
Patterson, Jerry 123, 124, 140
peers: influence on personality
traits 74–5
perfectionism 68–71; and
eating disorders 70; punitive

conscience 110; type of 69–70
performing arts 58; narcissism within 214
permissiveness, parental 125
personality: causes of 290–94; divorce 291; influences of childhood relationships 294
Personality Disorder 66, 200, 204–18, 274; addiction 207; advantages of 217–18, 242–3; Anne Robinson 260; As If personality 211–13, 233–7; borderline 205–7, 222; causes of 200–201, 219–21, 230–37, 296–7; complete absence of child care 219; denial 210; depersonalization 220–21; diminishing symptoms with age 267; dissociation 210, 220–21; fantasy 205; Frank Williams 215–17; genes 222; idealization 209–10; incidence with ageing 218; Jeffrey Archer 211–14; Julie Burchill 259; Kathryn Flett 257–9; low achievement 288–9; ludic love style 207; Michael Barrymore 211; multiple personalities 210–11; narcissism 205, 214; national variations in incidence of 238, 239, 242; omnipotence 204–5, 214–17; paranoia 209; parental maltreatment 228–9; Paula Yates 207–9; politicians 242; predictors of 222, 230; projection 210; psychopaths 214; self-revelation 256, 257–60; sexual dissatisfaction 206–7; splitting 209, 218; strategies for keeping reality at bay 209–10; success 214, 231–7, 242–3; Tony Blackburn 211; treatment of 267; unempathic care 241–2; weak sense of self 202; Wobbler pattern of attachment 210; Woody Allen 231–8
personality traits: heritability of 16–17, 20; lack of influence of peers on 74–5
Philip, Prince 74, 169–70; and upbringing of Prince Charles 48–50
Pioneer Fund of New York 313

play: importance of 271–3
Playing and Reality (Winnicott) 271
pleasure, loss of capacity for 202–4; see also anhedonia
Plomin, Robert 16, 321
Poe, Edgar Allan 58
Poisonous Pedagogy, doctrine of 115, 121, 171
political beliefs: and nature v. nurture debate 14; right wing xiv, 14, 302–5
politicians: and birth order 46; and Personality Disorder 215, 242
Politics of Experience, The (Laing) 2
postnatal depression 77, 223–8; see also depression
poverty 309; and child care 308; depression 23, 224; schizophrenia 27, 28, 83
power, the will to 57
pregnancy: stress during 227; subsequent child hyperactivity 203
prison population: mental illness among 8, 305
projection: and Personality Disorder 210; punitive conscience 113–14; sexual abuse 131
Prozac 157
psychoanalysis 262–4, 265–6; advantages of 263–4; shortcomings of 262–3; see also therapy
psycho-archaeology 196–9, 247
psychology: scepticism of 1
psychopaths 125–8, 214; national variations in incidence of 238–9
punishment: effect of erratic 123, 125
punitive conscience 93, 96; and achievement 110; Authoritarian Personality 115–17; characteristics of 110–11; fear of authority 110; fewer parents with 122; guilt 110; individuality 110; judgmentalism 110, 121; middle age rebellion 110–11; obsessiveness 110; panic attacks 111–12; perfectionism 110; projection 113–14; scripting 109–22; sex 111; underachievement 110; see also conscience

race: and schizophrenia 28, 83
rape 7; motivations of rapist 131

rats: effects of early environment 154–5, 158
Read, John 81–2, 83, 84
Reagan, Ronald 12
realism 273
reality: congenial versions of 9–10
rebellion: punitive conscience and middle age 110–11
re-experience: and insight 251–2
rejection: and Avoidant pattern of attachment 177–8
relationship patterns: causes of adult 290–94; development of benign conscience 109; effect of early years 5, 156–7, 158, 161; genes 186–9, 290; history of childhood 189–90; influence of childhood prototypes 89–90; insight from 254–7; ludic love style 207; parents as prototypes for 102–3; patterns of attachment 151–2, 176–83, 184; Personality Disorder 206–7; sexual 293–4; unempathic child care 290–91; use of Personality Disorder attributes 217–18; see also Attachment Theory
religious beliefs: and Avoidant pattern of attachment 177; Clinger pattern of attachment 179
repression: and childood sexuality 103–4
Repressors: and self-revelation 256
responsive child care: and Secure pattern of attachment 183
revolutionaries: and birth order 46–7; and childhood loss 57
Robertson, Jimmy 162
Robinson, Anne 260, 261
Roddick, Anita 56
Roeg, Nic xiv
role models, inappropriate 67
Rousseau, Jean-Jacques 56
rural environment: and schizophrenia 28

sanity 273
Sanity, Madness and the Family (Laing) 81
Sartre, Jean-Paul 58
schizophrenia 12, 21, 24–33, 231, 274; and abuse xiv, 82–3; and

anorexia 298; causes of xiv, 33, 81, 84, 300–301; dissociation as precursor of 220; drugs 29, 83–4; environmental explanation of 27, 79–88; genes xiii, 300–301; incidence in twins 27, 85–6; influence of birth order 47; insecurity 153; Julie 24–5, 84–5; parallels with child's thinking 274–5; parental negativity 79–86; race and incidence of 28, 83; Rufus May 29–33; smoking 97; social class and incidence of 28; treatment of 81, 83–4, 265, 268–9; weak sense of self 202
scientists: and birth order 46–7; and loss of parent 59
Secure pattern of attachment 152, 182–3
security: enduring effects of early child care 172–3
Seduction Theory 263
self, sense of 34, 200–201, 288: audit of 244–8; origins of 201; scripting 200–244; strong 221; weak 87–8, 200–204, 241, 288–9
self-esteem 62; low 61, 65, 67, 68
self-expression: and change 249
Selfish Capitalism xiv, xv, 22, 66
Selfish Gene, The (Dawkins) xiv
self-knowledge: and insight 249
self-revelation: as way of not knowing 254–61; exhibitionistic 246; generational differences 256; misuse of 257; narcissism 256; repressors 256; to intimates 254–7
Sensitive periods of development 157–8, 159
separation: maternal 155, 158; parental 153, 291–3
serotonin 157, 308–9
sex: and conscience 96–7; frequency of intercourse 111; patterns of attachment 293; punitive conscience 111; relationships 293–4; sexual conscience 102–6; roots of sexual personae 293
sex education 122
sexual abuse 87, 231; and depression 6–7, 62; and schizophrenia xiv,

82–3; and weak sexual conscience 130; *see also* abuse, child

sexual activity: in childhood 98–9; Avoidant pattern of attachment 176–7; Clinger pattern of attachment 179

sexual conscience: and childhood sexual abuse 130; development of benign 108; parental neglect 129–30, 134–5

sexual dissatisfaction: and Personality Disorder 206–7

sexual repression: and Authoritarian Personality 116; lovemaps 103–4

sexuality: childhood guilt 103; influence of family members 102; influence of parents on 48; parental expectations 42–3

shaming: of children 74

sibling rivalry: misdiagnosis of 1

siblings: differences between 7–8

Sinatra, Frank 4

Sincerity and Authenticity (Trilling) 256

Singapore: incidence of mental illness 22

smoking 251; and depression 97; heritability of addiction 21–2; schizophrenia 97

social class: and achievement 55, 288; and depression 67–8; and schizophrenia 28

social development: enduring effects of early care 154–61

social policy, right wing 302–5

society: encouragement of violence 297; obstacles to creating child-friendly 242; Personality Disorder and cultural trends 297; role in fostering Personality Disorder 237–44

speech disorder: symptom of schizophrenia 26–7, 82

splitting: and Personality Disorder 209, 218

Spock, Benjamin 121

Stalin, Josef 57, 61

State of the World's Children (UNICEF report) 8–9

Storr, Anthony 216, 217, 271

sub-personalities 231

success: and Personality Disorder 214, 231–7; *see also* achievement

suicide 82–3; and homicide 137–8; increase in rates of 71

Sukarno, Ahmed 57

Sulloway, F. 47

super ego: Freud on 138

Supernanny (Channel 4) xiv

SureStart scheme xv

Sweden: incidence of Personality Disorder 238

Swift, Jonathan 270

Taking a Stand (BBC Radio 4) 29

teenagers: proposed emotional audit of 310

Thatcher, Carol 73

Thatcher, Margaret 73, 303

Thatcher, Mark 73

therapy xiii; appropriate for problem 267–9; Cognitive Analytic Therapy 264; drugs 265; insight 250, 261–6; psychoanalysis 262–4, 265–6; reduction in insecurity 172; tackling effects of unresponsive early care 173–5

Tolstoy, Leo 56, 57, 269

Trilling, Lionel 256, 270

Trotsky, Leon 46

twins: Bouchard's study of 8; discordant 85–7; doubts over Minnesota Twins Reared-Apart Study 312–13; environmental influence on characteristics of 321–3; findings of twin studies 15–24, 27; Gayle and Gillian Blakeney 17–20; heritability of antisocial behaviour 139; incidence of schizophrenia 27; problems with studies of 315–19

Ullman, Tracey 58

unconscious, the 88

underachievement: and punitive conscience 110; reasons for 288; *see also* low achievement

unhappiness 66

UNICEF 8

United Kingdom: changes in parenting

practices 121; incidence of depression 67
United States: changes in parenting practices 116–17; incidence of depression 67; incidence of mental illness 22, 311; incidence of Personality Disorder 238; rise in anxiety levels 240; risks to weak-selfed 241; vulnerability to Personality Disorder 242
unresponsive child care 164: and Clinger pattern of attachment 180
urban environment: and schizophrenia 27–8, 83

Van den Boom, Daphna 187–8
Venter, Craig 15
violence 153; causes of 23, 297; childhood abuse 75, 137; childhood history 2; coercive families 124; increase in139–41; weak conscience 125–8

Washington, George 61
Wax, Ruby 270
weak conscience 93, 94, 96; and antisocial behaviour 122–3; authority 123; chaotic parenting 123–4; characteristics 122–3; erratic discipline 128–9; method of child discipline 123; repetition of abuse 130–38; scripting of 122–38; see also conscience

Willetts, David 303
Williams, Frank 242; and Personality Disorder 215–17
Williams, Serena and Venus 54
Williams, Virginia 215–17
Winnicott, Donald xvi, 201, 202, 271
Wobbler pattern of attachment 152, 181–2; and dissociation 220; low achievement 289; maltreating parents 228–9; Personality Disorder 210; and sex 293; see also Attachment Theory
Words, The (Sartre) 58
Wordsworth, William 55
work: and Avoidant pattern of attachment 177; Clinger pattern of attachment 179
working mothers 333 n.155; and insecurity in toddlers 162–70, 173
World Health Organization 311
Wright, Lawrence 31
writers: and depression 59; and loss of parent 55–6

Yates, Jess 208
Yates, Paula 2; and Personality Disorder 207–9
Your Child Can Be a Genius (Adams) 54

Zelig 234
Zola, Emile 56

A NOTE ON THE AUTHOR